# Geoheritage, Geoparks and Geotourism
Conservation and Management Series

Spectacular geo-morphological landscapes and regions with special geological features or mining sites are becoming increasingly recognized as critical areas to protect and conserve for the unique geoscientific aspects they represent and as places to enjoy and learn about the science and history of our planet. More and more national and international stakeholders are engaged in projects related to "Geoheritage", "Geo-conservation", "Geoparks" and "Geotourism"; and are positively influencing the general perception of modern Earth Sciences. Most notably, "Geoparks" have proven to be excellent tools to educate the public about Earth Sciences; and they are also important areas for recreation and significant sustainable economic development through geotourism. In order to develop further the understanding of Earth Sciences in general and to elucidate the importance of Earth Sciences for Society, the "Geoheritage, Geoparks and Geotourism Conservation and Management Series" has been launched together with its sister "GeoGuides" series. Projects developed in partnership with UNESCO, World Heritage and Global Geoparks Networks, IUGS and IGU, as well as with the 'Earth Science Matters' Foundation will be considered for publication. This series aims to provide a place for in-depth presentations of developmental and management issues related to Geoheritage and Geotourism in existing and potential Geoparks. Individually authored monographs as well as edited volumes and conference proceedings are welcome; and this book series is considered to be complementary to the Springer-Journal "Geoheritage".

Rosa María Ruiz Pedrosa

# Geomorphological Heritage as Didactic and Geotouristic Resources

## Case Studies from Castilla y León, Spain

Springer

Rosa María Ruiz Pedrosa
Department of Geography, Pangea Research Group
University of Salamanca
Salamanca, Spain

ISSN 2363-765X ISSN 2363-7668 (electronic)
Geoheritage, Geoparks and Geotourism
ISBN 978-3-032-06204-8 ISBN 978-3-032-06205-5 (eBook)
https://doi.org/10.1007/978-3-032-06205-5



This Springer imprint is published by the registered company Springer Nature Switzerland AG
The registered company address is: Gewerbestrasse 11, 6330 Cham, Switzerland

# Foreword

The study and assessment of geomorphosites has recently expanded, especially since it is no longer purely geological and gains strength as part of the abiotic nature intimately related to biotic, landscape, and territorial aspects. Its study began with the first inventories of geological heritage, which included geomorphological aspects as elements of knowledge of Earth's history. However, it is since the late 1990s, with the development of Applied Geomorphology and Cultural Geomorphology, driven by Mario Panizza and Sandra Piacente, and later in 2001 with the establishment of the Geomorphosites Working Group of the International Association of Geomorphologists, led by E. Reynard and P. Coratza, when a new integrated conception of geomorphosites allows rapid advances, first in inventory, cataloguing, and assessment, and immediately in their scientific and cultural appreciation. From this moment on, geomorphosites are appreciated scientifically, but also for their territorial values, their landscape imprint, and their close relationship with human activities and history. This new conception endows them with cultural content inherent to their location on the Earth's surface. Geomorphosites become important elements for the local population, not only as perennial but intangible physical components, but also as visible entities that are part of the landscape and offer a comprehensible view and therefore have a high educational and cultural potential. They become a resource available for current and future generations that fosters a relationship between the environment, the inhabitants, and visitors.

The progress towards a broad consideration of geomorphosites begins with the inventory and selection of outstanding elements based on various aspects that go beyond purely scientific ones. The selection of geomorphosites is based on three basic aspects: their scientific or intrinsic values, their additional values, where territorial and cultural aspects become important, and the values of use and management that make it an educational, territorial, geotouristic, or leisure resource. Currently, the inventory and assessment of geomorphosites in specific territories (municipalities, Natural Protected Areas, regions, etc.) consists of several stages. First, the evaluation of the element or site by scientists, territory managers, or the general public based on some of its intrinsic, additional, or use and management values. This stage is followed by the analysis and selection of geomorphosites, the designation of sites or places as legal or non-legal frameworks, and finally proposals for management actions aimed at their protection, positive management, and promotion. Thirdly, the evaluation requires a methodology for inventory and characterization of geomorphosites at different working scales, national, regional, or local (municipalities, Natural Protected Areas, geoparks, etc.). This phase requires detailed geomorphological studies, where geomorphological mapping plays an important role in framing geomorphosites in their geomorphological and territorial context and in deepening the knowledge of each inventoried geomorphosite.

As places and sites are better understood, it becomes necessary to manage the conservation of both the resources and the natural or cultural services it provides to society. In this sense, it is useful to ask who benefits from the existence of geomorphosites within specific territories and societies, as well as why the appreciation of landforms benefits. In regional and local-scale geomorphosites, or those included in Natural Protected Areas, the range of criteria to consider connected to the beneficiaries must be broader. Among them, we can highlight their educational values for formal studies or lifelong learning, their value for use by professional scientists, amateur naturalists, their possible historical value as an important element for understanding landforms and the advancement of Earth sciences, their historical or ethnological relationships with the societies that have inhabited them or with the infrastructures they have built (religious, civil, hydraulic, etc.), their aesthetic value and contribution to the landscape in the dual aspect of public awareness and appreciation of Geography and Earth Sciences, or as a cultural resource for residents and visitors; finally, criteria related to access and safety that promote or hinder their uses and exploitation. All of this is contained in multiple and numerous methodologies developed by different research groups from Italy, Switzerland, France, Spain, Poland, Iran, Czech Republic, Brazil, Argentina, Peru, and many others, which seek to grasp the diversity and richness of geomorphosites as elements of the territory that transcend mere scientific value.

The benefits of the assessment and conservation of landforms are wide-ranging, and as recently pointed out by J.E. Gordon, they affect numerous organizations, public bodies, societies, industries, communities, and individuals in three prominent aspects: research, management, and the use and dissemination of their values. In the first case, there are geographers and geologists seeking to understand the Earth, recent or past environmental changes, the landscape, and the relationships between the environment and humans through geomorphological research; in the second, technicians and managers of the natural environment along with owners, administrators, public services, or land planners who need to base their decisions and actions, as well as environmental organizations involved in nature conservation and management; and in the third, schools and universities in their fieldwork, naturalistic, geographical, or geological amateur societies that enjoy being in the field, nature interpreters and guides, and the general public, all aimed at recreational, educational, or cultural purposes, including local residents who thus better understand their territory and can benefit from it as a tourist or cultural resource.

We have been able to appreciate that geomorphological heritage is not only a scientific issue, but also concerns the territory, its cultural values, and the societies that inhabit or visit it. Therefore, sites and places of geomorphological interest must be managed for their correct use, improvement, and conservation. To do this, protection against exploitation must be addressed, on-site management that guarantees both their permanence when possible—because sometimes the geomorphological dynamics imply the ephemeral nature of landforms or processes—as well as potential uses related to interpretation, education, or geotourism. An approach to good management and protection practices developed by interested parties is necessary, which has recently expanded due to the increased interest in geomorphological heritage. Currently, there are numerous groups of stakeholders interested in geomorphosites: scientists, specialists in geography, tourism, and Earth sciences; people responsible for territorial policies at different scales; territory managers and planners; organizations and individuals involved in the conservation of both cultural and natural heritage; tourism professionals in active, ecotourism, or nature tourism, including geotourism; educators and teachers in various formal and regulated disciplines—geography, geology-; and finally, society in general including visitors and the local population. They all have different objectives, strategies, and perceptions of geomorphological heritage, and their visions and expectations must be coordinated through action plans, scientific monitoring, appropriate interpretation, good practices for protection and management, and appropriate uses of geomorphosites in their territorial context to enable orientations toward their sustainable use and management.

Recently, the concept of ecosystem services has been developed, which is fundamental for nature conservation and where abiotic elements are being incorporated very slowly. Ecosystem services are defined as the goods and functions of ecosystems and nature that benefit society in different aspects such as improving health, quality of life, or the economy. Landforms and modeling are also included in the wide range of ecosystem services, although they are not yet very well considered in assessments due to their abiotic condition. However, geomorphological forms and processes gain strength alongside geomorphosites due to their importance as cultural services, in addition to provisioning, knowledge, regulation, and support services, where terrestrial forms play a prominent role derived from their surface distribution and their direct relationship with human occupation, land use, and territory. Cultural services encompass a wide range of services that include recreational activities, aesthetic enjoyment, inspiration and artistic development in nature, environments of tranquility and relaxation, environmental and geographical education about the territory, interconnections with art and local, regional, national, or supranational history, as well as local identity, sense of place, social relationships, cultural diversity and meanings, spiritual or religious beliefs, and knowledge systems. All of these are important for human development and provide significant support for local populations and visitors to geomorphosites, who find in these services direct support for personal development and for the different groups that approach geomorphosites.

Rosa María Ruiz-Pedrosa's book is part of the work of inventorying, assessing, and considering geomorphosites in specific territories defined by depopulation and marginality as assets and resources applicable to school, university, or adult education, particularly in relation to local inhabitants. The author's work responds to a clear objective, where, without giving up basic research in geomorphology, aimed at explaining landforms in a specific local and regional context, it also seeks answers to concerns about the social projection of its study, with education as a fundamental pillar. It is oriented towards a "social utility" of Geography and particularly Geomorphology, disciplines in which this study is framed. The author focuses on the assessment of such elements and spaces regarding their use and conservation as heritage values, but also in relation to their utility. These are very sensitive elements in some cases that must be left to future generations, but we also know that, although conservation is ideal, if there is no utility or a way to use them, their deterioration can be more pronounced and not reach future generations. Among their potential uses are cultural services, where education again emerges as a service of maximum social utility, while geotourism is a resource that must be managed to fulfill the services of recreation, aesthetic enjoyment, or inspiration without compromising the places and geomorphological elements.

This book is a valuable contribution that connects with the knowledge that has been accumulating for decades about natural and geomorphological heritage. It incorporates this knowledge into the understanding of Castilla y León, where the author's work is located, as well as into national or international knowledge through innovations in methods applied to education or geotourism. Rosa María Ruiz-Pedrosa provides a detailed and precise description of the natural heritage. Her study includes fieldwork in areas closely linked to the author, where the sense of place is evident, as well as the intelligent application of both classic and innovative methods and techniques that form the basis of the research conducted for the book. The author's work is useful for those interested in and visiting the studied areas from a geographical, geomorphological, and educational perspective. It is also of particular interest to natural space managers, especially those where relief and landforms are essential to their landscapes and interact with territorial organization. Additionally, it is valuable for those who want to better understand the mountains of Ávila or Soria, as it delves into the entirety of their nature. The author presents a rigorous work, conducted with the necessary intelligence to unravel the reasons for the importance of relief and the most significant elements, and how their presence can be evaluated to provide current utility without disregarding their meanings or past uses. All of this can be found in Rosa María Ruiz-Pedrosa's well-organized book, the result of reflection, with theoretical development and its application in specific cases as an experiment that reveals the weaknesses of theoretical proposals and fieldwork. The goal is to enhance both geomorphological knowledge and the evaluation and proposed uses outlined in the book, particularly those focused on education at different educational levels, with a geographical perspective that combines the territory with the uses and exploitation of its components, in this case, geomorphological heritage.

Department of Geography, Pangea Research Group
University of Valladolid,
Valladolid, Spain

Enrique Serrano

# Acknowledgements

To University of Valladolid, which has funded my PhD with a Pre-doctoral Contract (Pre-doctoral Contracts of the University of Valladolid, 2020)

To my professors at the Department of Geography of my university, for teaching me the multiple and beautiful faces of Geography

To Jorge Rabassa and Andrea Brody-Barre for their encouragement and support in the publication process of this book

To Enrique Serrano for his invaluable support, both academically and personally, his endless generosity, and for instilling in me his passion for research and mountains

To my colleagues at the Pangea Research Group, from whom I always find support, inspiration, and from whom I always learn a lot

And most importantly,

To my mother and my brother

To my family from Ávila

To my friends

And to Mario

Rosa María Ruiz-Pedrosa

Valladolid

**Competing Interests** The author has no competing interests to declare that are relevant to the content of this manuscript.

# Contents

# 1 Introduction

This book focuses on three main concepts: geoheritage, geographic education, and geotourism. Geomorphosites, which are part of geoheritage, hold significant scientific and heritage value, necessitating their dissemination through geographic education and geotourism. The patrimonialization of the abiotic environment converges in Natural Protected Areas (NPAs), where protection entails planning and management of their uses, albeit not always fulfilled. This work, which is a modified translation of a doctoral thesis (Ruiz-Pedrosa, 2024), presents action proposals regarding geomorphological heritage and geomorphosites in two NPAs in Castilla y León based on geographic education (educational potential of geomorphosites and dissemination of geomorphological heritage and its relation to the territory and cultural heritage) and geotourism (diversification of tourism offerings through the design of georoutes and planning and management of public use around geomorphological heritage). It is necessary to review the state of the art on these topics, defining geoheritage and exploring it through formal and nonformal geographic education and geotourism. Additionally, there is also a review of the main methods for inventorying and assessing geodiversity, geosites, and geomorphosites.

## 1.1 Geoheritage

The term geoheritage is commonly used in academic and research works today, often linked to geodiversity, landscape, and conservation. We will define what we mean by geoheritage and geomorphosites, its relationship with geodiversity and landscape, the main conservation initiatives and regulations for abiotic Natural Heritage, as well as recent proposals for digitizing geoheritage. The approach is always integrative, as Geography itself is, highlighting the importance of geomorphology, which is often overlooked.

### 1.1.1 Geoheritage and Geodiversity

Some authors argue that the word and concept of geodiversity were introduced in 1993, shortly after the agreement on the Convention on Biological Diversity at the Earth Summit in Rio de Janeiro in 1992 (Brilha et al., 2018; Gray, 2018), as a way for geoscientists to assert themselves in response to the increasing focus on the concept of biodiversity. Geodiversity has been considered the abiotic equivalent of biodiversity since then. However, Serrano and Ruiz-Flaño (2007) pointed out that the term was previously coined by the Argentine geographer Federico Alberto Daus in the 1940s, referring to geographic diversity as the cultural diversities of geographical space and the territorial complexities at different scales related to human habitats.

Wilson introduced the scientific term biodiversity in 1988 to define the biological diversity of living organisms on Earth (Wilson, 1992). The definition of biodiversity shows a clear and precise hierarchy of levels (genes, species, and ecosystems), while geodiversity has shown a conceptual weakness that has left it adrift in various fields (Serrano & Ruiz-Flaño, 2007).

Since the 1990s, a naturalistic conception of geodiversity has emerged, distinct from the concept of biodiversity and independently coined by various geoscientists. This includes geodiversity as geological diversity (Sharples, 1995), geodiversity based on geological phenomena or environments (Nieto, 2001; Stanley, 2001), and a broader conceptual view of geodiversity (Gray, 2008; Kozlowski, 2004; Serrano, 2002), leading to the definition of geodiversity as "the variety of abiotic nature" in 2004.

According to Gray (2008), the components of geodiversity include Earth's history, tectonics, minerals, rocks, sediments, fossils, landforms, geomorphological processes, and soils. Kozlowski (2004) adds surface waters (springs, wetlands, lakes, rivers), and González-Trueba (2007) suggests that seas, oceans, and the physical elements and processes


R. M. Ruiz Pedrosa, *Geomorphological Heritage as Didactic and Geotouristic Resources*, Geoheritage, Geoparks and Geotourism, https://doi.org/10.1007/978-3-032-06205-5_1

within them should be included. In 2019, Gray presented a more comprehensive revision of the term geodiversity, defining it as "the natural range (diversity) of geological (rocks, minerals, fossils), geomorphological (landforms, landscapes, topography, physical processes), edaphic, and hydrological characteristics. It includes their assemblages, structures, systems, and contributions to landscapes" (Gray, 2019).

Natural diversity was introduced in the 1980s as a way to connect biotic and abiotic elements, leading to the spatial concept of "habitats." The European Union defines habitats as "terrestrial or aquatic areas differentiated by their geographic, abiotic, and biotic characteristics, whether entirely natural or semi-natural" (Habitats Directive, 92/43/EEC). Habitats include both abiotic and spatial components. The variety of abiotic elements that form habitats can also be referred to as geodiversity. Framing the concept of geodiversity in this context is of particular interest as it reflects an understanding of natural diversity and links the concept to the development of conservation policies and management of protected natural areas and Natural Heritage at national and supranational levels (Serrano & Ruiz-Flaño, 2007), which will be further explored in Sect. 1.2.

Today, the term geodiversity is widely used in Europe. In 2013, the International Union for Conservation of Nature (IUCN) established a Specialist Group on Geoheritage that "provides specialized advice on all aspects of geodiversity in relation to protected areas and their management."

Geodiversity is often discussed imprecisely, being used interchangeably with terms such as "geoheritage," "earth heritage," or "geoconservation" (Serrano & Ruiz-Flaño, 2007; Gray, 2018). However, we believe that geodiversity is not synonymous with geoheritage, and the aim here is to determine the relationship and difference between the two terms.

When it comes to heritage, UNESCO has been advocating for the internationalization of heritage since the end of World War II. In 1964, the Venice Charter created the first catalogue of World Heritage Monuments. It was only after the UNESCO Convention in Paris in 1972 that Natural Heritage began to be recognized alongside historical and artistic heritage. In Spain, Natural Heritage had a landscape perspective, focusing on "geological aesthetics" until the 1930s. In the 1960s and 1970s, more emphasis was placed on the biological aspect, and only recently has attention shifted toward abiotic elements and territories. In 2007, the Natural Heritage and Biodiversity Law (Law 42) was approved, which will be analyzed in Sect. 1.1.4 on Geoheritage and conservation.

A widely accepted stance among the scientific community working in geoheritage is that of Brilha (2016), who argued that geoheritage can be divided into geosites and elements of geoheritage. Geosites are places with in situ presence of geodiversity elements of high scientific value, while elements of geoheritage are those that, despite being displaced from their natural location, maintain high scientific value (e.g., minerals, fossils, and rocks available for research in museum collections). In addition to scientific value, both in situ and ex situ geoheritage can also have educational, aesthetic, and cultural value, which justifies their necessary use by society (teaching/learning, tourism, leisure, etc.) (Brilha, 2016).

Therefore, geoheritage could be defined as those places or elements with specific and recognized scientific value. Thus, geodiversity and geoheritage are not analogous terms, as a place with high geodiversity does not necessarily contain geoheritage, as the variety of abiotic elements does not necessarily imply that these elements have high scientific value. Similarly, geoheritage will always have recognized scientific value, which does not necessarily imply the diversity of abiotic elements. This geodiversity–geoheritage relationship has been advocated by various authors (Serrano & Ruiz-Flaño, 2007; Brilha, 2016; Brilha et al., 2018; Gray, 2018).

Currently, the way to assess the value of the natural environment is referred to as the ecosystem services approach. However, this terminology is considered unfortunate by various authors (Gordon et al., 2012; Gordon & Barron, 2013; Gray, 2011), as it focuses on biotic nature and overlooks abiotic elements. Ecosystem services are the benefits (goods and services) that society derives from nature and must be managed sustainably to ensure they remain available for future generations. In the case of nonliving nature—abiotic elements—these benefits can be referred to as "abiotic ecosystem services" (Gordon & Barron, 2011).

Gordon and Barron (2011) and Gray (2011) established a classification and evaluation system for the ways in which abiotic nature benefits society. Gray (2011) also identified an additional category called "knowledge services," which include benefits derived from understanding Earth's processes and history, the evolution of life, and the origin of landforms through research in geomorphology and geology (Gray, 2011). A more integrated view of Geography would also include cultural services.

Gray (2011) identified 25 distinct geodiversity services, classified into 5 categories defined by Brilha et al. (2018):

- *Regulation services*: natural processes that regulate the environment.
- *Support services*: processes that sustain natural environments.
- *Provisioning services*: natural materials used by society.
- *Cultural services*: intangible elements of the natural environment that benefit society spiritually or culturally.
- *Knowledge services*: benefits derived from understanding Earth processes, the evolution of life, and the origin of landforms through research in geomorphology and geology.

In some cases, the benefits of geodiversity are direct, while in others they are achieved through the influence that geological, soil, hydrogeological, geomorphological, or pedological factors and processes have on the landforms and biodiversity they host. Geodiversity is not only crucial for maintaining species and habitats but also plays a fundamental role in the health and well-being of people (Gordon et al., 2012).

The current status of abiotic services within the ecosystem services approach is inconsistent and confusing, as are ecosystem services themselves. Excluding abiotic services risks undervaluing the contribution of nature as a whole to human well-being (Brilha et al., 2018).

The real challenge in addressing these concepts is breaking down the barrier between geodiversity and biodiversity and between scientists from both disciplines, fostering collaboration to achieve a more holistic science of the Earth system. When assessing ecosystems, it is essential to analyze and evaluate the contribution of geodiversity in both monetary and nonmonetary terms to ensure that natural capital is not undervalued through omission. Demonstrating and communicating how investing in the natural environment can lead to increased provision of services and benefits for society is crucial (Gordon et al., 2012).

### 1.1.2 Geomorphological Heritage and Geomorphosites

Within geoheritage, geosites stand out, which are those places in the abiotic natural environment that have recognized scientific value in situ (visible in the field) and are of particular importance for understanding the Earth's evolution (Reynard & Coratza, 2013). Among geosites, we can distinguish several groups according to their scientific interest: structural, paleontological, hydrogeological, sedimentological, etc. When the geomorphological value is high, we can then talk about Geomorphosites, which can be defined as portions of the geosphere that are of particular importance for understanding Earth's history, are spatially delimited, and clearly distinguishable from their surroundings (Reynard, 2009). And when geosites and geomorphosites are identified and recognized, the desirable outcome is the process of heritage designation that turns them into part of the geoheritage (Fig. 1.1).

Geomorphosites can be individual objects or systems that showcase active or passive geomorphological processes. Active geomorphosites, such as river systems or active volcanoes, demonstrate ongoing geological processes, while passive geomorphosites serve as a record of past processes and have particular heritage value as a memory of Earth's evolution, landscape history, and climatic variations (Reynard & Panizza, 2005).

The recognition of the heritage value of geomorphology led to the establishment of the Geomorphosites Working Group within the International Association of Geomorphologists (IAG) in 2001 during the fifth IAG International Conference. Since then, the group has served as a key platform for the development of a specific research field on geomorphological heritage within the scientific community (Reynard & Coratza, 2013).

According to Reynard and Giusti (2018), geomorphological heritage can be considered as the set of landforms worthy of protection and transmission to future generations. Geomorphological heritage is part of geoheritage, which in turn is a component of Natural Heritage (Reynard et al., 2009). It is closely related to cultural heritage, influencing human uses and locations, and it is an important natural and cultural resource within the overall heritage of a territory (Serrano et al., 2020).

As highlighted by Serrano et al. (2020), the importance of geomorphology as Natural Heritage stems from its role as infrastructure for habitats and landscapes, supporting ways of life and cultural elements with total continuity on the Earth's surface. While relief, such as landscape, is part of a continuous whole, there are forms on the Earth's surface that are relevant for various reasons, such as landscape, cultural, religious, or simply for interpreting the landscape and enjoying nature due to their uniqueness. Geomorphology often explains the natural organization of the environment, as well

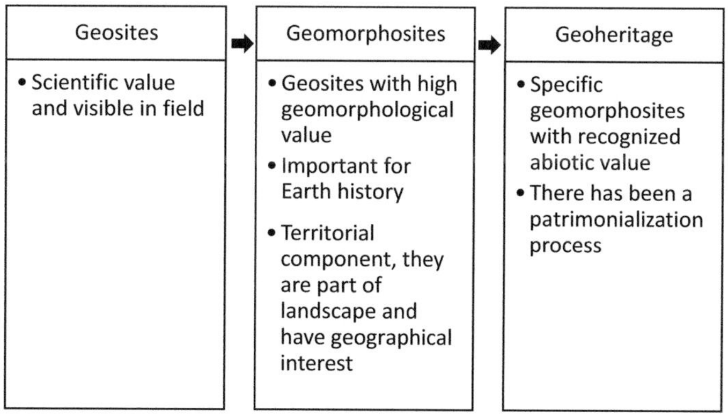


**Fig. 1.1** Conceptual relationship between geosites, geomorphosites, and geoheritage. Self-made

as the cultural organization through human struggle against unsuitable processes and forms for ways of life, and the human and cultural acceptance of relief from humanistic, identity, or aesthetic perspectives. Additionally, geomorphological elements and systems influence both natural and human processes and structures, are visible, have spatiality, and are at risk of degradation or disappearance due to human activity.

The value of geomorphological heritage comes from its inherited nature, the uniqueness of certain landforms, and the worldview of cultures that use, occupy, and interpret the terrain in a given territory. It can have value at different scales, from planetary to continental, regional, or local, and for different societies sharing the territory, so it varies greatly depending on the territorial scope in which it is located (Serrano et al., 2020).

Geomorphological heritage is often found within Natural Protected Areas (NPA). The first NPAs emerged in Spain and around the world from the idea of admiration and enjoyment of nature, a Natural Heritage that should be delimited, protected, and passed down from generation to generation. Thus, the existence of outstanding spaces that stand out for their beauty and value began to be recognized, and their preservation was acknowledged as a common societal goal. The result is the declaration of strictly delimited NPA to provide a legislative framework, restricting uses to ensure the conservation of the value of biotic and cultural elements, with greater emphasis than abiotic elements.

In NPAs, geomorphological elements have a territorial and landscape component that sets them apart from geological sites of interest. They are valued within the context of NPAs for their cultural and scenic significance, attracting travelers, hikers, and tourists seeking to understand the territory they are visiting. These elements become a valuable territorial resource. Local communities are also drawn to familiar forms that have been part of their lives yet may not be fully recognized for their cultural significance. The relief and landforms in NPAs define the essence of their landscape, shaping ecosystems, habitats, and territorial organization (Serrano et al., 2020).

### 1.1.3 Geoheritage and Landscape

The heritage is often divided into Natural Heritage and cultural heritage, although this distinction is erroneous and confusing, as argued by Martínez de Pisón (2012). This distinction is evident in Geomorphosites (Panizza & Piacente, 2009; Reynard & Coratza, 2016; Reynard & Panizza, 2005) and in the concept of geological or geomorphological landscape (Reynard & Giusti, 2018).

The term landscape has a broad polysemic field that is used in various research fields, from ecology and Geography to fine arts and architecture (Reynard, 2009; Martínez de Pisón, 2012). A global view of heritage leads us to the landscape as the meeting point of natural and cultural heritage (Martínez de Pisón, 2010). Therefore, we consider that every landscape has heritage value, as it serves as both a support (territory) and content (culture).

According to Reynard and Giusti (2018), a landscape is a mix of tangible and intangible elements. Tangible elements such as geodiversity, biodiversity, and human activity form the landscape as a complex object, both physical and anthropogenic. On the other hand, the intangible characteristics of the landscape, such as emotion, aesthetics, senses, symbols, phenomena, and individual feelings, form the landscape as a subject, perceived through the observer (Reynard et al., 2009).

Martínez de Pisón (2010) also referred to this duality of the landscape, stating that "the relationship with the landscape is aesthetic, scientific, sentimental, and moral." According to this author, one must first understand the geographical configuration of the landscape, its structure, and territorial morphology, without forgetting its subjective values, which constitute its content and are essential to understanding its meanings and its social systems and cultural references (Martínez de Pisón, 2012). The subjective aspect of the landscape is intertwined with the objective, shaped by it, and culturally reconfigured.

In the same vein, Serrano (2012) defined landscape as a space where different material, natural, and human elements come together, which is dynamic and manifested on the Earth's surface, diversifying the territory (Fig. 1.2). Serrano (2012) argued that landscapes reflect the territorial structures of their history up to current uses, or, in the case of abandonment, of natural elements. As shown in the author's proposed scheme, the landscape is a consolidator of both natural and cultural heritage, of very diverse nature, which requires an integrated management of landscapes to achieve their conservation, enrichment, and sustainable use and management (Serrano, 2012).

This inseparable relationship between geomorphology and cultural elements, as geomorphological features and cultural assets often integrate to form a single landscape, has led to the study of cultural geomorphology (Panizza & Piacente, 2009). These authors define cultural geomorphology as "the discipline that studies the geomorphological components of a territory that embody both a cultural feature of the landscape and its interactions with cultural heritage of archaeological, historical, architectural, etc. types," which can be studied in five phases:

1. Analysis of the physical environment where the cultural asset is located;
2. Analysis of the geomorphological factors that influence the location of the cultural asset in question;

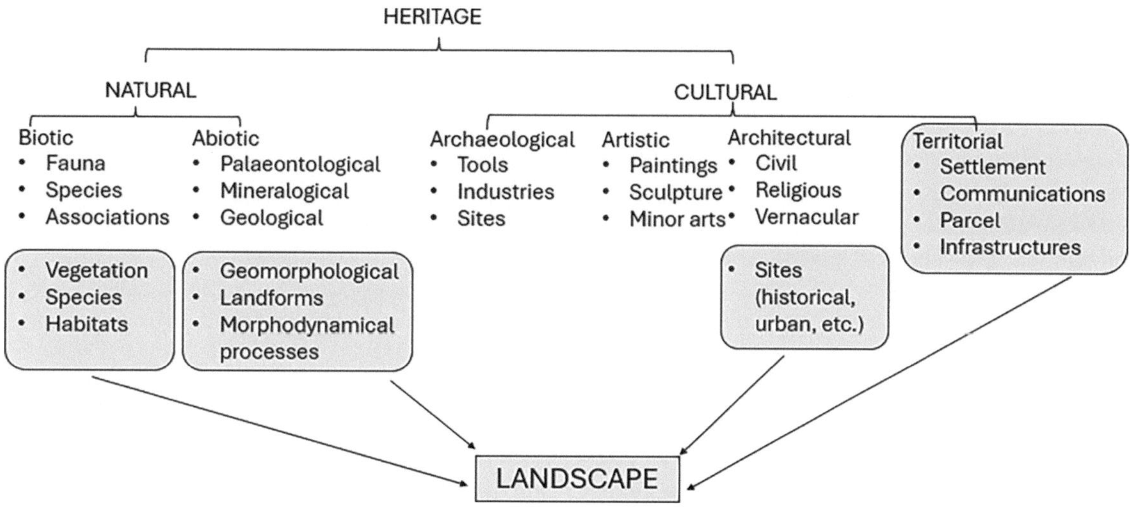


**Fig. 1.2** Relationship between landscape and natural and cultural heritage. Translated from: Serrano (2012)

3. Evaluation of geomorphological risks threatening the cultural asset;
4. Analysis of possible environmental impacts related to the use of the cultural asset, such as tourism;
5. Proposal of management measures.

It is undeniable that geomorphological heritage embodies a series of cultural values, with the landscape being the point of encounter and perception of the alterations that culture, especially historical human action, exerts on the territory, and how geomorphology influences the configuration of cultural elements, such as population settlements, economic activities, and so on. Therefore, one cannot be understood without the other, leading to the assertion that geomorphological heritage has a dual reading, between its intrinsic natural elements and the cultural uses and representations that have been given to it.

### 1.1.4 Geoheritage and Conservation

Geoheritage plays a crucial role in science and society, serving as a valuable resource for research and providing insights into Earth's history and human occupation. Therefore, it is essential to conserve and protect these sites as a global responsibility to ensure their preservation for future generations.

Ensuring the accessibility and proper conservation of geosites is vital for their use in scientific research, education, geotourism, and other purposes. These sites are often at risk from human activities and natural processes, highlighting the need for proactive conservation efforts.

Several authors (Crofts et al., 2015; Fuertes-Gutiérrez et al., 2016; Gray et al., 2013) have recently inventoried the threats, both natural and human, faced by geosites, including the following:

1. Civil engineering works, mainly construction of infrastructure and housing, which can lead to substrate loss.
2. Invasion and degradation caused by vegetation, resulting in loss of geological exposure or sensitive elements.
3. Quarrying or mining exploitation, which can lead to the destruction of caves, landforms, etc.
4. Damage to geomorphological features or processes as a result of, e.g., mineral extraction, river management plans, or forestry.
5. Damage to static or delicate geomorphological elements due to natural erosion.
6. Accidental losses or damages due to lack of awareness of the importance of geoheritage and the need for its conservation.

Many geosites have been and are being damaged or destroyed (Gray, 2019), particularly in developing countries but also in developed ones, due to lack of knowledge and awareness about geoheritage and the limited importance given to disciplines such as Geography. This lack of awareness is mainly due to the absence of a unified method that can be used internationally for inventorying geosites, with more emphasis placed on protecting biodiversity habitats rather than geodiversity. While geosites may seem more robust at first glance, many of them are subjected to multiple threats, as previously noted.

Gray et al. (2013) stated that "the goal is to enable local communities to take ownership of their geological and other heritage by protecting it, promoting it, and in doing so, deriving some sustainable economic benefit from it." Here, we believe it is more appropriate to talk about geoheritage, given its integrative nature, considering all the abiotic elements of the Earth and their relationship with the territory, in which human intervention plays a role, resulting in more effective geoconservation.

Regarding regulations related to the conservation of geoheritage, three levels of actions should be distinguished: international regulations, European Union community regulations, and national and specifically regional regulations of the NPAs under study.

#### 1.1.4.1 UNESCO: Global Geoparks and World Heritage List

The establishment of the **International Geoscience and Geoparks Programme by UNESCO** in November 2015 signifies global recognition of geological heritage by a prestigious international institution respected by the majority of countries (Brilha & Reynard, 2018). The Global Geoparks Network comprises 195 Geoparks in 48 countries (as of July 26, 2024). On the UNESCO Global Geoparks website, you can find the definition, values, and characteristics of these Geoparks:

> *UNESCO Global Geoparks are single, unified geographical areas where sites and landscapes of international geological significance are managed with a holistic concept of protection, education, and sustainable development. A UNESCO Global Geopark uses its geological heritage, in connection with all other aspects of the natural and cultural heritage of the area, to enhance awareness and understanding of key societal issues such as sustainable use of natural resources, mitigation of climate change effects, and reduction of risks related to natural disasters.*

Therefore, in Geoparks, geological elements are the central focus of declaration and protection, around which links can be established between that geological heritage and any potential natural, cultural (tangible and intangible) heritage that may exist in the area. The concept of a Geopark also includes protection, education, and sustainable development.

> *By raising awareness of the importance of the area's geological heritage in history and current society, UNESCO Global Geoparks instill a sense of pride in the local population and strengthen their identification with the area. This stimulates the creation of innovative local businesses, new jobs, and high-quality training courses, while generating new sources of income through geotourism, all while protecting the geological resources of the area.*

Geoparks are presented as a local development resource for their population through sustainable geotourism, following a bottom-up approach methodology that empowers local and regional stakeholders. However, despite the positive intention of declaring an area as a Geopark for conservation purposes, there is no formal legislative framework in place, leaving its protection to local, regional, or national laws. This lack of legal protection undermines its effectiveness.

In Spain, there are 16 UNESCO Geoparks (Table 1.1), with only one located in the Castilla y León region. The Geopark of Las Loras in the province of Burgos was declared in 2017 after 13 years of work by Jose Ángel Sánchez and his team at ARGEOL.

In 2004, Las Loras was declared a Biological Reserve, and guided georoutes, volunteer programs, and a variety of activities were initiated to strengthen the connection between the local population and their geoheritage. According to the director of the geopark (lecture by Jose Ángel Sánchez on November 9, 2021 at the University of Valladolid), this geopark serves as a space for socioeconomic development and social revitalization in a particularly depopulated area. This is evidenced by geopark's support for geotourism activities, intangible heritage recovery, agroecology, social par-

**Table 1.1** UNESCO Geoparks in Spain (as of July 26, 2024)

| Geoparks | Province | Declaration year | Surface (km$^2$) | Geological motivation |
|---|---|---|---|---|
| Sierras Subbéticas | Córdoba | 2006 | 320 | Ammonites, karst |
| Cabo de Gata-Nijar | Almería | 2006 | 472 | Vulcanism, coral reefs |
| Sobrarbe-Pirineos | Huesca | 2015 | 2202 | Pyrenees |
| Costa Vasca | Guipúzcoa | 2010 | 90 | Flysch, karst |
| Sierra Norte de Sevilla | Sevilla | 2011 | 1774 | Sierra Morena |
| Villuercas-Ibores-Jara | Cáceres | 2015 | 2500 | Appalachian relief |
| Cataluña Central | Barcelona | 2015 | 1250 | Monserrat mountain |
| Molina and the AltoTajo | Guadalajara | 2015 | 4300 | Karst |
| Island de El Hierro | El Hierro | 2015 | 268 | Vulcanism |
| Lanzarote and Archipiélago Chinijo | Lanzarote | 2015 | 2500 | Vulcanism |
| Las Loras | Burgos | 2017 | 960 | Folded relief |
| Origens | Lérida | 2018 | | Catalan Pyrenees, dinosaur remains |
| Montañas do Courel | Lugo | 2019 | 578 | Palaeozoic, glaciation |
| Maestrazgo | Teruel | 2020 | 2622 | Folded relief |
| Geoparque de Granda | Granada | 2020 | 4722 | Badlands |
| Cabo Ortegal | A Coruña | 2023 | 799 | Variscan Orogeny |

ticipation campaigns, and educational projects such as Erasmus+, in which it has participated with other geoparks. Being part of the UNESCO network allows Las Loras to engage in international projects, connect with other geoparks, and promote its development.

Geoheritage is also represented in another UNESCO initiative: the Convention concerning the Protection of the World Cultural and Natural Heritage, established in 1972, which **introduced the term "Natural Heritage" for the first time**. Currently, there are 1210 properties on the World Heritage List (as of July 2024), with 231 being natural and 40 being mixed (cultural and natural), compared to 939 cultural properties. Spain has three natural properties and two mixed properties. The natural properties include Garajonay National Park, Doñana National Park, and Teide National Park. The mixed properties are Pyrenees-Monte Perdido (spanning the border between France and Spain) and the Island of Ibiza.

The main international challenge is to raise awareness about geoconservation in all relevant international institutions, with the hope that this movement can then be implemented from top to bottom in national and local public administrations (Brilha & Reynard, 2018).

#### 1.1.4.2 European Union Regulations: Natura 2000 Network

The environmental regulations of the European Union are considered to be among the strictest in the world. However, when consulting European legislation on the environment, it can be noted that there is no specific regulation regarding geoheritage. The two main directives focus on the conservation and protection of biodiversity, specifically birds and habitats: Council Directive 79/409/EEC on the conservation of wild birds and Council Directive 92/43/EEC on the conservation of natural habitats and wild fauna and flora. This leaves abiotic elements of nature and natural spaces as a whole completely overlooked and unprotected, subject to national and regional regulations.

This simplistic and exclusionary perspective on the environment poses a challenge for geoconservation in European countries, as there is no available funding to protect sites or conduct research. Therefore, the challenge is to establish a European Union conservation policy that considers abiotic elements and geoconservation, promoting initiatives that may arise from its member countries and leading the way in protecting not only biodiversity but also geodiversity and, ultimately, geoheritage.

#### 1.1.4.3 National and Regional Regulations: Spain and the Castilla y León Region

The legislation regarding geoheritage and geoconservation at the national level varies greatly. Some countries have well-defined legal frameworks for the protection of geological heritage, with national inventories, while others do not consider abiotic elements at all. In general, conservation of flora and fauna, and therefore biodiversity, takes precedence over geodiversity in most countries. This lack of focus on geodiversity is not surprising given the absence of an international legislative framework as a reference. National environmental policies often prioritize other interests, such as economic ones, over conservation and protection efforts. In Spain, the basic regulatory framework for the conservation of Natural Heritage is defined by two laws: **Law 30/2014 on National Parks** and **Law 42/2007 on Natural Heritage and Biodiversity**.

The Law 30/2014, of December 3, on National Parks of 2014 established in its Title II that the objective of the law is to *"establish the basic legal framework to ensure the conservation of national parks and the network they form, as well as to establish collaboration and coordination instruments [...], returning capacity and protagonism to territorial actors"* and that the objective of National Parks is nothing other than the conservation of their natural and cultural values. The rest of the titles determine the corporate image of the network, the basic regulatory framework, planning instruments, intervention agents, territorial development, projection, and social participation, infractions, and sanctions. It is not the purpose here to analyze this Law in depth, as none of the selected study areas are affected by the National Park protection figure. In our autonomous community, the only National Park is Picos de Europa, whose socio-economic influence area includes the municipalities of Posada de Valdeón and Oseja de Sajambre in León.

On the other hand, Law 42/2007, of December 13, on Natural Heritage and Biodiversity of 2007 established the catalogue of habitats in danger of disappearance and the classification of protected areas into Parks, Nature Reserves, Marine Protected Areas, Natural Monuments, and Protected Landscapes, as well as the protected areas of the Natura 2000 Network.

Title III of this law is entirely dedicated to providing guidelines for the conservation of biodiversity; however, the same does not apply to geodiversity and its conservation, which is not considered. Once again, all efforts are focused on inventorying, defining, and protecting biotic elements, while abiotic elements are either sidelined at best or not considered at all. For example, in Article 2 of the Preliminary Title, the objectives of this law are established as *"the conservation and restoration of biodiversity and geodiversity through the conservation of natural habitats and wild fauna and flora."* Geodiversity is mentioned only to then discuss habitats and flora and fauna, without considering abiotic elements.

In fact, in this law, geodiversity is not understood as geographic diversity, as argued here, but is defined as *"Geodiversity or geological diversity: variety of geological elements, including rocks, minerals, fossils, soils, landforms, geological formations and units, and landscapes that are the product and record of Earth's evolution.*"

The Annex VIII of this law, titled Geodiversity of the Spanish territory, is particularly interesting as it lists the most representative geological units and globally relevant geological contexts in Spain. The term "geological" used in the annex is somewhat ambiguous, as it includes unique deposits and landforms of fluvial, lacustrine, and aeolian origin, as well as coastal and littoral deposits and landforms. Therefore, the term "geomorphological" should be included, as these features are not solely geological but also geographical, specifically in the field of geomorphology.

The environmental regulations of the two NPAs under study will be thoroughly analyzed later (see Sect. 3.2.1 Current Environmental Legislation in the Sierras de la Paramera and la Serrota and Sect. 4.2.1 Current Environmental Legislation for the P.N. Cañón del Río Lobos). However, since both are located in the region of Castilla y León and, therefore, subject to its environmental regulations, it is essential to refer to the regional legislation, specifically the Network of Natural Areas of Castilla y León and **Law 4/2015, of March 24, on the Natural Heritage of Castilla y León**.

Law 8/1991, of May 10, on Natural Areas of the Community of Castilla y León was pioneering at the time in creating the Network of Natural Areas, a concept that surpassed that of individual protected natural areas. This law establishes the Plan for NPAs of Castilla y León, which is purely indicative. The initial intention was to start the planning process for each area included in the Plan within 1 year of the law coming into effect, but in some cases, this process was never initiated.

Of the two study areas in this work, the Sierras de la Paramera and Serrota were proposed for designation in this law. However, there has been no progress in their environmental legislation since then, as they are only part of the Natura 2000 network and not the NPAs network of Castilla y León (see Sect. 4.2.1). The Cañón del río Lobos was already designated as a Natural Park in 1985 before this law, but it is also in a legislative process of stagnation, as its Natural Resources Management Plan has not yet been approved (see Sect. 4.2.1).

In the Law 4/2015, of March 24, on the Natural Heritage of Castilla y León, the Network of NPAs is established, following the previous law of 1991. It is comprised of three complementary networks (Fig. 1.3):

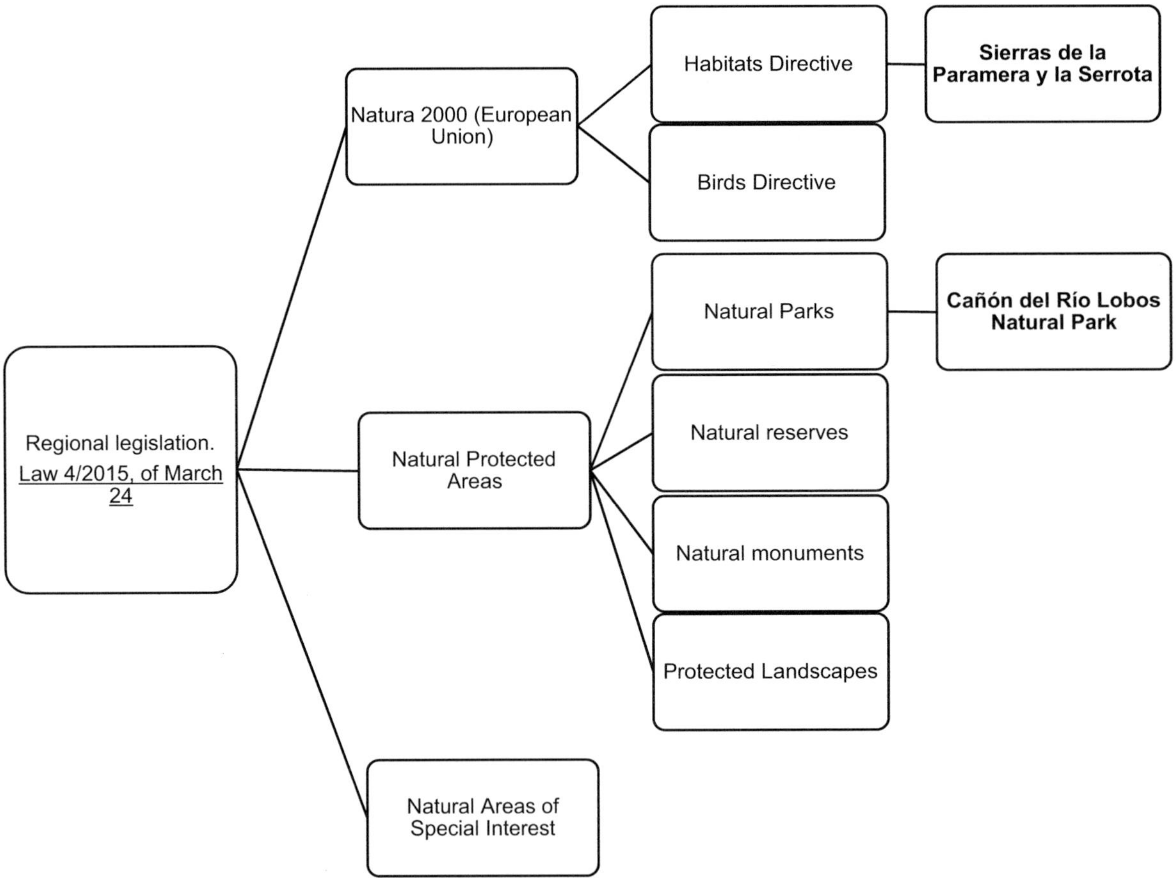


**Fig. 1.3** Environmental protection figures in the region of Castilla y León. (Prepared by the authors)

1. **Natura 2000** is the core network for biodiversity protection established by the EU Habitats Directive, consisting of Special Conservation Areas, Special Protection Areas for Birds, and Sites of Community Importance. **Sierras de la Paramera and la Serrota** belong to this network as a Site of Community Importance.
2. **Natural Protected Areas Network** includes NPAs, which align with Natura 2000 and go beyond the obligations of the Birds and Habitats Directives by implementing socioeconomic development systems based on conservation. This network follows the principles of the 1991 network and includes four categories:
   (a) Parks, which can be national, regional, or natural, such as the **Natural Park of the Cañón del Río Lobos**.
   (b) Nature reserves.
   (c) Natural monuments.
   (d) Protected landscapes.

   Planning instruments for these NPAs include Plans for the Management of Natural Resources, Management and Use Plans, and conservation regulations, which establish zoning, permitted activities, and management principles.
3. The Network of **Natural Areas of Special Interest** is designated to protect unique elements of Natural Heritage, regardless of location, with categories including:
   (a) Public utility forests.
   (b) Protective forests.
   (c) Wetlands of special interest.
   (d) Special interest livestock routes.
   (e) Natural recreational areas.
   (f) Micro-reserves for flora and fauna.
   (g) Notable trees.
   (h) Geological or paleontological sites of special interest. This category can be challenging as paleontological heritage is regulated by the Cultural Heritage Law under archaeological heritage.

   This regional law focuses all its efforts on the Natura 2000 Network and, therefore, on the protection of biodiversity, with a strong emphasis on habitat protection. Abiotic elements are once again relegated to the background, represented only by the Network of Natural Protected Areas and geologically significant sites. However, it is misleading to refer only to geological sites, as many of these locations are actually geomorphological.

### 1.1.5 Geoheritage Digitization

In recent decades, the development of new digital technologies has significantly influenced scientific practices in the field of geoheritage. Geographic Information Systems (GIS) have played a crucial role in the development of new visualization methods, evaluation, and mapping (Santos et al., 2018).

Numerous projects have been carried out in recent years using new technologies in the cultural heritage and/or geoheritage field, driven by various purposes such as enhancing accessibility to heritage sites, documenting changes in geosites due to climate variability, and studying the vulnerability of heritage sites to visitor influx (Cayla, 2014; Cayla & Martin, 2018; Cerrillo-Cuenca et al., 2021; Santos et al., 2018).

Geovisualization can be useful for managing geoheritage as it facilitates scientific study, monitoring, and management. It also allows new users, who were previously excluded from physical visits to these sites, to enjoy the geoheritage, such as people with physical limitations (Cayla & Martin, 2018). However, along with the growth of digital treatment of geoheritage, challenges arise, such as production costs, rapid technology obsolescence, the need for specialized personnel in data collection and processing, and in creating graphic outputs for dissemination. One of the main challenges facing these new solutions is network coverage, which is often irregular in natural spaces, and there may also be roaming conflicts if the visitor is in a foreign country.

One of the most relevant graphic outputs today is mobile applications. Taking advantage of recent advances in geolocation and web mapping, georeferencing and geoheritage cartography are increasingly using mobile solutions for data acquisition and visualization (Cayla, 2014). "Geovisualization" through real-time tracking or 3D visualization has not only benefited from advances in computer capabilities but also from the acceleration of digital data transfer through networks (Cayla, 2014).

The digitization of geoheritage is particularly useful for education and geotourism, as it helps bring scientific knowledge to the public in an intuitive, accessible, and visual way, making it of special interest in a society where the majority of the population uses a smartphone daily. However, there is a generational gap, as younger and adult populations are more adept at using new technologies, while older populations may struggle to engage with them. When it comes to education, the target audience is primarily young people of school age.

In this way, technology can serve as a support for teaching geographical knowledge, whether in the field (e.g., through scanning QR codes on a field trip) or in the classroom, using a web viewer or a 3D model to bring students closer to the territory when field trips are not feasible due to school resources, teacher training, or scheduling constraints (see Sect. 5.3.3 Geomorphology, Didactics and Field Trips: Case Study in Spanish High Schools). In a subject like Geography, where graphic and cartographic support is essential, these new tools based on Geographic Information Systems can enhance student understanding and motivation.

Following this approach, research projects focusing on virtual reality and field trips are being developed. One of the most relevant projects in Europe is the GeoVT project, aimed at providing a tool for distance learning of geomorphology, natural hazards associated with geomorphological processes, and geoheritage. The project involves five European universities: Stockholm University in Finland, National and Kapodistrian University of Athens in Greece, University of Modena and Reggio Emilia in Italy, University of Caen Normandy in France, and University of Wroclaw in Poland. A multidisciplinary team of geographers, geologists, and educators has created the GEOVT platform, which stores virtual field trips with 360° multimedia, audio, images, and texts. These virtual field trips have proven effective in participatory and interactive teaching of geomorphology, overcoming accessibility limitations, weather constraints, and economic barriers to travel, serving as auxiliary tools rather than substitutes for traditional field trips (Vandelli et al., 2024).

The same applies to geotourism, as geotourists often find it challenging to interpret and understand the landscapes and geomorphological processes in a territory. Traditional guided tours may not be of interest to certain audiences, so the option of using a web application in the field is presented as an alternative for self-taught individuals familiar with new technologies.

An example of successful digitization of geoheritage for tourism is the GeoGuide project carried out by the Institute of Geography and Sustainability at the University of Lausanne in 2013. This project involved creating a mobile geotourism application with educational routes illustrating different natural and urban environments in Switzerland and neighboring countries. The website guides users through the city of Lausanne, highlighting points of interest related to geomorphology, hydrology, history, and urban planning. To date, five guides have been developed: in Lausanne, Val d'Hérens in Valais, Vallon de Nant in the Alps of Vaud, Thonon-les-Bains in France, and finally in Rome (as of July 2024). A doctoral thesis on GeoGuide (Grangier, 2019) demonstrated that this initiative was well-received by the population, facilitating the dissemination and interpretation of natural and cultural heritage.

## 1.2 Geomorphosites and Their Didactic and Geotouristic Use in Natural Protected Areas

Geoheritage is a resource that is relatively new in terms of its economic, cultural, and educational value (Pellitero Ondicol, 2012). Currently, geomorphological heritage has significant potential as a cultural, educational, and socioeconomic resource associated with geoconservation, landscape, and territory (Serrano et al., 2020).

The intrinsic values of an NPA are considered a resource for the entire population, especially for the local population that inhabits them. We have discussed the abiotic services of nature in relation to the value of the natural environment, which can also be applied to NPAs as they are spaces with recognized natural value. NPAs should be resources for their population, providing natural benefits that can be used for territorial development, raising the level and quality of life in the area, while seeking a balance between conservation and traditional or new uses to ensure the settlement of the territory and decent living standards for the inhabitants (Serrano et al., 2020).

In this context, geomorphological heritage is a natural resource that can be used for educational and touristic purposes, as pointed by numerous touristic authors (Aryasa et al., 2017; Bollati et al., 2012; Clara Vasconcelos et al., 2020; Dowling & Newsome, 2017; Fernández Álvarez, 2020; Kubalíková, 2019, 2020; Kubalíková et al., 2021b; Newsome & Dowling, 2018; Orion & Fortner, n.d.; Reynard & Coratza, 2016; Stepišnik et al., 2017; Tormey, 2019).

Geoheritage and geomorphosites are considered the main resources for geotourism and geoeducation activities, which should be closely related and mutually supportive. Geoeducation can contribute to the recognition of geodiversity and geoheritage, impacting tourists' interests in NPAs. Proper development of geotourism can enhance accessibility and availability of geoeducational resources (Kubalíková et al., 2021a, 2021b).

Since the early 1990s, when geotourism began to be discussed, emphasis has been placed on education and interpretation as important tools that can increase awareness of geodiversity and conservation and contribute to the sustainable development of geotourism (Kubalíková et al., 2021a, 2021b). The educational aspect is integrated or reflected in numerous definitions and approaches to geotourism, from early to current holistic approaches (Newsome & Dowling, 2018).

Considering that the main goal of tourists is enjoyment, which is not incompatible with a desire to learn or with some currents of geotourism, and of educators to transmit knowledge, geotourism, and geoeducation share the stage in which they carry out their activities: the geographical space, whether natural, urban, rural, cultural, and so on, with natural spaces being the focus here. When discussing geotourism and geoeducation in NPAs, the space for their activities is limited to the natural environment, therefore, a desire to enjoy or learn about nature. Thus, since they operate in the same environment, they will share a series of common needs (Fig. 1.4):

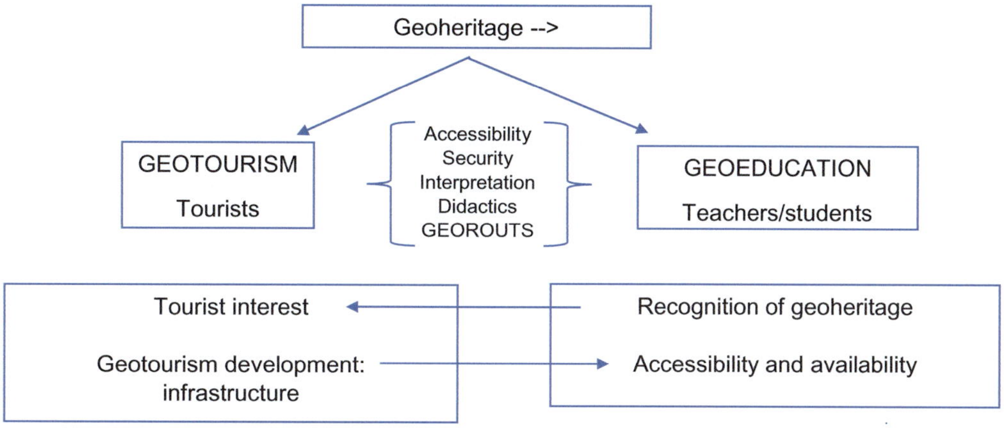


**Fig. 1.4** Relationship between geoheritage–geotourism–geodidactics. (Own elaboration)

- Accessibility: Places should be accessible to tourists, educators, and students to varying degrees.
- Safety is essential and should be considered alongside accessibility.
- There is a need for interpretation and educational application of geomorphosites.

Since the early twenty-first century, we have seen the rise of educational tourism, where visitors have a desire to learn. This trend is particularly evident in Geoparks, where knowledge transfer leads to increased environmental awareness and conservation efforts (Fernández Álvarez, 2020).

One example of this tourism–education union is geotrails, also known as geopaths, which are a way to present geoheritage to the public (Kubalíková et al., 2021a, 2021b). These combine the desire to learn, the experience of living on-site learning in the field, and physical exercise. In order to implement geotrails, a series of conditions must be met:

1. The geoheritage has significant scientific and landscape relevance.
2. The geomorphological significance can be easily understood by visitors without training in geomorphology, geology, or Earth Sciences.
3. There is a low risk of degradation as a result of human activities.
4. Basic accessibility and safety conditions must be met for visitors.

NPAs can be ideal locations for designing and implementing georoutes. Many of these areas already have georoutes, although this is not always the case in the natural areas we are studying. Protected natural areas have the advantage of attracting visitors and providing a space for exploration and continuous learning in the field (Tormey, 2019).

Geomorphological processes are sometimes considered dangerous and not seen as resources for geotourism and geoeducation activities. However, a rational use of these processes can help explain the origin and evolution of landforms, and proper interpretation can help understand the complex relationships between the process and the resulting relief. Understanding and explaining these processes can enhance awareness of potential geological risks and improve the planning of geoeducational, geotourism, and other beneficial local activities. Therefore, it is evident that geomorphic processes (including hazardous ones) should also be considered important resources for geotourism and should be taken into account when planning geotourism and geoeducational activities that are (or should be) closely linked (Kubalíková et al., 2021a, 2021b).

However, geomorphosites are not only resources for their geomorphological processes and dynamics, i.e., abiotic, but also their connections with civilization and culture are of interest for a comprehensive and integrative, essentially geographical understanding of the environment. Thus, once again, the nature–society connection becomes especially important in education and tourism, allowing for the discussion of humanized or anthropogenic landforms, which are unquestionably important for geotourism and geoeducation. This is evident in numerous tourist destinations, including the UNESCO World Heritage sites.

## 1.3 Geographical Education

The valorization of geomorphosites necessarily involves processes of geographic education, understood as the geographic literacy of individuals, equipping them to understand the physical and socioeconomic processes of the territory, the interrelation between them, and the resulting consequences. This can be done through two approaches: **formal education**, with a pedagogical-didactic focus aimed at students and teachers regulated by the educational system; and

**environmental education**, with a dissemination focus aimed at all audiences.

### 1.3.1 Formal Geographic Education: Geography Didactics

Changes in the objectives and methodologies used in formal geographical education have been closely linked to the evolution of geographical epistemology, or geographical thought. Geography has been a science of the Earth, of location and the situation of places, applied in service to humanity since classical Greek times. However, it was not until the nineteenth and twentieth centuries that modern Geography emerged with the founding fathers of Geography: Alexander von Humboldt and Carl Ritter. Geography became established as a science, leading to its institutionalization and the creation of the first national geographical schools and Geography departments in universities.

Geographical societies, such as those in Paris, Berlin, and London, while contributing to the dissemination of geographical knowledge and the incorporation of Geography as an academic discipline, were closely linked to imperialism (Solís Solís et al., 2019). Similarly, in Spain, since the late nineteenth century, Geography education was used for patriotic purposes, aiming to instill a sense of belonging to the state in students (García Álvarez and Marías Martínez, 2002). The Royal Geographical Society in Britain was one of the institutions that greatly contributed to the university study of geography, establishing departments at Oxford and Cambridge in the late nineteenth century.

Among the national schools, the German and French schools stand out. The German school promoted geographical determinism, which argued that the physical environment determined the way of life of societies, with Friedrich Ratzel being its main exponent. The French school promoted a possibilist Geography, in which the environment did not determine societies but conditioned them, offering different possibilities in the way they developed their activities. In 1903, Paul Vidal de la Blache published *Tableau de la géographie de la France* and is considered the beginning of human Geography.

In Spain, the teaching of Geography had a significant development in the early twentieth century, marked by the mandatory field trips (which we will discuss in detail later) and a predominantly naturalistic approach in schools. However, after the Civil War, it was relegated to history (Piñeiro-Peleteiro, 1997). It was not until the 1960s that a methodological renewal of Geography was resumed, with a regional focus following the lines of the French school.

After World War II, and particularly from the 1960s onwards, Geography and pedagogy underwent epistemological and methodological changes. Geography and its teaching had to adapt to the needs of a new world and the demands of a new society, leading to the reform of all study plans. In general, and with some exceptions, academic, ideological, and patriotic approaches were abandoned in favor of Geography as a discipline that prepares students to live in society, understanding the physical environment and its interrelation with humans, and being able to solve the problems that arise from it. Two historical events had a significant influence on this idea: the establishment of UNESCO in 1945 and the Universal Declaration of Human Rights in 1948. UNESCO has considered Geography, from its inception, as a discipline aligned with its objectives. As a result of the relationship between UNESCO and the well-established International Geographical Union, the UGI's Commission on Geography Teaching was created in 1952, renamed in 1968 as the Commission on Geographic Education. In 1965, the first work on methodology for teaching Geography was finally published by UNESCO, advocating for a Geography of globalization that could address the territorial problems of societies and promote mutual understanding among peoples, linking the teaching of Geography to universal ideals such as peace and international cooperation (de Miguel González et al., 2016).

In a political and epistemological context of Geography that is very different, the International Declaration on Geographic Education was published in 1992. Moving from political utopia to an attempt to strengthen Geography as a science, the focus now is on raising awareness about multiple socioeconomic and environmental issues. The Declaration emphasizes the right to education for all individuals and the powerful role Geography plays in international, environmental, and development education. It also advocates for Geography to be a subject taught by geography teachers, as a core and independent subject in primary and secondary education.

With the new challenges, proposals, and international guidelines of the twenty-first century, a constant review of Geography teaching methodology is necessary. Sustainable Development being one of the primary objectives at all levels of global action, the UN adopted the 2030 Agenda with the Sustainable Development Goals (SDGs) in 2015.

The approval of the Agenda 2030 has had a significant impact on the teaching of Geography. Many educators and researchers worldwide have advocated for the role of our discipline in teaching sustainable development (Brkic-Vejmelka et al., 2018; Crespo Castellanos & Rodríguez de Castro, 2020; Dube, 2017; Gress & Tschapka, 2017; Grindsted, 2015; Martínez-Hernández & Mínguez, 2023; Meadows, 2020). The multidisciplinarity of Geography and its curriculum content align perfectly with these goals.

In response to this new context, the UGI will proclaim a new International Declaration on Geographic Education in 2016, establishing an international action plan for quality geographic education accessible to all. This plan, aimed at policymakers, curriculum developers, and Geography teachers, is based on five premises:

- Increase the presence of Geography in the school curriculum.
- Establish minimum geographic knowledge requirements for Geography teachers.
- Promote the exchange of Geography teachers nationally and internationally.
- Facilitate research in geographic education.
- Create and maintain a strong professional network of Geography teachers and educators.

This idea has been reinforced by the National Geographic Society, which uses the term geoeducation to describe education about our world. The organization believes that a comprehensive geoeducation is essential preparation for life in the modern world, providing young people with the knowledge and skills they need to make important decisions in their personal, professional, and civic lives. The term geoliteracy is used to describe the goals of a complete geoeducation.

Despite being a fundamental subject for the education of citizens capable of understanding and navigating an increasingly changing planet, Geography does not have a significant weight in the educational system or in the students' mindset. A particularly interesting study conducted by the National Geographic Society focuses on the state of geographic education in the United States, with conclusions that could be extrapolated to most countries worldwide. The association states that subjects essential for geographic education have been neglected in educational reform efforts in recent decades, with subjects focusing on international and intercultural issues, such as Geography, global studies, and world history, being particularly overlooked. Even in the field of sciences, which has been the focus of educational reform in recent years, earth sciences, environmental sciences, and ecology have received much less attention than other traditional sciences such as biology, chemistry, and physics, and when they are taught, they are more likely to be taught by teachers without content expertise.

In Spain, students' interest in Geography-related subjects is minimal and continues to decline. Compared to science subjects—mathematics, physics, chemistry, biology—and technological branches, the field of social sciences is often relegated to the background, as seen in the subsequent demand for university studies by students who ultimately decide to attend university. Geography and history—which are still inseparable in compulsory secondary education studies—are often perceived as boring, useless, anchored in the past, and purely theoretical, with no practical application in daily life. However, when the teacher in charge of the subject makes an effort to carry out methodological practices such as classroom–laboratory activities, field trips, and project-based learning, making Geography a practical subject as it should be, interest increases, students can appreciate the usefulness of the subject, and the learning process is highly satisfactory. Some studies conducted in educational centers in our country already support these claims (Moreno, 2004; Muñoz et al., 2010).

Another challenge facing the teaching of Geography in Spain is that often the concepts specific to our science are studied in other subjects such as Contemporary World Sciences and Geology (see Sect. 5.3.1.1), so students often do not know what Geography really is or what its field and study objectives are. Therefore, studying the interest and knowledge of students in Spanish schools in Geography is quite complicated. Notari Llorens et al. (2016) conducted a study on knowledge of ecological awareness in compulsory secondary education in adolescents from schools in Castellón. This study, framed in the natural sciences field and published in the *International Journal of Learning in Science, Mathematics, and Technology*, confirms that many purely geographical concepts (such as environmental education, perceptions of the environment) are included in other subjects, with Geography linked to history and, therefore, to social sciences.

Outside of classrooms, experiences related to geographic education are often undervalued by society, which tends to focus on academic achievements. Cultural exchange, time in nature, travel, and scientific fieldwork are seen as enriching experiences for a few rather than essential for all.

In response to this, there is a growing movement within the Earth science education community, suggesting that the current emphasis on environmental awareness in many educational systems worldwide is insufficient. There should be a shift toward viewing the development of environmental consciousness as a central goal of scientific education (Orion, 2007). Orion (2007) defined two principles of environmental knowledge:

- Understanding that we live in a cyclical world built on a series of subsystems (geosphere, hydrosphere, biosphere, and atmosphere) that interact through the exchange of energy and materials.
- Understanding that humans are part of nature and therefore should act in harmony with it.

Orion and Fortner (2003) suggested that applying the Earth systems approach could be an effective learning tool for developing environmental awareness. Earth system science is the field of study that seeks to understand Earth as a system, including the individual systems (such as the geosphere, hydrosphere, and atmosphere) and how they interact and influence each other.

This work aims to emphasize geographic education and geotourism, which, as we have seen, are interconnected. It focuses on geomorphology, Natural Heritage, and fieldwork in Protected Natural Areas as the best way to directly engage with the territory in an integrated and engaging manner for students, educators, and tourists alike.

#### 1.3.1.1 Field Trips in Formal Geographic Education

The study of Geography focuses on the physical and human environment, making field trips an essential methodological tool for learning. Field trips to natural spaces are particularly important for teaching geomorphology through Geomorphosites.

The main advantage of field-based learning is the ability to address phenomena and processes that cannot be studied in a classroom. While the field can be a challenging learning environment with many distractions, it has the potential to stimulate students' learning instincts and help them see the relevance of classroom learning in their own surroundings. Therefore, one of the teacher's initial tasks is to identify and classify phenomena, processes, skills, and concepts that can be learned concretely and meaningfully only in the outdoor environment, as well as those that can be learned concretely in the classroom. Additionally, it is important to identify abstract processes and concepts that may not be easily understood in the field and require more sophisticated internal tools such as images, films, slides, and computer programs (Vasconcelos et al., 2020).

Orion (1993) suggested that direct experience with materials and concrete processes helps in forming abstract concepts and aids in memorization. This idea is supported by studies showing that students who participate in field activities have a greater ability to observe, memorize, and recall facts (Orion & Hofstein, 1994).

In Spain, the tradition of fieldwork or educational itineraries has its conceptual origins in the Institución de Libre Enseñanza, which advocated for learning as a result of research and discovery processes initiated by students and facilitated by teachers. After the hiatus caused by the Civil War, the promotion of fieldwork as the best methodological tool for teaching Geography was resumed, with numerous authors highlighting its advantages in the teaching–learning process, the interest and motivation it generates in students, and the importance of field trips in direct observation and the development of the geographical method (Crespo Castellanos, 2012; García Martín et al., 2019; García Ruiz, 1994; Morote Seguido, 2019; Sánchez Ogallar, 1995).

We must differentiate between field trips focused on the educational use of Natural Heritage for different educational levels. In this case, we need to distinguish between university and nonuniversity levels, as the purpose of field trips will be different:

- Fieldwork for nonuniversity students (primary, secondary, and high school) aims to work on concepts learned in the classroom, develop cognitive skills, and stimulate students' interest in Geography. The desired outcome is to instill respect and awareness toward Natural Heritage, as well as an understanding of Geography, the interaction between humans and the environment, and its consequences.
- On the other hand, field trips for university students (mainly Geography students) aim to train specialists in the field and promote research.

Despite the numerous advantages and the need for field trips, they are often overlooked, especially at nonuniversity levels. Reasons include insufficient teacher training or lack of interest, limited financial resources, lack of time in the school year schedule, the responsibility of taking minors to the field, and various other obstacles faced by teachers in schools. One alternative to physical field trips is virtual field trips, as previously mentioned (see Sect. 1.1.5), but they should be presented as a classroom supplement before the actual field trip, never as a replacement for direct contact with the territory.

On the other hand, some authors have pointed out that teachers face difficulties in teaching in the field as they do not feel prepared and struggle with geological/geomorphological content, compounded by a lack of supportive teaching materials (Orion & Hofstein, 1994). This is particularly significant in Spain, where there is a secondary education specialization in History, Geography, and Art History, where a historian or art historian may not have the necessary preparation to conduct field trips in natural spaces.

Closely related to field trips is the student's interest in the subject of Geography. Geography and History, which are still inseparable in compulsory secondary education studies, are often perceived as a boring, useless, outdated, and purely theoretical subject, with no practical application in daily life. However, when the teacher in charge of the subject makes an effort to implement methodological practices such as classroom–laboratory activities, field trips, project-based learning, and thus make Geography a practical subject as it should be, interest increases, students can appreciate the usefulness of the subject, and the learning process becomes highly satisfactory. Some studies conducted in educational centers in our country already support these claims (Moreno, 2004; Muñoz et al., 2010). The reasons for this denigrated Geography are multiple, starting with the constant modifications of the Education Law—as of the writing of this text, the Organic Law 3/2020, of December 29 (LOMLOE) is in force—with changing curricular content, especially when it comes to physical Geography. While the contents of human Geography—demography, urbanism, and economic Geography mainly—are covered in our own subject, when a student needs to learn about geomorphology and the environment, they find that these contents are included in the Geology subject. As we have discussed, this is not an exclusive situation in Spain but also occurs in neighboring countries such as Italy. Students do not know what Geography really is or what its field and study objectives are. They often associate it with memorizing rivers, mountains, and capitals.

In this context, it is worth highlighting the work that some Spanish geographers are doing in the field of teaching innovation related to Geography. The work of the Geography Teaching Group of the Spanish Geography Association (AGE) stands out, created in 1985 by professors from the University of Murcia. Its main objective is to support and promote methodological innovation in the teaching and learning processes of Geography. The variety of these works includes the use of Geographic Information Systems (Canovas García, 2021; García Paredes, 2022), urban planning (Hurtado Beltrán, 2024; Serrano Gil, 2021), landscape (Casas Jericó et al., 2018, 2019), and climate change (Sánchez Almodóvar & Olcina Cantos, 2024). Some authors have also focused their research on primary and secondary education (Fernández Álvarez, 2019; López Fernandez & Peral Velasco, 2018; Morón Monge et al., 2023; Zaragoza Sáez & Morote Seguido, 2024).

One of the objectives of this work, as mentioned earlier, is to evaluate the didactic potential of geomorphosites, i.e., their implementation in formal education. The best way to do this will always be through field trips, considering virtual reality, or virtual field trips, as a complementary support tool in the classroom, but never as a substitute for actual field trips. Knowing that field trips are well established at university levels and that the scientific content of Geomorphosites is too advanced for primary education, the goal would be to propose field trips in secondary education and high school. To achieve this, it was necessary to understand the situation regarding field trips to Natural Spaces in Spanish high schools where this teaching is provided. A survey was conducted in July 2023 among Geography and History and Biology and Geology teachers, and the results are presented in Sect. 5.3.3 Geomorphology, Geoeducation and Field Trips: Study Case in Spanish High Schools.

### 1.3.2 Nonformal Geographic Education: Environmental Education

Nonformal education refers to planned and structured education that is not regulated by educational institutions, targeting various population groups and not leading to an official accreditation or certificate. It is important to consider this type of education, as it focuses on professionals other than teachers and is aimed at populations excluded from formal educational institutions.

In Spain, environmental education emerged in response to the environmental crisis of the 1970s. Various international influences contributed to raising awareness of environmental degradation, leading to a need for education to protect the environment. The United Nations Conference on the Human Environment in Stockholm in 1972 recognized the social origins of environmental conflicts. The International Congress on Environmental Education and Training in Moscow in 1987 proposed a definition of environmental education as a process of raising awareness of the environment and the ability to solve environmental problems.

In the 1980s, with the Statute of Autonomy, environmental education facilities were established. The I and II Environmental Education Conferences were held in Sitges in 1983 and in Valsaín in 1987. In 1990, the new education law (LOGSE) included environmental education as a cross-cutting theme to be addressed in the classroom. By the 1990s, there was a new vision of environmental education as a tool to address environmental issues involving the entire society and aiming for sustainability.

The result was the approval of the White Paper on Environmental Education in 1999, which aimed to give more attention to social aspects, debate and prevention of problems, a shift in values and behaviors, and to elevate environmental education to all contexts and the entire population. Objectives included knowledge and understanding of environmental issues and their relationship with society, the economy, and culture; promoting new pro-environmental ethical values; engaging the entire society; and promoting sustainable behavior models.

Two decades later, the Environmental Education for Sustainability Action Plan (2021–2025) was approved, updating and addressing issues that have arisen since the publication of the White Paper and proposing solutions (Table 1.2). The main problems identified in the early decades of Environmental Education in Spain are related to the limited dissemination of its importance, the use of noninteractive methodologies, the existence of inadequate or outdated facilities, and the precarious situation of environmental educators. It is particularly interesting that one of the identified problems is that the recipients of environmental education are mainly students, although we have discussed in the previous section that environmental education and awareness about the environment are scarce in schools.

In Spain, environmental education facilities are defined as “centres with stable educational facilities whose main activity is to develop specific environmental education programs and projects, with a qualified team for their implementation” (Serantes et al., 2015). Three key elements are identified: facilities, educational programs, and the educational team. These facilities are considered part of nonformal education and are expected to be diverse to address a wide range of environmental issues (Serantes, 2011) (Fig. 1.5).

**Table 1.2** Problems and solutions regarding environmental education in Spain in the Environmental Education for Sustainability Action Plan (2021–2025)

| Problems | Solutions |
|---|---|
| Lack of communication to increase social, political, and economic support for protected areas | Explicitly include the ecosystem services provided by protected areas and their relationship with human well-being |
| Educational programs focused on heritage, not on its importance for humans and their well-being | |
| Environmental education: serving visitors instead of a tool for planning and managing protected areas | Integrate environmental education planning into the overall planning and management strategy of the protected area |
| Mainly targeted at school children | Ensure that environmental education programs target different audiences, including the local population and other social actors, and not just school groups. Promote active involvement of society in the conservation of natural spaces through volunteering |
| Less interactive methodologies | Diversify educational methodologies, including the use of new technologies and artistic expression |
| Inadequate or obsolete equipment | Analyze the needs for adaptation or updating of equipment, take advantage of existing constructions, traditional or of high heritage value. Promote and facilitate the use of equipment as places for work and activities by the local population and by associations, NGOs, companies, educational centers, and so on |
| Precarious situation of environmental education professionals | Recognize the figure of environmental educator as a professional category, through the establishment of standardized criteria |

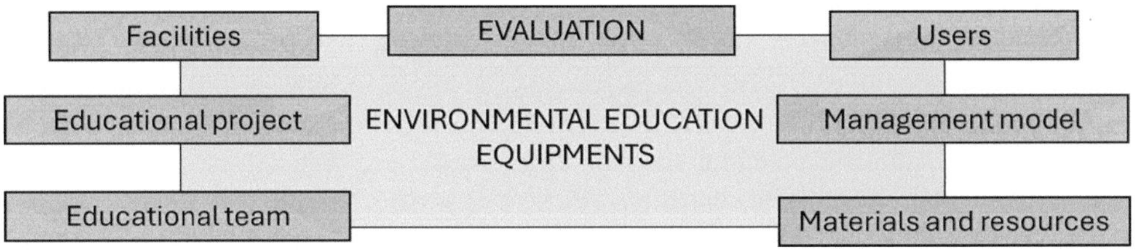


**Fig. 1.5** Elements that characterize environmental education equipment in Spain. Translated from Serantes (2011)

These facilities, for which public administrations do not commit or assume responsibilities, have self-regulation processes in terms of quality, aimed at ensuring quality education for the user. Serantes et al. (2015) proposed a classification of environmental education facilities based on their nature, content, activities, methodology, and purpose, establishing five categories:

- *Farm-schools*: located in rural areas, they aim to understand and experience the relationships between inhabitants and their environment, as well as with the vegetable garden and domestic animals, all based on sustainable agricultural practices.
- *Environmental classroom/center*: they carry out interpretation activities, workshops, and itineraries to promote sustainable behavior in environments (natural, rural, urban, etc.) whose resources or consequences of their use are of interest.
- *Interpretation/visitor centers*: associated with places of high heritage value, the goal is understanding and awareness through communicative strategies such as models, audiovisuals, interactive resources, and so on.
- *Reference centers*: responsible for training technicians, pedagogical advice on projects, support for working groups, and coordination of environmental education actions.
- *Museums for Environmental Education*: in addition to the usual functions of environmental education, they have their own projects.

The management, typology, and resources of environmental education facilities depend on each Autonomous Community. In the following section, we will analyze the Environmental Education Programs in the region of Castilla y León. According to a study by Serantes et al. (2015), it is the autonomous communities that promote most of the facilities, many of which are managed by private companies through concessions, leading to a lack of labor regulation characterized by precariousness in the sector: more than half of the workers have temporary contracts, there is no specific collective agreement, and salaries are below the qualifications of the workers.

Therefore, the sector demands a national agreement for its regulation and the need to create an attractive image of environmental education facilities.

#### 1.3.2.1 Programs of Environmental Education in the Region of Castilla y León

Environmental Education in Castilla y León was perfectly defined by Delgado Huertos (2007), who stated: "*Environmental education in Castilla y León, if we were to summarize its situation in an image, unfolds with the geological slowness of a sedimentary era.*" With three Environmental Education Strategies approved to date, with up to 9 years between each one, our community remains stuck in lethargy.

Without much change in the situation, the author already pointed out in 2008 that, in our community, the first Environmental Education Strategy led to the creation of a good network of facilities and qualified professionals, but was hindered by lack of political will, low social participation, and the negligible influence of nongovernmental organizations in environmental decision-making. Despite significant progress, the promises of this first strategy were not fulfilled, and it took 9 years for a new strategy to be approved in Castilla y León.

The Second Strategy was not implemented until June 2016 and was valid until 2020. It included an analysis of environmental education in the community, as well as guidelines, objectives, and action plans. It was highlighted that "*the rich natural, cultural, and social environment of Castilla y León, along with the expertise of environmental educators in the region, were key pillars for revitalizing future environmental education. The strategy aims to leverage the knowledge and experience gained from environmental education to promote sustainable development.*" The Second Strategy encompassed various environmental education initiatives in the community, such as the MYAS-MICOCYL program focusing on mycology, the European Mobility Week, and the "City, Without My Car" initiative. Additionally, there were environmental volunteer programs led by organizations such as the Duero River Basin Authority, the Caja de Burgos Foundation, and the Ministry of Development and Environment, covering different themes and formats. The Clear Sources for Sustainability in Small Municipalities Program included regional sustainability awards for small municipalities and training sessions for "Mayors' School." Some schools in the region independently implemented environmental education programs, while others received support from entities such as the provincial councils of Burgos, Palencia, and Zamora, as well as numerous municipalities. The VEEN School Visits to Natural Spaces program was also developed for the school environment.

Recently, the III Environmental Education Strategy of Castilla y León has been approved for the period from February 2024 to 2030, following a 4-year hiatus. The key themes include climate change, biodiversity, and circular economy. The strategy includes a SWOT analysis, highlighting weaknesses and threats in environmental education in our community:

- *Weaknesses*: environmental education overly focused on schools, inadequate and sporadic funding, precarious situation of environmental educators.
- *Threats*: professional intrusion, misinformation on environmental issues, reliance on public funds, limited willingness of the public to pay for environmental education services.

Key actions include strengthening environmental education in schools (particularly important given the identified weakness), enhancing the environmental educator sector, improving access to environmental information, innovating resources, promoting environmental education activities, expanding environmental education to society as a whole, and fostering environmental collaboration between public and private entities.

It appears that Environmental Education in our community, as well as nationally, has not progressed in recent years, facing the same issues highlighted in the 2021 Environmental Education Action Plan. It will be necessary to wait until at least 2030 to see if the commitments outlined in a mere 25-page Strategy translate into real support, and if our community will truly prioritize Environmental Education. Unfortunately, the precedents and historical background are not very promising.

We will focus on the VEEN program of School Visits as it is linked to formal education and takes place in Natural Spaces, making it of great interest in this work.

The VEEN program, or School Visits to Natural Spaces, offers educational visits to the network of park houses and thematic centers for schools in the community to learn about the natural and cultural richness of the Protected Natural Spaces of Castilla y León. The objectives of the program are as follows (Program dossier, accessed on 29/07/2024):

- Provide the opportunity for students to enjoy and learn about the natural and cultural values of the Protected Natural Spaces of Castilla y León.
- Encourage students to have direct experiences with the most important landscapes and places in their community.
- Provide valuable educational resources to teaching teams, who can incorporate the content of these visits into their educational practices.

Educational visits are available for children in preschool, primary, secondary, high school, and university, with a specific program tailored to each house and thematic center. These visits typically include guided tours of the exhibition, interactive workshops, and educational itineraries along interpretive trails. The program ensures that the activities align with the specific school curriculum of the visiting group and provides educational materials related to the Natural Space for both students and teachers.

Regarding the Natural Spaces studied in this book, the Sierra de la Paramera and La Serrota do not have the VEEN Program as they are not part of the Natural Spaces Network of Castilla y León and lack suitable facilities for such use. The Cañón del Río Lobos Natural Park is part of this program, which is conducted at the Park House in Ucero. A standard program is offered, but it can be customized in collaboration with the educational team based on the students' level and the teacher's interests. The standard program includes a guided tour of the Park House focusing on the karst landscape, various fauna and flora workshops, and an educational itinerary to the Ermita del Río, also focusing on flora and fauna. Additionally, participants can visit the River Classroom for a fishing art workshop and practice fishing techniques.

The Natural Heritage Foundation also offers a Golden Spaces program for individuals over 60 years old, aiming to connect participants with the natural environment and Natural Protected Areas of Castilla y León through guided tours led by environmental educators. There is no online dossier available for this program, so interested individuals need to contact each Park House directly.

## 1.4 Geotourism

Geomorphosites are part of the tourism system due to their high aesthetic and landscape value, as well as their historical and cultural connection to the territory, attracting tourists from around the world (Serrano et al., 2020). This book will focus on the geotouristic potential of geomorphosites, emphasizing nature tourism or geotourism, which goes beyond enjoyment, leisure, and beauty. Visitors seek to learn about geoheritage, leading to an awareness of the value and need for protection of Natural Heritage.

We will review the concept of geotourism and conservation, the importance of cultural services in geotourism, and the characterization of rural tourism in Spain and the Castilla y León region.

### 1.4.1 Nature Tourism, Rural Tourism, and Geotourism

The concept of geotourism is relatively new, experiencing a significant growth in the last decade, with its origins dating back to the late twentieth century, stemming from rural tourism and nature tourism. These types of tourism have their roots in the European Romantic movement from the mid-eighteenth to mid-nineteenth century and the development of landscape aesthetics. The Romantic movement changed the perception of "wild" landscapes, such as mountains and natural phenomena, from places to fear to landscapes to appreciate through a "romantic gaze." This shift inspired writers, artists, poets, and the social elites to seek and experience beautiful, sublime, and picturesque landscapes for recreational, scientific, or health reasons.

**Rural tourism** is defined as any leisure activity carried out in rural areas that is compatible with sustainable development, optimal resource utilization, integration of the local population, and consideration and improvement of the environment (Serrano et al., 2020). It emerged in Spain in the late 1980s as an alternative to beach tourism, aiming to diversify the economy in rural areas with limited opportunities, emphasizing the natural environment and landscape, and offering a unique experience based on the tranquility and familiarity of rural life (Grande Ibarra, 2006). Rural tourism and nature tourism are often confused as they share the rural setting, but not all rural tourism is nature-based, and not all nature tourism exclusively takes place in rural areas. Rural tourism encompasses a variety of nature-related activities, agriculture, fishing, and ways of life, while **nature tourism** focuses on recreational activities, nature interpretation, and sports in natural settings. It is argued that rural tourism is more closely related to cultural tourism than nature tourism, as its main motivation is to explore the rural environment and culture (Grande Ibarra, 2006).

Related to rural tourism, **sustainable tourism** has been part of the tourism industry for decades. It is more of a concept or characteristic of rural tourism, especially focused on travel operators, aiming for the conservation of nature as a tourist resource, respect for values, benefits for local communities, and harmony between tourists and the local population of the area (Serrano et al., 2020).

Within **nature tourism**, different modalities such as ecotourism and sustainable tourism emerge. Ecotourism is a type of tourism activity based on environmental conservation and benefiting the population, increasing awareness among the local population and visitors, and involving education on nature and culture values of the visited area (Wood, 2002) and management processes to minimize the negative impact of tourism and education.

The concept of geotourism is more recent, emerging as an evolution and combination of sustainable tourism and ecotourism. There have been multiple studies, and various authors have addressed the concept of geotourism in recent years (Bouzekraoui et al., 2018a, 2018b; Dowling, 2011; Escriche, 2011; Gordon, 2012; Kubalíková, 2013; Kubalíková et al., 2021a, 2021b; Marino Alfonso et al., 2021; Newsome & Dowling, 2018; Ólafsdóttir & Dowling, 2014; Dóniz-Páez et al., 2021), although the origin of the term can be traced back to the late 1990s.

The National Geographic Society has been publishing a series of geotourism maps since the late 1990s. J. Tourtellot from the National Geographic Foundation defines geotourism as "*tourism that sustains or enhances the geographical character of the place visited, including its environment, culture, aesthetics, heritage, and the well-being of its residents*." However, Hose (1996) refers to geotourism as geological tourism, where geological sites become attractive tourist resources. The author defines geotourism as "*the provision of interpretive resources and services to promote the value and social benefit of geological and geomorphological sites, and to ensure their preservation and use by students, tourists, or other visitors*." There is a duality in the meaning of the term between Geography and Geology, which share more similarities than differences, as argued in Serrano et al. (2020). Our stance is to advocate for a holistic view of geotourism, as it cannot be considered solely in terms of geological forms, but necessarily involves territorial and landscape implications. This is where Geography, Geology, Geomorphology, and Anthropology intersect, encompassing both natural and cultural heritage.

This holistic vision is reinforced by the creation of UNESCO Geoparks, which, as previously mentioned (see Sect. 1.1.4), aim at conservation and education for sustainable development, in harmony with tourism and local communities. The culmination of the diversity of opinions surrounding the debate on whether geotourism is geological tourism or not was the 2011 International Geotourism Congress held in the Arouca Geopark, where this topic was discussed and consensus was reached on the definition of the term: "*geotourism is tourism that sustains and enhances the identity of a territory, taking into account its geology, environment, culture, aesthetic values, heritage, and the well-being of its residents. Geological tourism is considered one of the various components of geotourism*". Therefore, all geological, geographical, historical, artistic, and ethnological resources of the territory are considered. In this work, we advocate and defend the holistic meaning of the term geotourism, always referring to the entirety of the abiotic heritage from an integrative geographical perspective.

Geotourism has experienced significant growth in recent decades worldwide and it is valued and accepted as a useful tool for promoting natural and cultural heritage, as well as fostering local and regional economic development, especially in rural areas (Kubalíková, 2019). Therefore, with the development of geotourism in other protected natural areas, it could contribute to the destigmatization and geographical deconcentration of those touristically overcrowded places, leading people to other nearby equally interesting but previously unknown spaces (Ruiz-Pedrosa et al., 2024).

According to Dowling (2011), geotourism can occur in a variety of environments, from natural and wild to built and planned, and includes both independent travelers and tourist groups. The key aspect of geotourism is that it involves all broader aspects of tourism activity, such as transportation, access, accommodation, services, planning, and management. Therefore, geotourism positively contributes to local development while also expanding the tourism sector as a whole.

Essential for the development of geotourism is understanding the identity or character of a region or territory. To achieve this, geotourism is based on the idea that the environment consists of abiotic, biotic, and cultural components. Three principles are fundamental to geotourism (Dowling & Newsome, 2017):

- It has a geological basis (based on the earth's geological heritage). Additionally, we should consider the rest of the abiotic heritage.
- It is sustainable, economically viable, improving the community and promoting geoconservation.
- It is educational, achieved through geointerpretation.

Geotourism aims to promote tourism development opportunities while ensuring the conservation and protection of geoheritage attributes. Stakeholders in geotourism include local communities, tourism industry, protected area managers, nongovernmental conservation organizations, and tourists themselves (Dowling & Newsome, 2010). Geotourists are interested in interacting with local communities and experiencing geological, geographical, cultural, and physical features. Local guides are valued for providing a better understanding of the surrounding environment (Mao et al., 2009).

The development of geotourism benefits local residents by generating income, employment, and skill development (Torabi Farsani et al., 2012). It can be seen as a way to conserve and manage Natural Heritage through the efforts of local populations and stakeholders, providing benefits, employment opportunities, and serving as a powerful tool for local development in depressed rural areas.

### 1.4.2 Geotourism and Conservation

Geotourism has evolved in response to the need to minimize the negative impacts of mass tourism in natural tourist environments, while also providing a tool for sustainable rural development. Sustainable tourism requires careful planning and management. Ólafsdóttir and Dowling (2014) propose a method for identifying opportunities for tourism development in natural areas by identifying significant features, critical areas, and compatible activities. Following this method, a series of planning zones are defined to protect conservation values while promoting development and tourism activities.

These zones are identified and described based on an approach where the terrestrial and aquatic areas of a region are classified according to their need for protection and compatibility with tourism. The following general zones and their main functions for geotourism development are proposed:

- *Sanctuary zones*: areas requiring special preservation;
- *Geoconservation zones*: areas maintaining a combination of protection and use, with an emphasis on the former;
- *Geotourism development zones*: small, concentrated areas of tourist attractions;
- *Outdoor recreation zones*: natural areas suitable for compatible outdoor recreation activities.

The involvement of local residents in geotourism planning is crucial for the sustainability of the process, ensuring that stakeholders have acceptance and empowerment in the development of geotourism. Appropriate eco-ethics for the participation of residents and tourists in the planning process includes developers taking into account the attitudes and feelings of the local community, including how an untouched local environment contributes to a sense of community belonging (Ólafsdóttir & Dowling, 2014).

Any geotourism development should not diminish the enjoyment of the environment by the local community and, whenever possible, should enhance it. Therefore, regional geotourism development should be carried out within the context of sustainable local, national, and international tourism development.

### 1.4.3 Geotourism and Cultural Services

Gordon (2018) explored how the connection between geoheritage and cultural heritage can enhance tourists' experience in natural areas and promote geoeducation and geoconservation. According to the author, the nature–culture symbiosis provides a way to enhance visitors' experience and engage with the geological—or geomorphological for us—heritage through various aspects of landscape appreciation (Serrano & González-Amuchastegui, 2020).

Cultural ecosystem services, a subset of ecosystem services, have emerged as a concept through which researchers and decision-makers can understand ecosystems in terms of their enriching and affirming contributions to human well-being (Gordon, 2018).

The Millennium Ecosystem Assessment (MEA) defined cultural ecosystem services as "non-material benefits that people obtain from ecosystems through spiritual enrichment, cognitive development, reflection, recreation, and aesthetic experiences." It identified ten categories: cultural diversity, spiritual and religious values, knowledge systems, educational values, inspiration, aesthetic values, social relationships, sense of place, cultural heritage values, recreation, and ecotourism. However, we believe that ecotourism is a use, not a service, and should not be included as a cultural ecosystem service.

This concept is interesting for studying the relationships between geoheritage, geotourism, and cultural landscapes, especially within Natural Spaces. In addition to providing economic benefits from tourism activities, they also lead to a range of relational benefits: shaping people's identities (belonging, sense of place, spirituality); experiences (tranquility, inspiration) that enhance well-being, mental and physical health; and knowledge, skills, and capacities (Gordon, 2018). Kubalíková (2020) further emphasizes the importance of cultural ecosystem services based on their link to geodiversity, as it offers multiple services for both nature and people, and states that cultural systems should be an important basis for conservation, landscape planning, and responsible and sustainable use of natural resources. Concepts that geographers have been using and implementing for decades in Territorial Planning.

Therefore, it is evident that geoheritage significantly contributes to cultural ecosystem services, benefiting the population. This establishes a relationship between the natural environment, people, and culture within the geographical space. According to Gordon (2018), cultural services can enhance geotouristic experiences by:

- The diversity of the physical environment influences cultural diversity, cultural identity, and people's sense of belonging.
- Abiotic and biotic characteristics shape the landscape, where physical and cultural elements converge, as advocated by geographers for decades (Martínez de Pisón, 1983, 2010, 2012; Serrano, 2012).
- Geotourism can promote understanding of geomorphological processes in relation to current environmental issues.

### 1.4.4 Nature Tourism in Spain and the Region of Castilla y León

Since rural tourism became an alternative to mass tourism in the late 1980s, it has continued to gain importance in our country, particularly in the region of Castilla y León, the leading autonomous community in rural tourism in Spain. With 85% of the territory belonging to rural areas and 28% of the land being part of protected natural spaces, it is not surprising that rural tourism numbers are thriving.

The first numbers on travelers accommodations and overnight stays in rural tourism accommodations correspond to the beginning of the century, between 2001 and

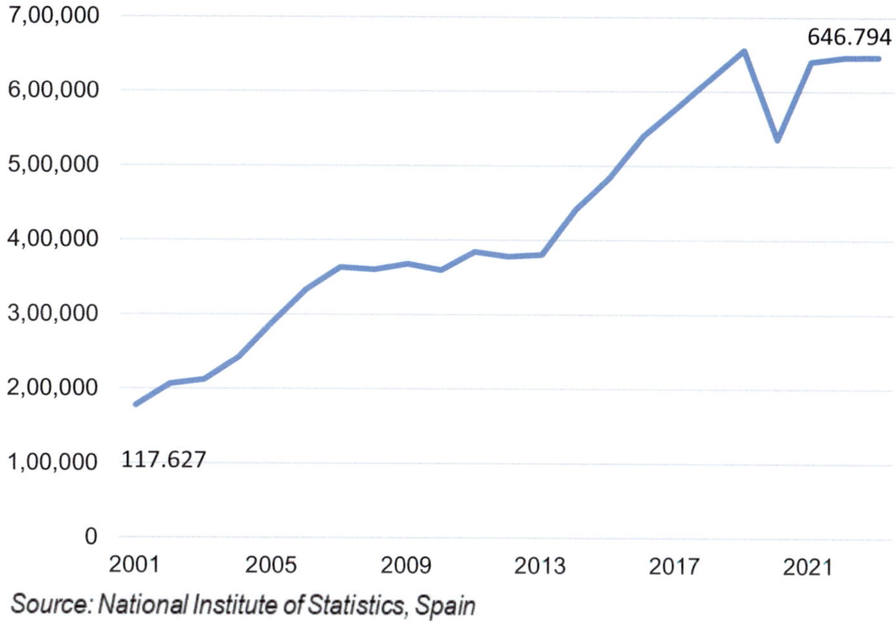


**Fig. 1.6** Number of travelers in rural tourism accommodation in Spain in August (2001–2021)

2005. The first period of growth will be between 2003 and 2007, during which the number of travelers almost doubles, from 200,000 to almost 400,000 (Fig. 1.6). Travelers are defined as those who stay overnight in rural tourism accommodations. In 2007, the Law on Natural Heritage and Biodiversity was approved, regulating strategic and sectoral plans, establishing conservation objectives and actions for the use of Protected Natural Areas (NPA). The number of travelers will remain stable, below 400,000, until 2014–2015, when the Sectoral Plan for Nature and Biodiversity Tourism was approved, which was in effect until 2020. The objective of this sectoral plan was to make conservation compatible with the tourist use of Natural Heritage, a model that, as the plan itself defines, "*contributes to the enhancement of Spain's natural wealth for nature tourism, promoting balanced socio-economic development and boosting income and employment generation, without compromising biodiversity and improving its management and conservation.*" At this point, a significant effort is made in the configuration of sustainable nature tourism destinations and products, and a multitude of rural tourism infrastructures are created, financed with LEADER and PRODER funds. Thus, in the last decade, rural tourism has grown exponentially both in the number of travelers (reaching almost 700,000 before COVID, Fig. 1.6), as well as in the number of accommodations (increasing by 40,000 until 2019, Fig. 1.7), and in overnight stays (increasing from 7 to 8 million in 2019, Fig. 1.8).

After the COVID-induced hiatus, nature tourism is not only resuming but also gaining momentum, signaling a potential shift in the tourism sector. Following the strict restrictions imposed during the health crisis, many experts have noted a growing trend toward rural and nature tourism, which served as an alternative to urban crowds during the reopening phase. This change in tourism typology is not limited to Spain (Gago-García et al., 2021; Agudo Segura et al., 2022; Dot Jutglà et al., 2022; Méndez Gutiérrez del Valle, 2022; Ríos Rodríguez et al., 2022; Capdepón Frías, 2023) but it is observed worldwide, including countries such as Ecuador (Herrera Anangonó et al., 2021), the Czech Republic (Vaishar & Šťastná, 2022), Portugal (da Silva Lopes et al., 2021), Russia (Polukhina et al., 2021), and Turkey (Kürüm Varolgüneş et al., 2022), among others. The desire for freedom, health, and relaxation has led people around the globe to seek refuge in natural spaces, driven by the proximity and ease of travel in a context where international trips were restricted.

In the case of Spain, this new paradigm is reflected in the data, with a record number of overnight stays in 2022, surpassing 12 million annually. Rural houses dominate, accounting for 70% of the total, followed by rural hotels at 15%, with the remaining distributed among other types of rural accommodations (hostels, active tourism lodgings, etc.).

Regarding the destinations chosen by tourists, the INE provides data on overnight stays by tourist destination, which we have shown in Table 1.3. Two destinations stand out well above the rest: the Pyrenees and the Island of Mallorca, with 19% and 16%, respectively. The INE distinguishes several areas for the Pyrenees: the Catalan, Navarre, Aragonese, and Basque Pyrenees (the list of municipalities that make up each tourist area can be found at https://www.ine.es/daco/daco42/ocuptr/zonas_eotr.xlsx, date of consultation 31/07/2024), which we have grouped together. It is undeniable that the Pyrenees are preceded by their natural and cul-

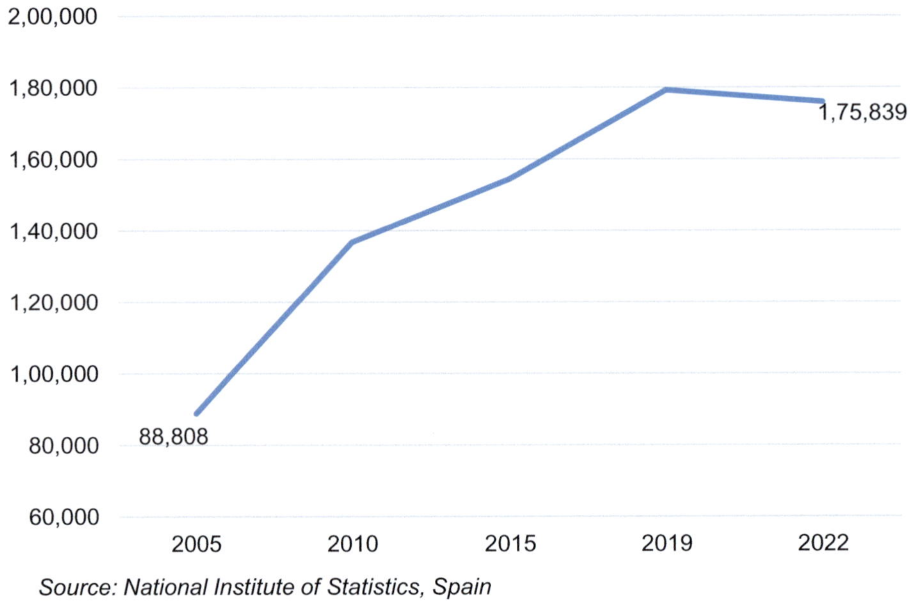


**Fig. 1.7** Estimated number of bed places in rural tourism in Spain for the month of August (2001–2022)

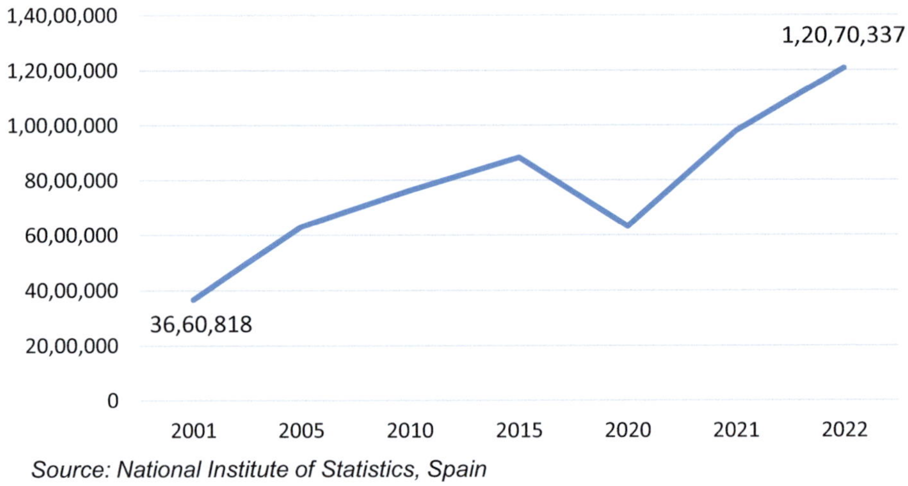


**Fig. 1.8** Overnight stays in rural tourism accommodation in Spain (2001–2022)

tural heritage, in Spain and worldwide, which has attracted explorers, mountaineers, naturalists, geographers, and artists for centuries (Serrano, 2023).

We must question the data for Mallorca, which includes the capital, Palma de Mallorca, Manacor, Alcudia, Andratx, and Calviá among the municipalities. These municipalities focus their tourist offer on mass tourism, sun and beach, and as much as the accommodations may have a rural category, it is at least implausible to think that the tourists staying there visit the Island for purposes compatible with rural tourism. Beach tourism is so massive in Mallorca that these tourists occupy accommodations throughout the Island, regardless of their category. The same can be said for the Costa Brava, Costa del Sol, Costa de Almería, and the rest of the Balearic and Canary Islands, although in the Canary archipelago, we can recognize a tourism linked to volcanism, although, equally, as a complement to sun and beach tourism. Already with very low percentages, there are other destinies that count upon with an environmental protection and declaration and in which it is actually articulated around Nature, as it is the case of the Picos de Europa National Park, the Sierras de Tejeda, Almijara, and Alhama National Parks.

It should be highlighted the exclusive presence of the Fuentes Carrionas and Fuente Cobre Natural Park in Palencia (now known as Montaña Palentina) in Castilla y León, with nearly 50,000 annual overnight stays. Additionally, the Las Batuecas-Sierra de Francia Natural Park and the Arribes del Duero Natural Park in Salamanca together account for over 55,000 overnight stays.

The Spanish government announced a new Sustainable Tourism Strategy 2030 in June 2020. Although it has not been approved yet, the draft and main objectives of this strat-

**Table 1.3** Overnight stays in rural accommodation by tourist destination (total and percentage of annual total by 2022)

| | | |
|---|---|---|
| Pirineos (Catalán, Navarro, Aragonés y Vasco) | 1,326,490 | 19% |
| Mallorca Island | 1,150,784 | 16% |
| Costa Verde | 426,238 | 6% |
| Costa Brava | 320,933 | 5% |
| Extremadura Norte | 300,834 | 4% |
| National Park Picos de Europa | 255,504 | 4% |
| Natural Park Sierras de Tejeda, Almijara and Alhama | 181,645 | 3% |
| Paisatges de Barcelona | 175,669 | 2% |
| Menorca Island | 172,569 | 2% |
| Costa del Sol (Málaga) | 160,328 | 2% |
| National Park Sierra de Guadarrama | 133,836 | 2% |
| Tenerife Island | 132,741 | 2% |
| La Palma Island | 122,376 | 2% |
| Rías Baixas | 102,691 | 1% |
| Natural Park Corona Forestal | 100,940 | 1% |
| Costa Barcelona | 95,738 | 1% |
| Costa de Gipuzkoa | 93,811 | 1% |
| National Park Teide | 91,444 | 1% |
| Terres de Lleida | 85,516 | 1% |
| Costa Daurada | 83,427 | 1% |
| Natural Park Sierra de Aracena and Picos de Aroche | 82,887 | 1% |
| Ibiza-Formentera Islands | 78,671 | 1% |
| Natural Park Sierra de Grazalema | 75,324 | 1% |
| Natural Park Oyambre | 73,364 | 1% |
| Natural Park Sierra and Cañones de Guara | 73,030 | 1% |
| Natural Park Calares and Cabeceras de los Ríos Mundo, Tus and Guadalimar | 71,168 | 1% |
| National Park Taburiente | 62,810 | 1% |
| Natural Park Cumbre Vieja | 61,150 | 1% |
| Natural Park Sierras de Cazorla, Segura and Las Villas | 48,378 | 1% |
| Natural Park Fuentes Carrionas and Fuente Cobre | 47,885 | 1% |
| Natural Park Sierras Subbéticas | 46,130 | 1% |
| Costa de Almería | 44,747 | 1% |
| Gran Canaria Island | 44,507 | 1% |
| National Park Ordesa | 43,883 | 1% |
| Natural Park Los Alcornocales | 43,494 | 1% |
| Natural Park Sierra Norte de Sevilla | 43,304 | 1% |
| Natural Park Saja-Besaya | 41,264 | 1% |
| Island de La Gomera | 41,225 | 1% |
| National Park Garajonay | 41,225 | 1% |
| Natural Park Las Batuecas-Sierra de Francia | 36,161 | 1% |
| Natural Park Sierra Nevada | 34,665 | 0% |
| Natural Park Alto Tajo | 32,014 | 0% |
| Natural Park Alt Pirineu | 30,442 | 0% |
| Natural Park Pagoeta | 28,616 | 0% |
| Natural Park del lago de Sanabria | 27,866 | 0% |
| Natural Park Aiako Harria | 25,576 | 0% |
| Mariña Lucense | 25,151 | 0% |
| National Park Sierra Nevada | 24,182 | 0% |
| Costa de Bizkaia | 23,332 | 0% |
| Rías Highs | 22,443 | 0% |
| National Park Aigüestortes | 21,392 | 0% |
| National Park Cabañeros | 21,019 | 0% |
| Costa Da Morte | 20,062 | 0% |
| Natural Park Arribes del Duero | 19,199 | 0% |
| Natural Park Posets-Maladeta | 14,078 | 0% |
| Hierro Island | 13,624 | 0% |
| Natural Park Majona | 13,612 | 0% |

Source: National Institute of Statistics, Spain

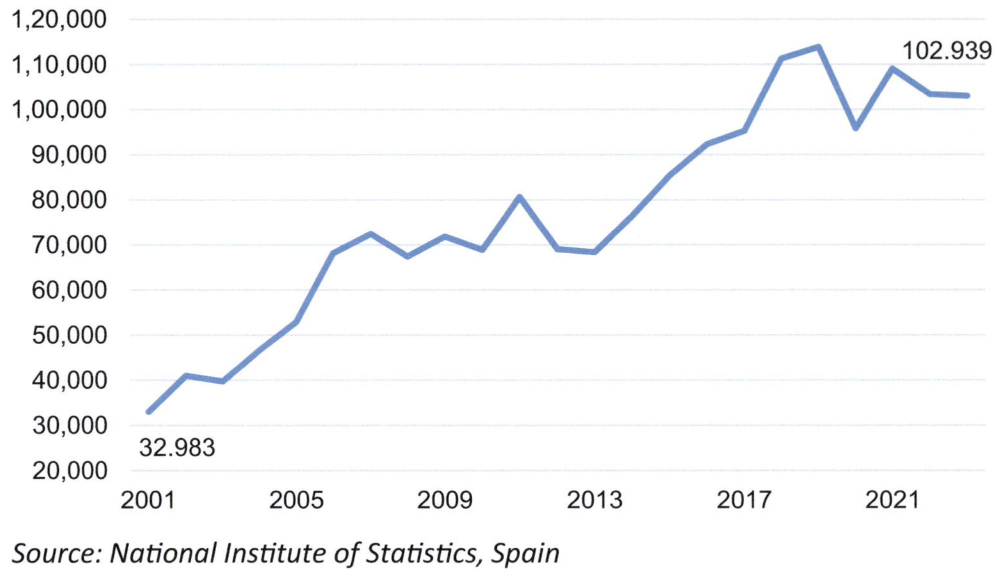


**Fig. 1.9** Visitors to rural tourism accommodation in Castilla y León (2001–2023)

egy can be consulted. The goal is to transform Spanish tourism toward a sustainable model, focusing on socioeconomic growth, preservation of natural and cultural values, social benefits, participation and governance, and continuous adaptation. It will take a few years to analyze the consequences of this new Strategy (if approved) and how this new paradigm around natural tourism evolves. Currently, the impacts of this "nature boom" can already be observed in some Protected Natural Areas: overcrowding, pollution, and inconvenience for local residents and other tourists due to uncivil behavior. Not all visitors are aware of the value of Natural Heritage or the need for conservation. The larger the audience, the greater the risk of degradation of these fragile Protected Natural Areas.

Regarding the region of Castilla y León, it has become the leading autonomous community for rural tourism in recent decades. Since the early 1990s, LEADER funds have provided a territorial promotion opportunity for the region's villages. In 1994, thanks to the Regional Plan for the Tourism Development of Castilla y León, there was a significant growth in rural accommodation offerings, with a predominance of Rural Houses (Alario Trigeros, 2018). It is not surprising that the vast natural and cultural heritage of the region attracts visitors from all over Spain and the world, with 33 Protected Natural Areas, 120 Special Conservation Areas, 70 Special Protection Areas for Birds of the Natura 2000 Network, a UNESCO Global Geopark, 9 World Heritage Sites, 8 Biosphere Reserves, 2625 Cultural Interest Assets, 13 Protected Designations of Origin, and 18 Protected Geographical Indications.

The region of Castilla y León leads in all rural tourism statistics. With over 30,000 accommodations, it has the highest number of touristic lodging spots in the country, accounting for 18%, well above the next highest percentages in Andalucía and Cataluña at 11%, and Castilla-La Mancha at 10% of the national total. The region also tops the list in terms of visitors (defined by the INE as those who stay overnight), with over 100,000 visitors, making up 16% of the total in 2023, followed by Asturias and the Balearic Islands, both at 10%.

Rural tourism in Castilla y León began to flourish in the early 2000s: by 2007, the number of visitors had more than doubled compared to 2001 (Fig. 1.9), reaching 72,000 visitors. Tourism then stagnated until 2013 due to the economic crisis, but from that year until the onset of COVID, there was significant growth, with 113,854 visitors in 2019. The COVID-19 pandemic significantly impacted social and economic activities, and our region has not been able to fully recover to pre-pandemic levels. The latest data from 2023 shows 102,939 visitors.

The Statistical Information System of the Junta de Castilla y León provides us with provincial-level information on overnight stays, number of rural accommodation places, and the number of accommodations, campsites, hostels, and active tourism companies (Table 1.4). We have been able to compare, based on the variable, the most recent data (some for 2022, others for 2024) with the data from the beginning of the century, in 2003. Regarding overnight stays, the most recent provincial-level data is for 2022, with Ávila leading the region with over 300,000 overnight stays per year, followed by Segovia with over 260,000. This leadership can be attributed to the proximity of both provinces to Madrid, and in the case of Segovia, to the Guadarrama National Park, which, as mentioned, is one of the most visited destinations in the country, mainly by residents of Madrid for weekend tourism. They are also leaders in estimated number of accom-

**Table 1.4** Infrastructure and services of the rural tourism system in Castilla y León (A: overnight stays, B: estimated number of bed places (August 2022), C: rural tourism accommodation, D: campsites, E: hostels, F: active tourism companies)

| | A | B | C | | D | | E | F |
|---|---|---|---|---|---|---|---|---|
| Province | 2022 | 2022 | 2003 | 2024 | 2003 | 2024 | 2024 | 2024 |
| Ávila | 334,053 | 5751 | 284 | 996 | 13 | 15 | 16 | 26 |
| Burgos | 252,906 | 4162 | 157 | 471 | 17 | 19 | 78 | 13 |
| León | 245,712 | 1973 | 248 | 534 | 35 | 35 | 122 | 36 |
| Palencia | 111,665 | 1808 | 141 | 262 | 5 | 4 | 34 | 14 |
| Salamanca | 186,579 | 4344 | 188 | 585 | 18 | 19 | 20 | 19 |
| Segovia | 260,508 | 4375 | 94 | 462 | 5 | 6 | 22 | 48 |
| Soria | 179,797 | 4042 | 79 | 386 | 9 | 8 | 17 | 16 |
| Valladolid | 99,476 | 2520 | 47 | 211 | 5 | 3 | 21 | 14 |
| Zamora | 149,204 | 3072 | 70 | 269 | 8 | 4 | 9 | 6 |
| Total | 1,819,900 | 32,047 | 1308 | 4176 | 115 | 113 | 339 | 192 |

Source: INE y SIE

modation places. Below them, with slightly lower overnight stay numbers, are Burgos and León, with over 245,000 overnight stays. León also stands out as a leader, well above the other provinces, in the number of hostels and campsites, and second in the number of active tourism companies. This indicates a specialization of the province in sports and youth tourism, as these are the users who typically frequent campsites and hostels, possibly influenced by the Camino de Santiago passing through the province and its proximity to Galicia. Burgos also has one of the highest numbers of hostels. While the rest of the provinces maintain average data, Valladolid lags far behind, which is not surprising given that it is the most urbanized province with fewer Natural Protected Areas.

Despite these encouraging statistics, specific planning for rural and/or natural tourism has only recently been developed. Two Strategic Tourism Plans have been approved in the region, the first covering the period 2014–2018 and the second for 2019–2023, both general for the entire tourism sector. The first Strategic Tourism Plan of 2014 aimed to strengthen the business fabric, professionalize the human resources of the sector, enhance competitiveness, and internationalize the offer, with no focus on the tourism in question. The second Plan analyzes the results of the first one and identifies key issues to be addressed in this new plan, including two of interest. The first is the promotion of responsible and sustainable tourism, fostering respect for the tourism heritage, specifically the Natural Heritage. The second is the promotion of nature tourism and active tourism in nature for sports activities. Only 35% of visitors in our region engage in natural activities, compared to 64.50% visiting cultural monuments, 54% enjoying gastronomy, and 51.30% exploring art and history (2017 data, Junta de Castilla y León). The Junta de Castilla y León is proposing a new model of responsible and sustainable tourism development, with five general objectives and four lines of action, none of which are focused on natural tourism.

Following the second Tourism Strategic Plan, the Smart Growth Plan for Rural Tourism in Castilla y León was published in 2021. The plan aims to position rural tourism as a strategic sector, modernize its productive model, and promote development in depopulated areas of the region. An analysis of the rural tourism context in the region over the past 5 years is presented, highlighting the leadership in national tourism data and identifying weaknesses. The plan emphasizes the need to improve productivity, competitiveness, and profitability of rural tourism businesses through specialization, an area where the region lags behind other autonomous communities. A SWOT analysis is presented (Fig. 1.10), focusing on economic aspects, and four specific objectives are established, each with multiple strategic lines: governance, competitive specialization, tourism sustainability, and innovation and digitalization. Under the sustainability objective, one strategic line addresses environmental sustainability, emphasizing the importance of preserving natural integrity, promoting resource efficiency, and reducing pollution.

All these objectives are intended to be implemented through Destination Tourism Sustainability Plans, which were recently awarded in April 2024, with a budget of 23 million euros. The beneficiaries can be found in Table 1.5. The selected projects can be classified based on the type of rural tourism:

- Natural tourism, with projects in the Tietar Valley, El Bierzo, the Geopark of Las Loras, and Medinaceli Natural Heritage.
- Wine tourism, with the Rueda Wine Route and wine landscape in the Palencia Cerrato.

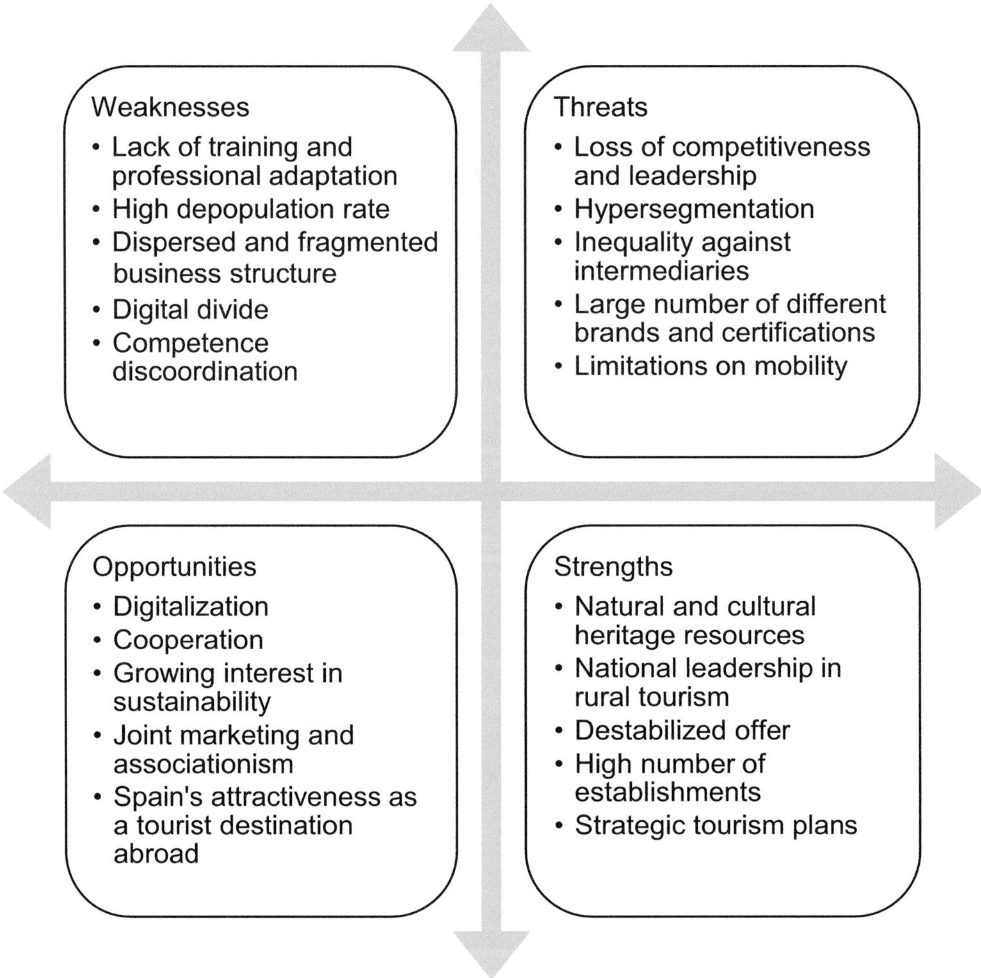


**Fig. 1.10** SWOT analysis of rural tourism in Castilla y León (2016–2021). Translated from: Smart Growth Plan for Rural Tourism in Castilla y León. Own elaboration

**Table 1.5** Tourism sustainability plans awarded in Castilla y León (2024)

| Plan | Beneficiary entity | Amount |
|---|---|---|
| Tourism Sustainability Plan in the Destination of La Adrada: Ecomobility for a sustainable tourism 360–365 in the Upper Tiétar Valley | Ayuntamiento de La Adrada | 1,367,000 € |
| Tourism Sustainability Plan in the Destination of UNESCO Global Geopark Las Loras (Municipalities in the province of Burgos) | Asociación para el Desarrollo Rural Integral de las Comarcas Circundantes al Camino de Santiago | 1,800,000 € |
| Tourism Sustainability Plan in the Destination "Conventus Iurídicus Cluniensis" | Diputación de Burgos | 2,400,000 € |
| Tourism Sustainability Plan in the Destination of the municipality of Carucedo | Ayuntamiento de Carucedo | 1,100,000 € |
| Tourism Sustainability Plan in the Destination Canal de Castilla | Consorcio para la Gestión Turística del Canal de Castilla | 2,000,000 € |
| Tourism Sustainability Plan in the Destination of the Palentino Cerrato, cultural landscape of wine | Diputación Provincial de Palencia | 2,000,000 € |
| Tourism Sustainability Plan in the Destination of Ciudad Rodrigo, a reference tourist destination | Ayuntamiento de Ciudad Rodrigo | 1,800,000 € |
| Tourism Sustainability Plan in the Destination, creation of the archaeological park Roman villa of Saelices El Chico toward cultural and rural tourism | Ayuntamiento de Saelices El Chico | 1,209,000 € |
| Tourism Sustainability Plan in the Destination Renacere | Ayuntamiento de Berlanga de Duero | 1,500,000 € |
| Tourism Sustainability Plan in the Destination of Medinaceli Natural Heritage | Ayuntamiento de Medinaceli | 1,500,000 € |
| Tourism Sustainability Plan in the Destination of Peñafiel | Ayuntamiento de Peñafiel | 1,500,000 € |
| Tourism Sustainability Plan in the Destination consolidation of the heritage and historical offer of Tordesillas | Ayuntamiento de Tordesillas | 1,500,000 € |
| Tourism Sustainability Plan of the Rueda Wine Route | Consorcio de la ruta del vino de Rueda | 1,580,000 € |
| Tourism Sustainability Plan in the Destination Castles of Valladolid: observe, enjoy, and share | Diputación Provincial de Valladolid | 2,100,000 € |

Source: Junta de Castilla y León. Date of consultation: 1/08/2024

- Cultural tourism (the category with the highest representation), with initiatives in Tordesillas, Ciudad Rodrigo, Canal de Castilla, Cluniac monasteries in Burgos, the archaeological park of the Roman villa of Saelices-El Chico, and Castles of Valladolid.

Thus, the Junta de Castilla y León has opted for cultural rural tourism, with projects that are not only more numerous but also have a higher economic allocation (Table 1.5).

The coming years and political actions will shape the future of natural tourism in Castilla y León. The focus has always been on the economic and territorial development benefits of a region heavily affected by depopulation, presenting rural tourism as a salvation tool. However, some authors have pointed out that while rural tourism provides additional income and maintains some jobs, it lacks the strength to create a viable economic alternative in rural areas that can attract population (Alario Trigueros et al., 2018). For now, we should focus on geotourism and geographic education instead of the overcrowding of Protected Natural Areas. In a context of climate change, pollution, and historical environmental degradation, awareness is more necessary than ever, and the only way to achieve the conservation of this heritage that must be protected and passed on to future generations, regardless of the economic value it may generate.

## 1.5 Geoheritage Inventory and Assessment Methods Review

In recent years, numerous methods for inventorying and evaluating geoheritage have been developed to study not only their scientific and geomorphological value but also their additional and/or cultural values, including some that assess their use and management (e.g., Brilha, 2016; Herrera-Franco et al., 2022; Mucivuna et al., 2019; Németh et al., 2021; Pasquaré Mariotto et al., 2023). While some European authors have emphasized the geotouristic and educational applications of geosites or geomorphosites, such studies are still scarce in Spain, particularly in the autonomous community of Castilla y León.

Recognizing the importance of geomorphosites for education, it is essential to determine how they can be integrated into formal education. The didactic potential refers to the suitability of geomorphosites for teaching Earth Sciences disciplines within formal education. This potential will vary depending on the educational systems of each country and the curriculum established for each educational level in subjects related to Earth Sciences (e.g., Geology, Geography, geomorphology, natural hazards, hydrology).

One of the objectives of this book is to design a novel and standardized methodological proposal that can be applied in Natural Spaces to assess their didactic and geotouristic potential. A brief overview of the most relevant methods used in recent decades for evaluating geomorphosites and geosites for tourism and educational purposes will be provided (Table 1.6).

Serrano and González-Trueba (2005) proposed a method for identifying Geomorphosites in Protected Natural Areas in Spain, known as the Valladolid Method. This method is based on a triple assessment:

- Scientific values (or intrinsic), purely geomorphological: genesis, morphology, dynamics, chronology, lithology, and structure.
- Cultural values (or added), which add value to natural elements: landscape and aesthetics, heritage elements, ecological, economic.
- Use and management (territorial components and potential use): ensuring their conservation, with appropriate management, impossibility of use without proper management, and possible serious deterioration.

The final assessment, reflecting the three elements evaluated, allows for an immediate comparison of the dominant value (natural or added) and therefore the context in which its management, use, and conservation are framed. The authors aim to provide managers of Natural Areas with a useful tool, in addition to information on the intrinsic, added, and use and management values of the places of interest to facilitate decision-making. For this purpose, geomorphosites mapping and assessment are developed.

Reynard et al. (2007) from the Institute of Geography at the University of Lausanne in Switzerland developed a method for evaluating geomorphosites that combines the scientific value of the sites with their specific (cultural and added) values, in the context of promoting geoheritage in the creation of Geoparks and the development of geotourism.

The method consists of a form divided into six parts, with the evaluation of scientific and additional value in the third and fourth parts, all using quantitative values.

The form includes:

1. General site information: location, type, and ownership.
2. Descriptive data: description and morphogenesis.
3. Evaluation of scientific value: rarity, representativeness, integrity, and paleogeographic value.
4. Evaluation of additional values: ecological, landscape, cultural, and economic value.
5. Synthesis: overall value, educational value, threats, and management measures.

Bollati et al. (2012) developed a method for assessing Geomorphosites for tourism and educational purposes. In addition to evaluating their scientific and additional characteristics, the third step involves assessing their potential for use, taking into account variables such as accessibility, avail-

**Table 1.6** Geoheritage inventory and assessment methods review

| Authors | Application | Scientific values | Additional values | Use and management | Educational/tourism application |
|---|---|---|---|---|---|
| Serrano and González Trueba (2005) | Geomorphosites in NPAs | Genesis, morphology, dynamics, chronology, lithology, structure | Landscape, aesthetics, heritage elements, didactics, science, and tourism | YES (appropriate management, impossibility of use without appropriate management, and possible serious deterioration) | X |
| Reynard et al. (2007) | Geomorphosites | Description, morphogenesis, rarity, representativeness, integrity, and palaeogeographical value | Ecological, landscape, cultural, economic, and educational value | YES (threats and management measures) | X |
| Bollati et al. (2012) | Geomorphosites for tourist and didactic use | Geomorphological evolution, educational exemplarity, spatial extent, geodiversity, geohistorical significance, ecological supporting role, integrity, rarity, other geological interests | Cultural, aesthetic, socioeconomic | X | Potential for tourism/ educational use: accessibility, services, visibility, number of tourists, sporting activities, legal restrictions, geomorphological, and additional interest |
| Kubalikova (2013) | Geomorphosites for tourism purposes (considering the didactic potential) | Integrity, rarity, scientific knowledge of site, morphology, genesis, age, diversity | Cultural, ecological, aesthetic | Threats and risks: conservation activities, risks, and threats to the site, current state of degradation of the site | Educational potential: exemplarity and representativeness, clarity, visibility of features and processes, educational facilities<br>Tourism: quality of tourism services, local producers, accessibility |
| Brilha (2016) | Geological and geodiversity sites | Representativeness, integrity, rarity, scientific knowledge | X | X | Educational: educational potential, geological diversity, accessibility, safety<br>Touristic: landscape, interpretative potential, accessibility, and safety |
| Stepišnik et al. (2017) | Geodiversity for educational purposes | Identification of areas with high geodiversity value | Additional values | X | Educational: geodiversity, accessibility, security |
| Bussard et al. (2022) | Geomorphological landscapes | Integrity, rarity, representativeness, and palaeogeographical interest | Aesthetic, ecological, and cultural | X | Educational: visibility and level of complexity |

ability of services, visibility, number of tourists, sports activities, legal restrictions, geomorphological interest, additional interest, and the presence of geomorphosites in the surrounding area.

Kubalíková (2013) proposed a method for evaluating geomorphosites for tourism purposes, while also considering their educational potential. According to the author, an appropriate method for assessing geotouristic potential should consider the following criteria:

1. Evaluation of scientific and intrinsic values.
   (a) Assessing the integrity, rarity, and importance of the Earth Sciences of the site.
   (b) Evaluating the scientific knowledge of the site.
   (c) Assessing the morphology, genesis, age, and diversity of the site.
2. Evaluation of exemplarity and educational potential.
   (a) Assessing the exemplarity and representativeness of the site, clarity, and visibility of features and processes.
   (b) Evaluating the presence of educational facilities (brochures, websites, informational panels, guided tours).
   (c) Assessing the level of educational use of the site.
3. Evaluation of site accessibility and visibility and the presence of tourist infrastructure.
   (a) Evaluating the number, distance, and quality of tourist services.

   (b) Assessing the presence of local products that support and/or promote the site.
   (c) Evaluating accessibility.
4. Evaluation of existing threats and risks, assessing conservation activities or existing legislative protection of the site.
   (a) Evaluating conservation activities (legal protection, other types of protection).
   (b) Assessing risks and threats to the site.
   (c) Evaluating the current state of the site, the level of disturbance or degradation.
5. Evaluation of added values.
   (a) Assessing cultural values (historical/religious/archaeological).
   (b) Evaluating ecological value (relationships with living nature).
   (c) Assessing aesthetic/scenic/landscape value.

Brilha (2016) developed a method for inventorying and evaluating geodiversity sites, proposing a quantitative assessment of their educational and touristic potential based on the following criteria:

- For educational purposes:
  - Didactic potential
  - Geological diversity
  - Accessibility
  - Safety
- For tourism:
  - Landscape
  - Interpretive potential
  - Accessibility
  - Safety

Brilha emphasizes that accessibility and safety are equally important for both tourism and education, giving more weight to diversity for education and to landscape for tourism. After fieldwork, a final characterization of the sites is prepared, and quantitative assessments of their educational or touristic potential and risk of degradation are applied.

Stepišnik and Trenchovska (2018) developed a method to assess geodiversity for educational purposes. They argue that geodiversity is valuable for teaching due to its diverse landscapes. To evaluate the educational potential, the authors consider criteria such as accessibility, safety, alignment with curriculum content, interdisciplinary integration, broader context integration, fieldwork techniques, and availability of materials for teachers.

Martin (2012) and Bussard et al. (2022) emphasize the pedagogical potential of geomorphosites, suggesting that their educational value depends on how a teacher utilizes them as educational resources. They prefer to discuss pedagogical potential as a characteristic of use rather than an inherent scientific or additional value. According to Martin (2012), every geosite has some pedagogical value, and educational potential can be realized through interpretation, allowing for the evaluation of the educational value of an interpretation product (Bussard et al., 2022). Building on this concept, Bussard et al. (2022) propose a method to assess the pedagogical potential of geomorphological landscapes, involving three stages: selection of geomorphological landscapes, description of the sites and evaluation of their heritage values, and assessment of their educational potential. Two criteria are considered to determine the pedagogical potential of geomorphological landscapes: visibility, indicating the presence or absence of unobstructed views to observe geomorphological features, and complexity level, which varies based on the number of objects and concepts needed to understand the site's morphogenesis.

The review of methods concludes that while early methods focused on analyzing the scientific and additional values of Geomorphosites, in recent years there has been an effort to assess the tourism and educational potential of these sites. However, there is a proliferation of tourism interest, with the educational potential often being subordinate to tourism or linked to it, rather than valued in its own right, and always focused on awareness, dissemination, and environmental education, rather than formal education.

It is important to clarify that the choice of the term "didactic" (didactic potential of geomorphosites) means linking geomorphosites to the field of formal education, specifically Geography, while in the field of geointerpretation, the term "educational" (educational potential of Geomorphosites) is used to raise public awareness about the heritage value of Geomorphosites. Therefore, the difference between didactic potential and educational potential lies in the objective: didactic implies disciplinary learning in the context of formal education, while educational aims at dissemination and awareness.

The didactic potential of geomorphosites can be evaluated based on the degree of alignment between geomorphosites and established didactic concepts for teaching Geography, specifically Geomorphology. To do this, a detailed analysis of the educational legislation of the relevant autonomous community is necessary, as done in this book (see Sect. 5.3.1.1).

# 2 Objectives and Methodology

The objectives of this research are aligned with the developed methodology.

The first objective is the geomorphological study and inventory of geomorphosites in Natural Protected Areas (NPAs) in Castilla y León region. Cañón del Río Lobos Natural Park and Sierras de la Paramera y la Serrota have been selected. Cañón del Río Lobos Natural Park, declared in 1985, was the first NPA in Castilla y León. The Pangea Research Group has conducted a comprehensive study and description of its geomorphological heritage, inventory and geotouristic evaluation of geomorphosites, and its socioeconomic environment, culminating in the publication of the book *Natural Heritage and Geomorphology: the Geomorphological Sites of the Natural Park Cañón del Río Lobos* by Enrique Serrano, Mª José González Amuchástegui, and Rosa María Ruiz Pedrosa, 2020. Therefore, this book does not aim to provide a detailed description of the physical environment, although brief introductions to the geomorphology of the Park are included. The objectives in the Río Lobos area are to provide a brief introduction to its physical environment, update socioeconomic data, and adapt the geomorphosites inventory to the new methodology designed in this work. Sierras de la Paramera and la Serrota NPA, although it has been part of the Natura 2000 network since 1998 as a Special Conservation Site, has not been included in the Natural Protected Areas network of Castilla y León. It lacks any recognizable protection status, heritage management, or environmental protection initiatives. Therefore, the objectives in Paramera and la Serrota have been the comprehensive analysis and description of the physical and socioeconomic environment, including an inventory of geomorphosites and the development of geomorphological mapping.

After the geomorphological study, the second objective is to design a method for assessing geomorphosites and their didactic potential, understood as their suitability for didactic use in formal education for teaching Earth Sciences. And, after the methodological design, it must be applied to assess the geomorphosites potential use, geotourism potential (already developed in Serrano et al., 2020), and didactic potential proposed in this book to the inventoried geomorphosites in both NPAs.

Finally, the aim is to propose dissemination, planning, and management actions tailored to the specific needs and potential of each study area in order to promote the valorization, appreciation, and conservation of geomorphological heritage. In Sierras de la Paramera and la Serrota, there are three main goals. The first one is to develop a geomorphological map and establish connections with cultural heritage.

In Paramera and Serrota, it is also necessary to explore didactic approaches to geomorphosites. On the other hand, in Cañón del Río Lobos Natural Park, the objective is to design georoutes to diversify geotourism offerings, additionally to create an inventory of geomorphosites and develop a geomorphological map for planning, management, and public use.

The methodology of this research outlines the following research phases: (1) study of geomorphological heritage and inventory of geomorphosites, (2) development of geomorphological maps, (3) implementation of the educational potential assessment method, and (4) evaluation of tourism potential and design of a geotourism map.



R. M. Ruiz Pedrosa, *Geomorphological Heritage as Didactic and Geotouristic Resources*, Geoheritage, Geoparks and Geotourism, https://doi.org/10.1007/978-3-032-06205-5_2

## 2.1 Geomorphological Analysis and Geomorphosites Inventory

### 2.1.1 Geomorphological Maps

Geomorphological maps are efficient graphic inventories of landforms that represent surface and subsurface materials. The availability of high-resolution remote sensing data (aerial and satellite images, digital elevation data) allows for competent interpretation and spatial representation of geographic features (Otto & Smith, 2013; Seijmonsbergen, 2013). Geomorphological mapping is the first step in developing thematic maps, such as hazard, risk, geoheritage, and geotourism maps, but they need to be simplified for different purposes such as management and education (Regolini-Bissig, 2008, 2010, 2011; Bollati et al., 2017a, 2017b; Bouzekraoui et al., 2018a).

Geomorphological mapping is useful for understanding the heritage elements or sites of a specific territory, but it also allows for establishing interrelationships with other geomorphological or cultural elements. In the case of Geomorphosites classified as sites, geomorphological schemes allow for analyzing their internal characteristics and their structural or dynamic relationships with cultural elements (Serrano et al., 2020; Migoń, 2021).

Geomorphological mapping is a useful communication tool that, through a cartographic system understandable by experts, becomes a preliminary tool for environmental management and geomorphological risk (Cendrero, 1980; Coratza & Regolini-Bissig, 2009; Otto & Smith, 2013; Coratza et al., 2021), particularly in its application to protected natural and cultural spaces. Published works use geomorphological mapping and its derivatives for research, management, or education, with necessarily differentiated criteria. Thus, geomorphological mapping has been oriented toward different proposals with significant advances in the last 20 years (Regolini-Bissig, 2012; Fuertes-Gutiérrez et al., 2016; Coratza et al., 2021). Scientific cartographies have been proposed exploring cartographic methodology and techniques used—such as GIS and geomatics (Carton et al., 2005; Coratza & Regolini-Bissig, 2009).

All the works show the need to incorporate geomorphological cartography and its derivatives for the knowledge and dissemination of Geomorphosites, but also the difficulties in representing the elements and their characteristics (Serrano & González Trueba, 2011; Coratza & Hobléa, 2018; Coratza et al., 2021). Even knowing that the map is just an interpretation of the territory where what the author wants to show and explain is expressed, the geomorphological map must be as precise, clear, and objective as possible, at the same time as an interpretative synthesis useful for users and therefore oriented to different audiences, sometimes specialized, such as managers, educators, and interpreters of the territory, and other times nonspecialized, such as tourists, hikers, and visitors. Undoubtedly, the maps will be different if they are oriented toward environmental and territorial management or toward educational or geotouristic uses. It is in this place where the main challenges are found.

The development of the geomorphological map of the Ulaca Vettonian hillfort at a scale of 1:20,000 and of the Serrota was based on a review of literature and maps, including the georeferencing of existing geomorphological sketches (Garzón et al., 1981; Moreno, 2008), and the interpretation of remote sensing data. Remote sensing data include digital elevation models and slope derived from LiDAR terrain models, digital orthophotomaps, and Google Earth images (Table 2.1). In the case of Ulaca, images obtained by UAV (Maté-González et al., 2022) have been

**Table 2.1** Products used to compile the geomorphological map

| Product | Source | Characteristics |
|---|---|---|
| Topographic Map 1/25,000 | Instituto Geográfico Nacional (IGN) (http://centrodedescargas.cnig.es/) | Contours line, interval original 10 m, in the map 50 m |
| Aerial photography | Agrarian Research Institute of Castilla y León (ITACYL) | 25 cm resolution aerial photographs |
| Orthophotomaps | PNOA Plan Nacional de Ortofotografia Aérea (https://mirame.chduero.es/chduero/viewer)<br>Instituto Geográfico Nacional (IGN) | Variable resolution (25–45 cm/pixel)<br>Terrain characteristics and landforms recognition (outcrops, deposits, landforms, etc.) |
| MDT | Confederación Hidrográfica del Duero (https://mirame.chduero.es/chduero/viewer) | 5 m/pixel<br>Terrain characteristics and landforms |
| Stereoscopic viewer | Instituto Geográfico Nacional (IGN) (http://www.ign.es/3d-stereo/)<br>Photograms of the PNOA | Variable resolution (22–45 cm/pixel)<br>Characteristics of the ground surface<br>Morphology |
| 3D models satellite images | Browser of Google Earth (https://earth.google.com/web/)<br>IGN imagery (large scale) and Landsat imagery | Validation of the ground and terrain characteristics and a broader framework of geomorphological features |
| 3D models, UAV images | Ulaca Virtual Tour<br>USAL, Diputación de Ávila, IGDA (https://tidop.usal.es/Ulaca/) | High resolution images<br>Micro-landforms and detailed features recognition |

used to analyze small-scale elements and granitic microforms.

On the other hand, the geomorphological mapping of the Cañón del Río Lobos National Park, carried out at scales of 1:10,000 and 1:25,000, has been done through the interpretation of aerial photographs and orthophotographs from the National Aerial Orthophotography Plan (PNOA), analysis of the digital elevation model (DEM), and fieldwork (Serrano et al., 2020). The map presents four layers of information: planimetry, altimetry, toponymy, and geomorphology, organized in layers as well. Twenty-eight geomorphological elements have been represented, using a combination copatterns and symbols to represent structural forms, erosion surfaces, karstic, fluvial, and slope forms. The colors and shapes follow established and classic guidelines (Tricart, 1970; Peña Monné, 1997; Smith et al., 2011), adapted in some cases to the conditions of the Natural Park. In this case, mapping has been carried out for the knowledge and management of the Natural Park, as a geoscientific document for managers and users (Verstappen, 2011).

### 2.1.2 Geomorphosites Inventory

It consists of three steps: desk work, field work, and geomorphosites assessment.

1. **Desk work**. For an initial geomorphological contextualization, a study of the physical environment and geomorphology of the study area is carried out through the consultation of maps and existing works, accompanied by a preliminary photointerpretation before fieldwork. Based on this desk work, an initial list of potential Geomorphosites is prepared, which is later contrasted with fieldwork.
2. **Fieldwork**, in which the NPAs of study are visited and, after a thorough geomorphological and geoheritage analysis in various field days, the final inventory of potential geomorphosites is prepared, accompanied by a detailed description of each of them in which their scientific, added, and accessibility values are described and analyzed. This is a modification of the method developed by Serrano and González Trueba in 2005, with more detail added on accessibility and added values (Table 2.2). The geomorphosite is classified according to whether it is an Element (an isolated landform) or a Site (a set of landforms with the same geomorphological genesis), and whether it is Representative in its geomorphological context (there is more than one similar example) or Exceptional (it is the only example of a particular landform). This method has already been applied in previous works in protected natural areas and mountain landscapes (Serrano & González-Trueba, 2005; Pellitero Ondicol, 2012; González Amuchastegui et al., 2014; Serrano et al., 2020). It involves a systematic collection of basic identification data, location, and Altitude, followed by the description of their geomorphological characteristics. Especially important for the subsequent evaluation of didactic and tourist potential is the description of additional values and accessibility that has been added to the table developed by the authors, i.e., those characteristics that add value to the geomorphosites and will condition their uses, such as landscape value, cultural and economic elements, and their impacts and infrastructures.
3. **Geomorphosites assessment**. The previously identified geomorphosites are numerically evaluated, giving values from 0 to 4. The evaluation consists of an expert judgment based on three steps (see Table 2.3):
   (a) Factors that condition their use, considering their accessibility (the safety of the path to access the geomorphosites), fragility of the geomorphosites in the face of visitor influx, seasonality (the months that can be visited per year, depending on the weather), intensity of physical activity necessary to access the site (characterized by distance and elevation, measured in kilometers and altitude, respectively), visibility, and current uses and legislation (level of environmental protection).
   (b) Physical elements assessment, including its geological richness (number of geological eras present in the geomorphosites and characterization of lithology),

**Table 2.2** Geomorphosites identification sheet

| Identification | **Name:** | **Site:** | **N°** |
|---|---|---|---|
| | | | Altitude: |
| Situation | Municipality: | Coordinates: | |
| Geomorphological values | TYPE (Place/Site) | | |
| | Genesis | | |
| | Morphology: description, morphostructures, erosion | | |
| | Dynamics | | |
| | Chronology | | |
| | Primary interest | | |
| | Secondary interest | | |
| | Representativeness | | |
| | Geomorphosite attribution | | |
| Additional values | Landscape | | |
| | Cultural elements | | |
| | Economic | | |
| | Social | | |
| | Conservation status | | |
| | Current uses | | |
| | Infrastructure | | |
| | Impacts | | |
| | Legal situation | | |
| Accessibility | Walking distance | | |
| | Slope | | |
| | Security | | |

**Table 2.3** Geomorphosites assessment method

| | | |
|---|---|---|
| 1. Conditioning factors for its use | Accessibility | 0: no path; 2: loose rock path; 4: well-defined path |
| | Fragility (risk of degradation) | 0: high (the geomorphosite is already degraded); 2: moderate (the geomorphosite is in process of degradation); 4: low (no degradation) |
| | Seasonality | 0: can be visited less than 3 months per year; 2: can be visited 3–6 months per year; 4: can be visited all the year |
| | Intensity of physical activity (distance and slope) | 0: >15 km and +1.000 m; 2: <10 km and +500 m; 4: <5 km and <500 m |
| | Visibility of the geomorphosite | 0: covered visibility; 2: medium visibility; 4: full visibility |
| | Current uses | 0: there is not; 2: occasional use; 4: frequent use |
| | Legislation | 0: there is not; 2: in development; 4: strong protection |
| | **Total (up to 28)** | |
| 2. Physical elements | Geology | 0: one geological era; 2: two geological eras; 4: three or more geological eras |
| | Rocks (igneous, sedimentary, and metamorphic) | 0: there is not; 2: up to four; 4: five or more |
| | Superficial landforms | 0: there is not; 2: one or two; 4: three or more |
| | Sedimentological cascade | 0: no evidence; 2: one example; 4: two or more examples |
| | Weathering (chemical, physical, or biological) | 0: no evidence; 2: one example; 4: two or more examples |
| | Active processes | 0: no active processes; 2: one example; 4: two or more examples |
| | Biodiversity | 0: no evidence; 2: one example; 4: two or more examples |
| | Hydrology | 0: no liquid or solid water; 2: one example; 4: two or more examples |
| | **Total (up to 32)** | |
| 3. Additional values | Landscape and aesthetic | 0: no landscape view, landscape entirely covered by natural or anthropogenic features; 2: landscape view enclosed, e.g., by vegetation or other elements of the natural environment; 4: wide landscape view, no obstacles to observation |
| | Cultural elements | 0: no existence; 2: one example; 4: more than one example |
| | Anthropic influence | 0: dominance of nature, space little, or not at all modified by humans; 2: rural/natural space partially modified by man; 4: anthropogenic influence dominates over natural influence |
| | Economic value | 0: no economic activities; 2: one economic activity; 4: more than two economic activities |
| | **Total (up to 16)** | |

distinction of landforms, the presence of exogenous forces shaping the relief (weathering, erosion, transportation, and sedimentation processes), active processes, degree of biodiversity, and evaluation of the hydrology.

(c) Assessment of its additional values, considering landscape and aesthetic (the view of the landscape from the geomorphosite depending on the degree of visibility), cultural elements (the number of examples of both tangible and intangible cultural heritage), anthropic influence (evaluates the degree of anthropic transformation of the territory in which the geomorphosite is framed), and economic values of the site (the presence of economic activities related to the geomorphosite or the territory in which the geomorphosite is located). Example: active tourism companies, livestock farming, forestry.

Based on the assessment results, a final list of geomorphosites is designed to compare and identify the key attributes of each site, and assess their educational and geotouristic potential.

## 2.2 Didactic Potential Assessment

The method for assessing the didactic potential of geomorphosites is designed for application in NPAs in the region of Castilla y León, but it can also be applied in Natural Spaces (protected or not) in any other Spanish regions and even in other countries, as it has been done in this book.

The objective of the method is for geomorphosites to be resources for the formal teaching of Geography, Earth Sciences, Environmental Sciences, or Geology at different educational levels (primary, secondary, and higher education). To determine the suitability for each educational level, it is necessary to consult the curriculum content related to Geography, Geology, and Geomorphology of each subject linked to these disciplines (Table 2.4). Geomorphosites represent physical processes, their relationship with the territory, human activity, and cultural heritage, so depending on the intrinsic scientific value and added values of each geomorphosite, they may be more or less suitable for different educational levels. A geomorphosite with a very complex geomorphological content may only be suitable for secondary and higher education stages (e.g., a hanging syncline), while simpler landforms with the presence of cultural elements may be suitable for teaching from the primary stage (e.g., a river valley). The method also evaluates the availability of didactic materials, such as activity sheets for students or geoheritage guides for teachers, and the possibility of applying field techniques (e.g., creating topographic profiles, vegetation pyramids, etc.) and the relationship with other school subjects.

**Table 2.4** Geomorphosites didactic suitability assessment

| áers | Criteria | Scores |
|---|---|---|
| Didactic suitability assessment | Primary education suitability | 0: the values represented by the geomorphosite are not included in curriculum; 2: considered in a cross-cutting manner; 4: compulsory in the curriculum |
| | Secondary education suitability | 0: the values represented by the geomorphosite are not included in curriculum; 2: considered in a cross-cutting manner; 4: compulsory in the curriculum |
| | Higher education suitability | 0: the values represented by the geomorphosite are not included in curriculum; 2: considered in a cross-cutting manner; 4: compulsory in the curriculum |
| | Relation with other disciplines | 0: no relationship; 2: related with social science disciplines; 4: related to other science disciplines |
| | Suitable material for formal education | 0: no existence; 2: presence of didactic material for formal education (e.g., activity sheets for students, geoheritage teaching guides for teachers); 4: didactic projects are under development |
| | Fieldwork techniques | 0: no field techniques; 2: possibility to apply some field techniques; 4: possibility to apply several techniques |
| | **Total (over 24 examples)** | |

Likewise, each criteria is scored from 0 to 4, obtaining a numerical result that allows the comparison between the didactic potential of the geomorphosites, and to select those with more potential for the design of didactic activities or routes.

## 2.3 Touristic Potential Assessment: Geotouristic Maps

Geotourism assessment focuses on managers and users of geomorphosites, aiming to provide elements for enjoyment, leisure, and learning about geomorphology. The emphasis is on added values and management rather than intrinsic values (Pralong, 2005). The outcome of this assessment is the development of a geotourism map as an applied tool.

Serrano et al. (2020) provide a detailed evaluation of criteria considered by other authors (Pralong, 2005; Kubalikoba, 2013; Brilha, 2016), combining them to obtain a comprehensive view of the natural, cultural, and additional value of geomorphosites (Table 2.5). Criteria such as scenic, scientific, cultural, educational, conservation (vulnerability, usage limitations), and added values have been taken into account.

Following Brilha (2016), weights have been assigned to each criterion, giving more importance to values appreciated by tourism such as scenic, cultural, conservation, accessibility, and safety, than to other attractions (Table 2.6). Cultural values are considered in relation to existing cultural assets that complement or enhance the assessment of geomorphosites through the mutual interrelations between geomorphology and historical-artistic or cultural elements.

After assessing the tourist potential of Geomorphosites, which is medium to low in the Sierras de la Paramera and la Serrota, only the geotourism map of the Cañón del Río Lobos National Park has been developed. This map is the result of simplifying the geomorphological map and incorporating tourist information, taking into account criteria such as target audience, objectives, graphic design choices, and practical aspects to enhance its usefulness and applicability (Regolini-Bissig, 2011). Among the various types and proposals of maps for geotourists, the geodidactic map is considered the most suitable tool for education and recreation. These maps are large-scale, reproducible in brochures and leaflets, with schematic backgrounds, figurative symbols, and relevant additional information for tourists (Regolini-Bissig, 2008, 2011). The map is aimed at tourists who are not familiar with geomorphology or Earth Sciences, making it easier for them to understand the visible geomorphological or geological phenomena in the geomorphosites.

**Table 2.5** Assessing tourism potential methods review

| Criteria | | Authors | | | |
|---|---|---|---|---|---|
| | | Pralong (2005) | Kubalikoba (2013) | Brilha (2016) | Serrano et al. (2020) |
| Scenic | | Yes | No | Yes | Yes |
| Scientific | | Yes | Yes | No | Yes |
| Cultural | | Yes | No | No | Yes |
| Economic | | Yes | Yes | Yes | No |
| Educational | | No | Yes | No | Yes |
| Conservation | Vulnerability | No | Yes | Yes | Yes |
| | Use limitations | No | Yes | Yes | Yes |
| | Uniqueness | No | Yes | Yes | No |
| Additional | Accessibility | No | Yes | Yes | Yes |
| | Security | No | Yes | Yes | Yes |
| | Logistics | No | Yes | Yes | No |
| | Population | No | Yes | Yes | No |
| | Observation conditions | No | Yes | Yes | Yes |
| | Proximity to recreational areas | No | Yes | Yes | Yes |

**Table 2.6** Tourism potential assessment and assigned values

| Criteria | | Importance | Numerical assessment proposal |
|---|---|---|---|
| Scenic: panoramic view, size of panoramic view, geographic diversity, natural diversity | | | 0–5–10–15 |
| Scientific: integrity, rarity, geodiversity, scientific knowledge | | | 0–5–10 |
| Cultural: presence of cultural values, value of cultural elements, number of elements, historical diversity | | | 0–5–10–15 |
| Educational: representativeness and clarity of forms or processes, pedagogical exemplarity, didactic documentation available, current educational use | | | 0–5 |
| Conservation | Vulnerability: risk of degradation and fragility | 10 | 0–5–10 |
| | Limitations on use: legislation for its protection | 5 | 0–5 |
| Added | | 10 | 0–5–10 |
| | Accessibility | 10 | 0–5–10 |
| | Security | 15 | 0–5–10–15 |
| | Observation conditions | 5 | 0–5 |

The basic principles of geotourism maps (Castaldini et al., 2005, 2009; Carton et al., 2005; Coratza & Regolini-Bissig, 2009; Regolini-Bissig, 2010) emphasize highlighting only recognizable landscape features and being simple, clear, and practical in the field. These principles are synthesized to provide a useful map for visitors to discover and understand abiotic elements while maintaining scientific rigor as a document for dissemination and scientific knowledge. Geoconservation should be a key objective in preparing detailed geotourism maps with interpretive routes or paths. It involves selecting key elements from the geomorphological map, simplifying reading levels, and spatially representing the most significant elements (Castaldini et al., 2005, 2009; Coratza & Regolini-Bissig, 2009; Regolini-Bissig, 2010, 2012).

# 3 Sierras de la Paramera y la Serrota

## 3.1 Physical Environment Study

The Paramera and Serrota mountain ranges are located southwest of the city of Ávila (Fig. 3.1), within the western sector of the Sistema Central and are part of one of the three major structural complexes of the Sistema Central: the Gredos Complex, according to Capote et al. (1982). This mountain range serves as the watershed between the Duero and Tajo river basins, as it borders the Valle de Amblés to the north, with the Adaja River as its main watercourse belonging to the Duero basin, and the Alberche River to the south, belonging to the Tajo basin. Altitudes in the study area range from 850 m in the valley bottoms to 2294 m, the highest point at La Serrota peak. The Paramera and Serrota mountain ranges are separated by the Mengamuñoz mountain pass. To the east of Serrota is the Piedrahita mountain range, while on the other side of the Valle de Amblés is the Ávila mountain range. To the south, the Alberche River separates our study area from the Gredos mountain range.

### 3.1.1 Geological Context

From a geological point of view, the area is located in the Central-Iberian Zone of the Hesperian Massif. Most of the study area is composed of a large Hercynian basement of granites and metasediments of Paleozoic origin.

The materials in the study area are grouped into two main sets (Fernández Ruiz et al., 2006): one consisting of igneous and metamorphic rocks belonging to the Hercynian Massif, and the other consisting of Tertiary and Quaternary sediments corresponding to the western sector of the Amblés Valley.

The largest spatial extent is occupied by biotitic granites and granodiorites whose age ranges from the Permian to the Carboniferous, with later intrusions of other granitic masses, usually within the previous granodiorites, which have a smaller extent, lower degree of fracturing, and higher acidity, making them more resistant to erosion. Their presence in the relief is characterized by the abundance of in situ rock, with prominent rocky outcrops (Bullón Mata et al., 1988).

At the end of the Mesozoic, when the Central System had not yet risen, there was a single erosion surface, on which a significant weathering profile would develop (Fernández Ruiz et al., 2006). It is during the Alpine orogeny, especially in the Castilian and Neocastilian phases, when the relief of the Central System is generated, acquiring a morphology of stepped blocks, caused by compartmentalization and vertical unevenness of the initial basement. Thus, it gives rise to a Germanic-style relief (Fig. 3.2), leading to the dismantling of the elevated blocks, with their materials being deposited in the depressed areas, filling the tectonic basins. Therefore, this sierra-valley–sierra-valley–sierra structure (Sierra de Ávila–Amblés Valley–Paramera and Serrota Sierras–Alberche Valley–Gredos Sierra) corresponds to a Germanic relief of elevated blocks and sunken blocks: horst (the sierras) and graben (the tectonic basins, the valleys) typical of the Central System, with the blocks uneven from ancient fracture lines.

### 3.1.2 Geomorphology: Structural Relief and Landforms

The structural relief of the Paramera and Serrota, as in the entire Central System, is defined by the fractures that limit the tectonic blocks, thus conditioning their Altitude and orientation or crushing and breaking the rocks. The Central System, located in the province of Ávila, was part of the internal zones of the Hercynian chain, where the deformation and metamorphism of the rocks were more intense and early, leading to extensive crystallizations of granitoids. Once the orogenic effort was completed, the entire sector achieved great stability, and the granites became more rigid,


R. M. Ruiz Pedrosa, *Geomorphological Heritage as Didactic and Geotouristic Resources*, Geoheritage, Geoparks and Geotourism, https://doi.org/10.1007/978-3-032-06205-5_3

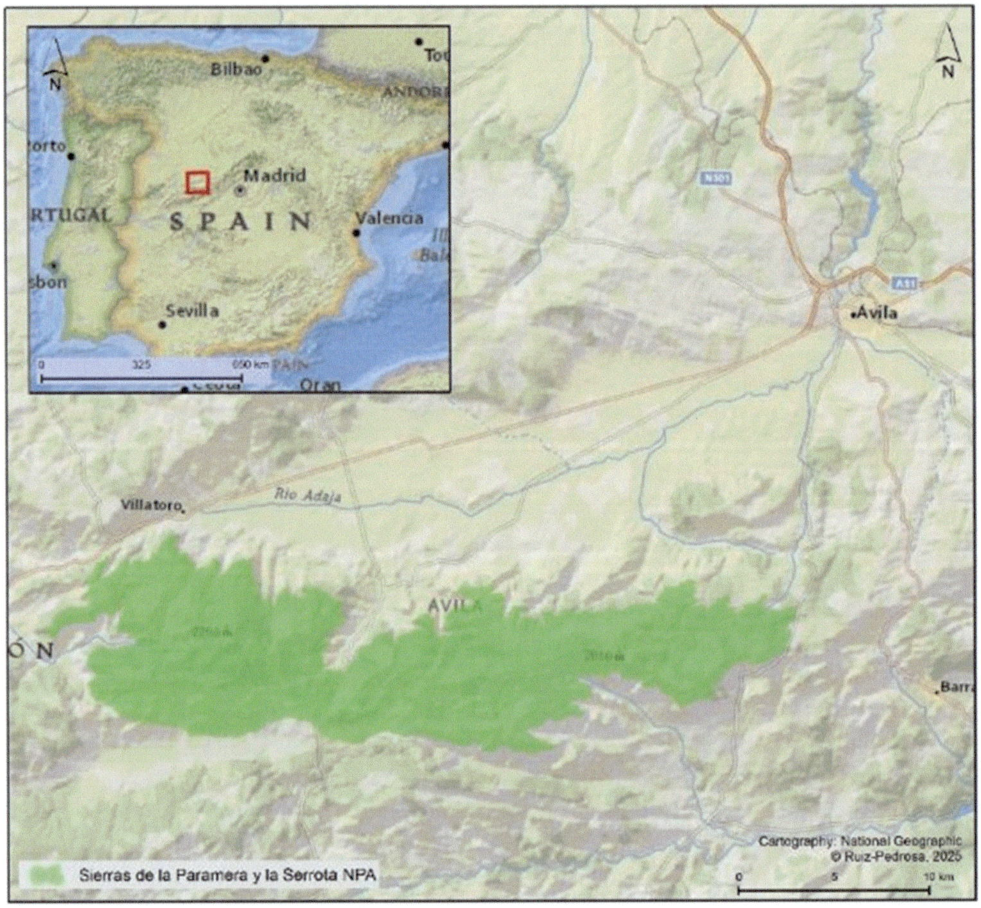


**Fig. 3.1** Location of Sierras de la Paramera y la Serrota, 25 km southwest of the city of Ávila (central Spain)

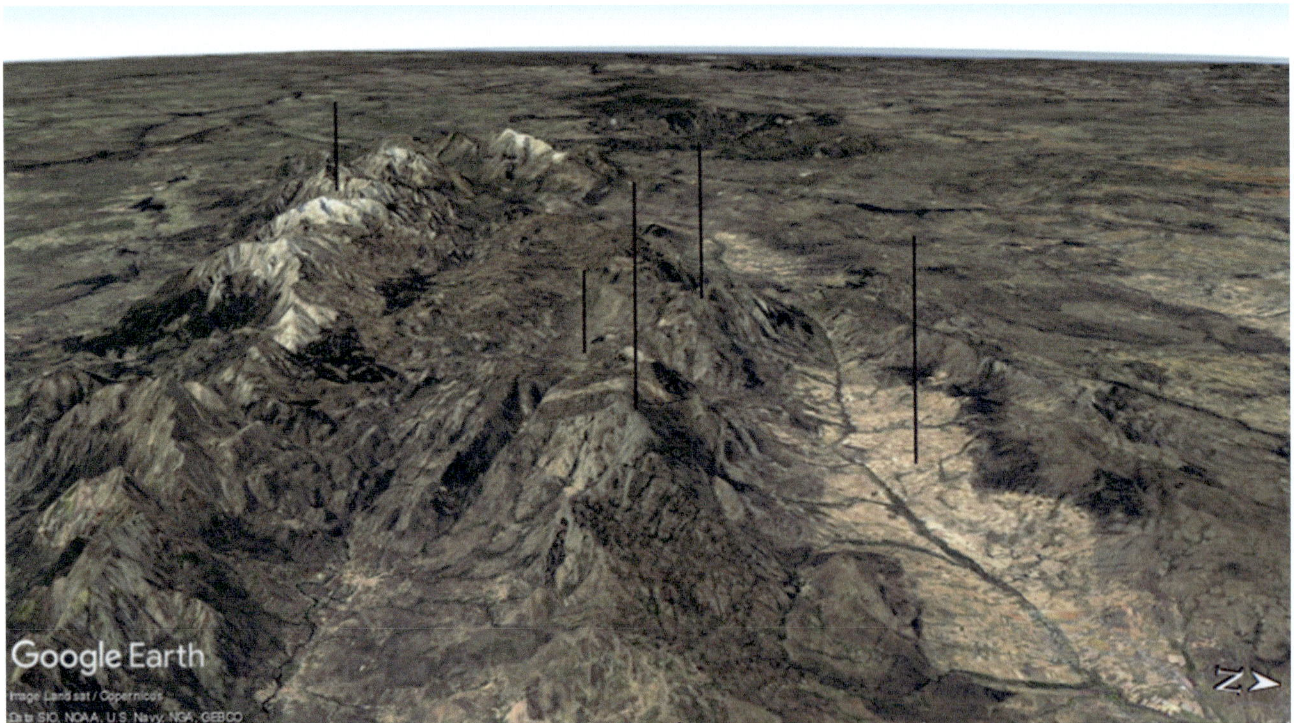


**Fig. 3.2** Aerial view of the study area. (Source: Google Earth)

so the response to tectonic thrusts from this point on would be fracturing (Garzón et al., 1981; Bullón Mata et al., 1988; Pedraza, 1989; Fernández Ruiz et al., 2006; Martín-Parra et al., 2008).

The fractures in the Sierras of the Paramera and Serrota belong to the same families present throughout the Hesperian massif, with their layout, dynamic history, directionality, and origin being the same as extensively described in the Central System (Pedraza et al., 1989). For the province of Ávila, Bullón Mata et al. (1988) recognized a series of main faults with NE–SW and NW–SE directions formed at the end of the Primary era, in the late Hercynian period, from which block mobilizations occur. With this structural framework, tertiary reactivation acts selectively, mobilizing mainly on the NE–SW direction fractures, although the E–W fractures also become important in Ávila. Together, they give rise to the general direction of the Central System in this province, and to the most important mountain alignments and basins (Bullón Mata et al., 1988; de Vicente et al., 2018). In the Sierra de la Paramera and Serrota, the Puerto del Pico Fault stands out, dividing this southern mountain alignment of the Province of Ávila into two—La Paramera and La Serrota—through the Mengamuñoz mountain pass.

In the Central System, three large morphostructural units are easily delimited (Sanz Herráiz, 1988), which are recognized in our study area:

- The ramps are half horsts that stand out above the relatively sunken blocks of the basins as they have been raised by important faults. The internal structure of these ramps would have been erased by the development of piedmont slopes. They act as channels for runoff from the mountains. An example is the Ulaca half horst.
- The mountains, being the horsts that form the most prominent reliefs, such as Picos Zapateros and Pico de la Serrota.
- The valleys, grabens, or intramountain basins that articulate the elevated blocks and their contact with the ramps. In the study area, these are the Amblés Valley, the Picuezo River valley, and the Alberche Valley.

The Paramera and Serrota mountain ranges are the elevated horsts that stand out between the Amblés Valley to the north and the Alberche River to the south (Fig. 3.3). It constitutes a mountain range developed in an east–west direction, with elevations exceeding 2000 m, highlighting the Zapatero Peak in the Paramera, at 2158 m, and reaching 2294 m in the Serrota. To the north, they are limited by the Amblés Valley basin, with steep cliffs toward it, softer in the case of the Serrota than in the Zapatero Peak, as a flat surface, the Paramera de Ávila, with an average Altitude of about 1500 m, gently connects the mountain slopes and hangs over the Amblés Valley.

The Paramera, which gives its name to the mountain range, is defined by Bullón Mata et al. (1988) as a set of blocks delineated from fractures, resulting in a tectonic block with uniform relief, shaped by very continuous erosion surfaces, partially dissected by alteration alveoli. These alteration mantles make the soils relatively suitable for agricultural exploitation, which in the case of the Paramera in Ávila and its climatic conditions, it is limited to livestock use.

The connection of these parameras with the Picos Zapateros and La Serrota occurs gradually, with the slopes gradually losing their inclination and merging gently with the surface of the paramera. In those areas with significant river incisions, which occur on the slopes in contact with the lower blocks, especially if they are tectonic basins, the edges of the paramera are dissected abruptly, leaving the paramera

**Fig. 3.3** Amblés valley and Sierra de la Paramera in the background

**Fig. 3.4** Picuezo Valley, seen from the Gavilán peak. In the background, the Zapateros Peaks

fragmented or surrounded by steep slopes, as in the valley of the Río Picuezo (Fig. 3.4).

It is also necessary to consider the broad blocks hanging over the bottoms of the tectonic pits, which, due to their lower Altitude, have relief characteristics different from the previous ones. At their culmination, the rocks are covered with significant alteration masses, which have a considerable influence on the development of soils and vegetation. Fluvial dissection affects the entire Paramera region, which is therefore compartmentalized into a large number of fragments over valleys with strong incision and diverse directions, indicating the structural subdivision of the entire ensemble.

In La Paramera, we find a granitic landform that, although unknown, is outstanding, as deduced from the fieldwork carried out in this work and whose most spectacular geomorphosites will be explained later (see Sect. 3.3).

On the other hand, in La Serrota and according to Fernández Ruiz et al. (2006), there is evidence of glacial landforms in the surroundings of its southern and Majalespino peaks, in the form of degraded glacial cirques and morainic deposits. The periglacial model stands out in areas of slope creep and endorheic areas with a tendency to develop peat bogs.

To the north, the Adaja River has shaped the Amblés Valley (Fig. 3.5), with an area of about 750 km$^2$ and an elongated shape in an ENE–WSW direction, following the Hercynian guidelines of the Hesperian Massif (Martín-Parra et al., 2008), and whose boundaries are strongly broken by the fault scarps perpendicular to its main direction. Two lithological sets outcrop in it: the basement, which occupies more than two-thirds of its area, and the Tertiary fill of the pit, which shows partial and little active emptying in recent times.

Garzón et al. (1981) study the morphostructure of the valley, differentiating two sedimentary cycles: one characterized by siliceous detrital deposits and the second of arkose character. The first is characterized by the outcrop of coarse-grained sandstones and conglomerates of small pebbles and quartz clasts, with even materials from granite disintegration. The sedimentary cycle of the arkose occupies most of the valley's surface, distinguishing between polygenic conglomerate facies, with a strong concentration of calcite, and facies of the arkose set, formed by rubble and mud flows. The eastern zone, which is the widest and most arid, has been used for agricultural and livestock purposes, while in the northwestern end, there are holm oak groves with scrubland. In the south, the Alberche Valley (Fig. 3.6) has a West–East development, in which two well-differentiated parts are distinguished (Fernández Ruiz et al., 2006). In the westernmost part, the headwaters of the Alberche River are located, with a morphology of a wide valley filled with Quaternary sedimentary materials, as well as deposit morphologies such as scree slopes, degraded scree slopes, and alluvial fans. In the eastern part, on the other hand, the river is entrenched, with a notable tectonic control of the entire network that results in sections with straight alignments.

In the Sierras de la Paramera and la Serrota, we find remnants of erosion surfaces as a result of the tectonic leveling of a single original surface, which causes compartmentalization. When this leveling activity ceases, erosion surfaces are finally developed, usually dated as late Tertiary, which shape, above all, the broad and massive blocks of the highlands or

**Fig. 3.5** Amblés Valley from Manqueospese Castle (Mironcillo municipality). In the background, Sierra de Ávila

**Fig. 3.6** Alberche Valley seen from the Pico Zapatero pass

the basal platforms. These erosion surfaces, which define the morphology of a large part of the Central System in Ávila, when preserved, are characterized by the presence of numerous rocky outcrops that rise slightly above the ground surface, with signs of alteration and sanding of the granite. Around these outcrops, rounded blocks detached from the rock appear, slightly displaced from their in situ position, semi-buried, so that sometimes it is not clear if they are rooted or not. For these reasons, better-preserved and evident rocky ensembles are formed as the rock is less alterable (Bullón Mata et al., 1988).

According to Bullón Mata et al., another phase of leveling could follow this one, which produces strong alterations of the rocks in preferred locations, giving rise to variable-sized hollows and alteration corridors, which also appear on mountain slopes in the form of basins, where river headwaters and valleys are subsequently established.

Thus, the general arrangement of the erosion surface presents a distribution in horsts and grabens according to the general Germanic relief of the Central System, defining the relief of the Sierras de la Paramera and la Serrota, and is the starting point for the granitic and glacial processes that characterize the geomorphology of this area, being its main tourist and educational attraction.

In the Sistema Central, granite forms and processes have been studied since the nineteenth century, both from genetic

**Fig. 3.7** Las Chinitas granite dome

and morphological perspectives, and the most characteristic forms and emplacement processes have been established (de Prado, 1864; Sanz Herráiz, 1988; Pedraza, 1989; Pedraza et al., 1989; Molina, 1991; Sequeira et al., 2002; Migoń & Vieira, 2014). Examples of all these forms can be found in the Sierra de la Paramera.

Landforms are conditioned by the type of granite, fracture patterns, climatic conditions, and emplacement in relation to the topography (Sanz Herráiz, 1988; Migoń & Vieira, 2014). Classifications of granite forms are very diverse, either by their genesis, location, or size (Twidale, 1982; Sanz Herráiz, 1988; Twidale & Vidal-Romaní, 2005; Migón, 2006, 2021).

Granite is a plutonic rock that forms in the upper levels of the crust after the cooling and crystallization of its minerals, being subsequently fractured by tectonic stresses and exhumed as a result of erosive processes to which it is exposed, mainly by the action of the atmosphere, hydrosphere, and biosphere. Once exhumed and therefore exposed on the surface, weathering begins, a process by which the true shaping that will give rise to the most characteristic granite forms begins.

The weathering of granite is given by the conditions of pressure, temperature, and humidity, which can decompose the rock—chemical weathering—and/or disintegrate it—physical weathering. In the study area and throughout the Sistema Central, hydrolysis is common, i.e., the chemical alteration of granite caused by water, breaking down the rock into clays (formed by feldspars and micas), and leaving loose quartz grains, with a granular appearance known as “grus” or sandy granite, a process known as granite arenization.

Granitic landforms can be classified into major and minor, with examples of all of them found in the Paramera and Serrota:

- Major granite landforms:
  - *Granite domes*: These are vaulted forms in relief, bounded by smooth and curvilinear walls that give rise to large rock platforms. These are initial forms of granitic modeling, as they are barely degraded. In the Paramera, the modeling is highly degraded and nubbins and boulder fields are more common, although vertical rock platforms of some domes that have not yet been fully exhumed can still be seen (Fig. 3.7).
  - *Granite ridges*: These are irregular forms, generally with a prismatic or pyramidal tendency, whose origin is due to the action of ice and water through frost weathering or cryoclastics, taking advantage of the vertical joints present in the granitic block. A granite ridge is referred to when it appears individually, or as a crest when several associated and continuous ridges are present, as is the case in the so-called Picos Zapateros in the Sierra de la Paramera (Figs. 3.8 and 3.9), which from east to west are: Risco del Sol (2113 m), Pico Zapatero (2158 m), and Risco Redondo (2054 m). Another example can be found in Los Gavilanes (Fig. 3.10), with Gavilán chico (1527 m) and Gavilán grande (1523 m), and in Peña Cabrera (Fig. 3.11).
  - *Nubbins and boulder fields* (Figs. 3.12 and 3.13): These are the most representative forms of granite modeling, as they are the result of the grouping of other forms. It can be defined as the chaotic association of boulders of different shapes and sizes that have remained in place, as a result of the alteration or weathering of the granite bodies. When the boulders do not remain in place but show clear signs of displacement, they are called boulder fields, thus being a more mature physiognomy, the result of the evolution of the nubbin.

**Fig. 3.8** Zapatero peak (left) and Risco Redondo (right)

**Fig. 3.9** Picos Zapateros: (a) Zapatero pass, (b) Zapatero peak, (c) Risco del Sol, (d) Alberche valley

**Fig. 3.10** Gavilanes peaks

**Fig. 3.11** Peña Cabrera

**Fig. 3.12** Nubbins and Boulder fields in Majadas Highs

 - *Tors* (Figs. 3.14 and 3.15): A pile of granitic rocks on which weathering has acted along joint planes, resulting in a pile of blocks that have not yet detached from the main massif.
 - *Rock platforms* (Fig. 3.16): These are flat and smoothed surfaces that appear when the granite dome has been completely degraded or dismantled, although they can also correspond to the top of a dome that has not yet emerged. Every dome-shaped form is made up of rock platforms, so we can find them on walls or summits, forming horizontal or subvertical planes.
- Minor granite landforms:
  - *Gnammas* (Fig. 3.17): These are circular depressions in the granite, with little depth and formed on horizontal surfaces. The action of water causes the rock to disintegrate and become sandy.
  - *Tafone* (Figs. 3.18 and 3.19): They are perforations or holes that appear in granite in sloped areas, usually vertical walls, formed by water runoff and/or generated moisture. Sometimes, a group of connected tafone can be generated.
  - *Pedestal rocks* (Figs. 3.20 and 3.21): They are isolated rocks that act as support for the elevation of the pedestal rocks, which is usually large and stands out on the surrounding surface. They are usually found in rocky areas, scree slopes, or tors, as a result of highly degraded relief.
  - *Platforms* (Fig. 3.22): It is about smooth surfaces compartmentalized by the intrusion of vegetation due to the presence of joints.
  - *Cracked blocks* (Figs. 3.23 and 3.24): These are blocks that appear vertically open due to weathering.

**Fig. 3.13** Picuezo nubbin

**Fig. 3.14** Tor in the Sierra de la Paramera (ascent to Zapatero). In the background, the Amblés Valley and the Sierra de Ávila

**Fig. 3.15** Tors in El Concho area

**Fig. 3.16** Unexhumed vertical dome rock platforms on the northern slope of the Ulaca semihorst

**Fig. 3.17** Gnammas in Ulaca

**Fig. 3.18** Tafone in Ulaca surroundings

**Fig. 3.19** Chinitas tafone

**Fig. 3.20** Pedestal rock in Ulaca

**Fig. 3.21** Pedestal rock in Las Chinitas

**Fig. 3.22** Platform in Ulaca

**Fig. 3.23** Cracked block in Castro de Ulaca

**Fig. 3.24** Cracked block in Gavilán grande peak

Glacial modeling occurs in the study area in the Sierra de la Serrota, and its general characteristics are well known (Schmieder, 1915; Hernández Pacheco & Vidal Box, 1934; Arenillas & Martinez de Pison, 1976). The interpretation of glaciation in La Serrota has varied, from a culminating ice dome with outlets to the valleys to the presence of cirque glaciers adapted to the topoclimatic conditions that generated the cirques and moraines in three glacial periods or phases. In the Sierra, there are five glacial cirques that define a cirque glacier morphology: Cerradillas, La Honda, Medialuna, Belesar, and Hornillos (Fig. 3.25).

### 3.1.2.1 Cerradillas Glacier

Glacial cirque oriented to the southeast that develops ~20 m below the ridge with a length of 1.1 km (Table 3.1). The cirque is carved on the substrate, where erosion forms can be seen in the granite. It consists of two overdeepened basins filled with accumulations of peat. Both basins are separated by a non-visible rock bare and morainic deposits.

The cirque has three moraine complexes: the outer one is formed by very deteriorated moraines that form block fields, and on top of it is the largest and best-preserved one, which completely closes the cirque, and in which at least three smaller recessional moraine arcs are preserved. Inside the cirque, a split moraine closure is preserved in two arcs that close the upper peat platform. Finally, a small discontinuous arc is located on the molded granite, indicating a small glacier lodged in the cirque wall.

The accumulation forms indicate the complexity of glacial evolution, with a deteriorated expansion phase (FI); a well-preserved advancing and equilibrium phase with up to three minor recessional pulses (FII); an internal cirque glacier phase (FIII); and finally a small residual glacier (FIV).

### 3.1.2.2 La Honda Glacier

Glacial cirque oriented to the southeast that develops ~50 m below the summit of La Serrota and extends 1.1 km in length to the outermost moraines (Table 3.1, Fig. 3.26). The cirque does not have outcrops of the substrate and is covered by a continuous debris talus that connects with peat bog depressions. It is formed by three overdeepened basin infilled with peat bogs separated by small thresholds. In the southern portion, there are outcrops of the substrate of limited extent, and for the most part, it is covered with debris composed of large blocks that form block streams.

The set contains three moraine complexes: the outer one, formed by deteriorated moraines, reaches 1810 m above sea level. Inside, the best-preserved and most voluminous arch that completely closes the cirque is preserved. At least four smaller recessional arches can be seen. Inside the cirque, 210 m higher, two very close and discontinuous moraines are preserved, indicating a period during which the southern portion of the cirque was not occupied by the glacier, with intense periglacial processes that generated the covering of scree. Finally, similar to Las Cerradillas, a small discontinuous arch is located at the foot of the cirque, formed by a very small glacier.

The accumulation forms indicate a glacial evolution very similar to that of Las Cerradillas, with a phase of expansion (FI), deteriorated; a well-preserved advancing and equilib-

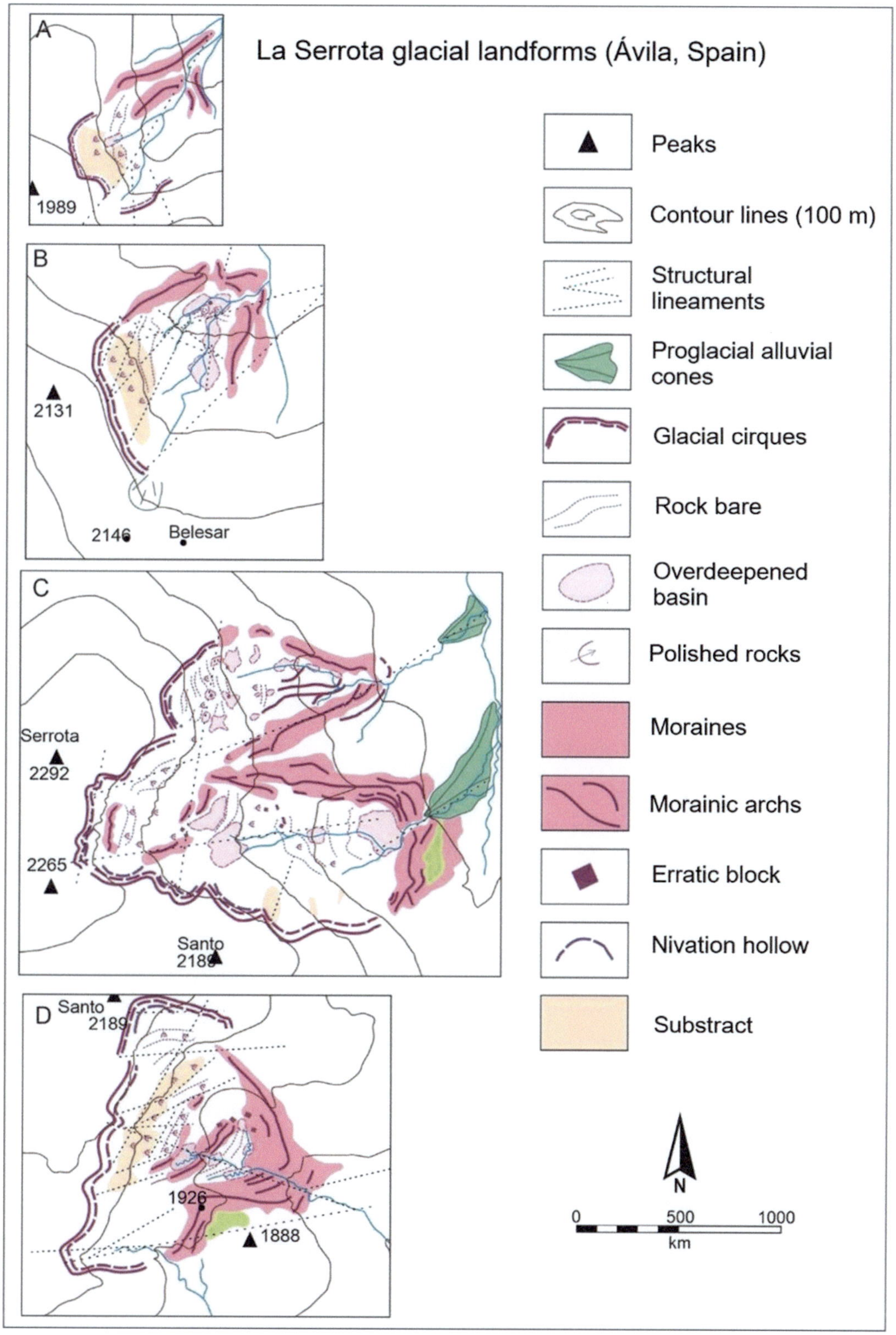


**Fig. 3.25** Detail of the glacial landforms of La Serrota. Glaciers: (**a**) Hornillos, (**b**) Belesar, (**c**) Media Luna and La Honda, (**d**) Cerradillas

rium phase with at least four minor recession pulses (FII); an internal cirque glacier phase (FIII); and finally a small residual glacier (FIV).

During phases I, FII, and FIII, this glacier was connected to the Medialuna glacier, forming a single glacier split into two tongues descending toward the Canto Moreno stream. Thus, during the maximum expansion, it would be the largest glacier in La Serrota, with an extension of 163 ha at maximum expansion and 144 ha at the peak ice volume, with tongues over 1 km in length.

**Table 3.1** Morphometric characteristics of the Serrota glacial morphology

| Glacier | | Area (ha) | Length (m) | Width (m) | Orientation | Cirque height (m) | Front height (m) | Peak height (m) |
|---|---|---|---|---|---|---|---|---|
| Cerradillas | FI | 93 | 1180 | 750 | SE | 2180 | 1780 | 2199 |
| | FII | 82 | 1140 | 745 | SE | | 1800 | |
| | FIII | 42.5 | 750 | 673 | SE | | 1870 | |
| | FIV | 17.5 | 410 | 315 | E | 2100 | 1910 | 2150 |
| La Honda | FI | 113.5 | 1178 | 830 | ESE | 2200 | 1810 | 2292 |
| | FII | 103.5 | 1610 | 750 | ESE | | 1830 | |
| | FIII | 24.5 | 520 | 460 | ESE | | 2040 | |
| | FIV | 5.5 | 165 | 285 | E | | 2170 | 2260 |
| Media Luna | FI | 50 | 1070 | 600 | E | 2180 | 1790 | 2279 |
| | FII | 41 | 1010 | 565 | E | | 1800 | |
| Belesar | FI | 68 | 1060 | 645 | NE | 2100 | 1840 | 2145 |
| | FII | 54 | 960 | 600 | NE | | 1850 | |
| Los Hornillos | FI | 42 | 950 | 460 | NE | 1900 | 1560 | 1988 |
| | FII | 28.5 | 870 | 350 | NE | | 1580 | |

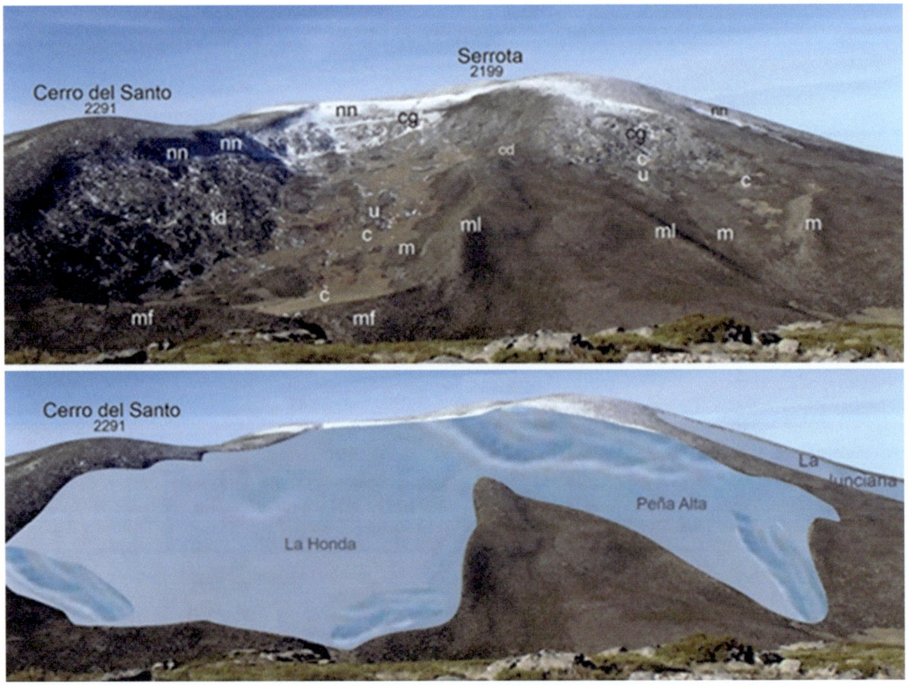


**Fig. 3.26** Palaeoglacial reconstruction of La Honda glacier during the Last Pleistocene Glacial Maximum. Letters means: cg: glacial cirque, nn: nivation hollow, u: rock bare, c: overdeepened basin, m: moraines, td: debris talus

### 3.1.2.3 Medialuna Glacier

This glacial cirque is a continuation of La Honda's cirque, with which it formed a single glacial cirque although the tongue was individualized. The cirque in this portion is oriented to the east, also with the elaborated cirque that extends about 110 m below the summit and 90 from the ridge, with a length of 1 km (Table 3.1). At the top, there are alternating peat plateaus and scattered moraine material accumulations.

In this area, three very close moraine complexes are located, both in Altitude and spatially. The outer one is also formed by very deteriorated moraines that cover the slope with blocks, and on top of it is once again the largest and best-preserved one, in which three smaller recessional arcs can be seen. Inside the cirque, a very distorted moraine closure is preserved with at least two alignments, all reaching the intermediate moraine complex.

The moraine complexes show a simpler glacial evolution than in the previous ones, where a phase of expansion (FI) is followed by one of advance and equilibrium, with well-preserved morphologies and minor recession pulses (FII). Finally, an internal phase that would still occupy the majority of the cirque (FIII).

It should be noted, as in the previous case, that in the first three glacial phases, the ice that occupied this cirque and basins was connected to the La Honda glacier and was part of the larger La Serrota glacier.

### 3.1.2.4 Belesar Glacier

It is a NE-oriented glacial cirque that develops ~45 m below the ridge and had a length of 1 km (Table 3.1). The cirque is carved on the substrate, with abundant abrasion forms in the granite that reach to form sheepback rocks and polishings. It consists of two overdeepened basins separated by

40 m of elevation difference, both filled and with peat accumulations. In the upper basin, solifluction forms occur on the fines.

The cirque has two moraine complexes. The outer one is formed by very deteriorated moraines affected by intense solifluction processes. On top of it, there is a well-preserved complex divided into its fronts that closes the cirque and generates the overdeepened basin, where there is also a large glacial erratic block.

The accumulation forms indicate a simpler glacial evolution than in the previous ones. A phase of expansion (FI), also with deteriorated moraines, is followed by the advance and equilibrium phase, well preserved (FII). After that, there are no accumulation forms, showing the disappearance of the ice.

#### 3.1.2.5 Los Hornillos Glacier

It is the smallest glacial cirque, which also feeds the smallest complex in the range. The glacial cirque is oriented to the NE, also more than 80 m below the ridge, which according to the map (MTN E. 1/25,000) does not reach 2000 m (1988 m). The glacier had a maximum length of 960 m (Table 3.1). The cirque is carved on the substrate for most of its extent, and between hollows, the substrate is shaped in sheep rock, glacial polish, and asymmetrical rocks emerges. It is the most extensive area in the range with representation of erosion mesoforms.

The cirque has two moraine complexes that close at the valley bottom at an Altitude of 1560 m. It is the lowest Altitude reached by glaciers in La Serrota. The outer complex is formed again by very disfigured moraines that are seen as piles of linear blocks, better preserving their shape to the SE of the cirque. Inside, again a well-preserved frontal–lateral moraine is the largest of this complex.

The accumulation forms indicate a very simple glacial evolution, marked by a phase of expansion (FI), a well-preserved advance and equilibrium phase (FII) followed by the complete deglaciation of the cirque.

In the group of the four existing glaciers, divided into five along the deglaciation process, the preferred orientations in 100% of the cases are east, with variations between Southeast and Northeast components. In the latest glacial phases or in the Half Moon one, the component is directly to the east.

The size of the Pleistocene glaciers was very small, with only 365 ha of ice at their maximum extent, and only 23 ha in the last phase.

The profiles of the glacial cirques show poorly fitting cirques, stepped in basins and slightly pronounced bare rock, often covered with moraine material. The thicknesses are moderate, between 30 and 50 m, indicating the limited capacity to generate well-configured erosion forms, although they were able to deposit voluminous moraines during the second phase, the longest in time. The most significant aspect is that the heads of the glaciers did not reach the ridges and divides, and accumulation occurred between 30 and 90 m below the divides. This fact indicates that in the divides and high slopes, there were no thermal conditions for the maintenance of snow and its transformation into ice. The genesis of the glaciers was not the low radiation that allowed the maintenance of snow and ice, which occurs in predominantly N, NE, and NW orientations, or in all of them if the isotherms drop below 0 °C. The most significant topoclimatic elements would be two. On the one hand, the snow drifted from the west, southwest, and northwest, the most frequent directions during the Pleistocene, and accumulated on the lee side, on the eastern slopes (E, NE, and SE). On the other hand, the effect of insolation on the eastern slopes, when the sun shines in the morning, the period with the coldest atmosphere, when radiation warms the atmosphere but weakens on the ground and is therefore less effective, while in other orientations the sun shines when the atmosphere is warmer and radiation is more capable of melting the snow. The combination of both factors, well known in the mountains of the Central System, implies the overaccumulation of snow on the lee side and its subsequent preservation, favoring its transformation into ice. This marginality of the conditions explains the small size of the glaciers and the moderation in the erosion forms shaped by the ice. The exception is the Los Hornillos glacier, with the cirque also to the NE, but the tongue sheltered to the north, conditions that favored the genesis of a glacier at a very low Altitude, under a summit of 1988 m in Altitude with the cirque at 1910 m.

Glaciers adapted to the structure, following the lines of fracturing, as pointed out by Arenillas and Martinez de Pison (1976), which gives them the linearity of the small tongues that advanced from the cirques. If the cirques are oriented to the NE, SE, ESE, and E, the body adapts to the structure and takes directions N, NE, or SE according to the fracturing.

As we have pointed out, these are glaciers adapted to different factors unrelated to average temperatures and Pleistocene isotherms. We can establish that they were marginal glaciers, developed not so much by climatic and morphoclimatic conditions (average summer temperatures, Altitude of annual isotherms, preferred orientations by temperature), but by site factors, where the structure, orientation, and snow overaccumulation by the wind determined the development of glaciers in the Pleistocene. The glaciers were not located in the highest sectors. All this indicates that they did not have optimal climatic conditions for glacier development, so the Equilibrium Line Altitude (ELA) would be located above 2200 m. However, snow accumulation and its transformation into ice and glaciers would occur in orientations directed by the wind, not by solar radiation in shady areas, adapted to the preglacial torrential headwaters directed by fracturing, without glacierism transcending this influence, except very locally.

In the recent Pleistocene, the Sierras de La Paramera and La Serrota are located in a colder environment, although their Altitude is below the glacial equilibrium line, the Altitude above which snow accumulation dominates on glaciers, while below it, melting dominates. This fact implies an important morphological asymmetry; in La Serrota (2294 m), with a N–S oriented ridge above 2000 m, a group of glaciers is located in favorable orientations determined by the wind and overaccumulation, retaining snow in predominantly eastern orientations, under peaks over 2100 m. However, in La Paramera, with its highest point at 2137 m, most of its surface at 2000 m Altitude or below and an E–W-oriented ridge, there were no glaciers due to the lack of overaccumulations on the leeward side and the retention of snow in eastern orientations.

The marginality of glaciation implies that structural elements prevail over the colder climatic conditions typical of glaciation. The key factors in glacial morphology are the preglacial elements that guide the shaping (fractures, preglacial topography, granite alteration) and topoclimatic factors (wind, leeward overaccumulation, moderate radiation in the east component). Below 2000 m, these small glaciers began to melt, so the materials removed in the cirques were deposited, forming moraine arcs that extend up to 1750–1800 m in Altitude, where the glacier fronts would be located. The moraines have voluminous arched ridges that allow the reconstruction of the contours of the glaciers, which experienced phases of retreat, up to three well-marked ones in the Cerradillas, before their disappearance at the end of the Pleistocene.

During this period, while in the culmination of this period while in the culminating zones the glaciers were exerting their forces, the culminating zones were carved by the glaciers, in the zones without glaciers of La Serrota and the zonas culminantes ejercían su labor los glaciares, in the zones without ice had intense they have intense processes of ice and lack of ice, solifluxion, linked to fusion of the snow care, and frozen soils, that presented a wide set of periglacial land forms today inactive. There are at least six periglacial landforms related to different periglacial processes (Table 3.2), such as tors, boulder fields, debris talus, organized soils, lóbes and flows and nivation hollows. All of them appear in the extensive summit plains, and the slopes of both sierras, and characterized by the model, inherited from the colder periods of the Recent Pleistocene.

However, the asymmetry between both sectors is not morphoclimatic. In La Serrota the glacial forms give energy to the relief and slimness to the summit forms, while in La Paramera it is the granitic forms derived from weathering and exhumation of the granitic structures that characterize large portions of the summit and slope areas. The textural and mineralogical combinations and the fissuring of the granites direct the main forms of modeling in granites at all scales. The granitic forms are inscribed in the reliefs at very variable scales, so the range of forms is very wide, as some are inscribed within others. The granitic modeling forms are the product of a long morphological evolution, both morphostructural, through the emplacement, fracturing, and exhumation of intrusive bodies, and, above all, by the evolution of relief forms.

**Table 3.2** Periglacial landforms and processes of La Serrota

| Forms | Process | Conditioning processes | Activity |
|---|---|---|---|
| Tor | Cryoclastics<br>Thermoclastics | Jointing | No |
| Boulder fields | Cryoclastics | Granite outcrops<br>Granito | No |
| Debris talus | Cryoclastics | Slopes | No |
| Organized soils | Frozen soils<br>Cryoclastics | Filled with finer materials | No |
| Lóbes and flows | Solifluxion<br>Nivation | Slopes<br>Water availability | Yes |
| Nivation hollow | Nivacion | Slopes with over snow accumulation | Yes |

The chronology of the glaciation of La Serrota is not studied here. Initially, it was established that the moraine complexes belonged to different Quaternary glaciations, but in the 1970s and according to monoglacialist theories, it was established that all glaciations belonged to the last glacial cycle, the Würm epoch. With the arrival of new dating technologies, the glaciers of Gredos have been studied, which, due to proximity, it is logical to think that they have had a similar paleoclimatic evolution.

The work carried out in the Alto Gredos, the highest massif with an E–W orientation of the ridges, where cirques and glacial tongues aligned to the north and south, with a strong asymmetry, as well as the datings of Béjar and La Covacha, shows a complex evolution with four glacial phases (Table 3.3) attributable to the last glacial cycle, after 26,000 years (Carrasco et al., 2022).

The Local Maximum Glacier (FI) is represented in both Gredos and La Serrota by the longest glaciers that existed, although they are the thinnest. In Gredos, they are attributed to a glacial advance between 29 and 26 ka, which has been detected in both Alto Gredos and Béjar and La Covacha. This period of thin but extensive glaciers can be attributed to a first phase related to the dominance of fronts coming from the SW that brought a lot of snowfall through very humid masses, although not very cold and during a brief period that allowed for an initial glacial advance (Serrano et al., 2013) in the entire northern Iberian Peninsula, confirmed in Guadarrama, Gredos, and the Cantabrian Mountains.

A second phase (F-II) corresponds to the Last Global Glacial Maximum, a very cold and dry period between 29 and 16 ka that generated the formation of shorter glaciers than the previous ones, which were more typical of humid conditions, but thicker, as befits drier conditions and a colder

**Table 3.3** Correlation of glacial phases from moraine complexes at La Serrota

| | Cerradillas | | La Honda | | Media Luna | | Belesar | Hornillos | Characteristics |
|---|---|---|---|---|---|---|---|---|---|
| 1 | FI | | FI | | FI | | FI | FI | Larger glaciers but reduced volume. Very deteriorated moraines |
| 2a | FII | a | FII | a | FII | a | FII | FII | More voluminous glaciers, with the greatest thicknesses. Very well-preserved moraines |
| 2b | | b | | b | | b | | | Small, discontinuous moraines. Indicate equilibrium processes in retreat |
| 2c | | c | | c | | | | | |
| 2d | | | | d | | | | | |
| ¿? | – | | – | | FIII | c | Total deglaciation | | Moraine very close to the previous phase |
| 3a | FIII | a | FIII | a | Total deglaciation | | | | Well-preserved internal moraines, arched, on bare rock. Deglaciation: disappearance of ice in the circuses La Honda, Belesar, and Hornillos |
| 3b | | a | | b | | | | | |
| 4 | FIV | | FIV | | | | | | Very small glaciers, with poorly preserved and small moraines. Glaciers Cerradillas and Honda |
| 5 | Total deglaciation | | Total deglaciation | | | | | | Disappearance of ice in La Serrota |

**Fig. 3.27** Navalacruz wildfire. (Source: David Castro)

period of longer duration. The most important glaciation is followed by a deglaciation period with phases of equilibrium and minor advances between 17.5 and 14.6 ka that have generated intermediate moraine complexes composed of successive moraines.

A new advance (F-III) of glaciers, of lesser development, occurs in all the Gredos massifs, including La Serrota, with moraines inside the cirques. Immediately after, around 14 ka, a glacial advance occurs in Alto Gredos (F-IV), which correlates with the last glacial phase of La Serrota (F-IV). This is housed in the cirques of Gredos, but in La Serrota it is manifested in very small glaciers nestled in the walls of the cirques with elongated moraines above 1910 m in Cerradillas and 2170 m in Las Hondas.

Subsequently, deglaciation begins, with a rapid glacial retreat leading to the disappearance of glaciers in La Serrota, although in Alto Gredos there is still a balance and a very brief advance in time, during the Recent Dryas, around 11 ka, with very small glaciers. However, in La Serrota between 14 and 12 ka, the mountain is free of ice although there is intense cold that generates energetic periglacial processes. These processes reshape the slopes, particularly those facing north, through coverings of large clasts and block streams, as well as the glacial cirques already devoid of ice.

### 3.1.3 Navalacruz Wildfire

The so-called Navalacruz fire (Fig. 3.27), named after the place where it originated, although some municipalities, such as Sotalbo, saw 90% of their municipal territory burned, devastated the Sierra de la Paramera almost entirely in August 2021, burning more than 22,000 ha of this Natural Space. It became the fourth most important in the history of Spain and the second largest in the Castilla y León region, being surpassed by the Sierra de la Culebra fire in 2022.

The perimeter of the fire affected 30% of the surface area of the Sierra de la Paramera and La Serrota Red Natura area. According to data provided by the Junta de Castilla y León, of the 22,037.85 ha burned, 12.4% corresponded to wooded areas, 83.5% to non-wooded areas, and around 4.1% to non-forest areas.

Lasting for 13 days, with a thousand people evacuated from up to nine population centers and a sense of abandonment and negligence among the residents, this event marked a before and after in the heritage of these mountains and the lives of their inhabitants. In the midst of a heatwave, with temperatures of 38 °C, winds of up to 60 km/h, and a lack of firefighting equipment, were the triggers of this fierce fire, which started from a burning car on the shoulder of the N-502 road and ended up ravaging the entire Paramera.

The Copernicus Emergency Management Service's viewer allows observation, based on NASA's VIIRS (Visible Infrared Imaging Radiometer Suite) data, of the mapping of burned areas and the perimeters of active fires. As of the consultation date of 31/03/2022, the burned area of the Navalacruz fire can be observed (Fig. 3.28).

On August 24, 2021, the Government declared the area affected by the Navalacruz fire as a catastrophic zone in order to access assistance not only to alleviate material damages but also personal, infrastructural, residential, industrial, livestock, and agricultural damages. Additionally, in September of the same year, the Junta de Castilla y León presented the Recovery Plan for the area affected by the Navalacruz fire, with a budget of 25 million euros and a 10-year execution period. Among the planned actions, those focused on infrastructure recovery, hydrological protection and forest recovery, replacement of fences, and repair of rural roads stand out. Furthermore, the sale of the burned commercial wood is planned, with the revenues initially allocated to the municipalities of the Ávila Association.

In April 2022, the Junta de Castilla y León published the Restoration Plan for the August 2021 fire in the Sierra de la Paramera, detailing the effects of the fire, zoning analysis, and prioritization of the burned area, and proposing actions:

- Emergency actions, including population supply, hydrological–forest correction, extraction of burned wood, compatibility and support for livestock activity, and protection of cultural assets.
- Restoration actions for infrastructure, the forest environment, and biodiversity.
- Protection and defense actions against future forest fires, as well as the improvement of the natural environment through the organization of livestock use, maintenance of pastures, periodic clearing, and plantations in riverbanks and wetlands.

This Plan, with a 10-year action framework, proposes a monitoring program for the actions using GIS tools and updatable information tables (Fig. 3.29).

Among the actions already carried out, it is worth mentioning the improvement of tracks and water crossings, hydrological–forestry correction, repair of breakwaters and roads, and the reforestation of a total of 50 ha in Navalmoral and 10 ha in Sotalbo. There are ongoing actions of hydrological and forestry correction, improvement and restoration works in the Duero and Tajo rivers, and the planting of another 10 ha in Sotalbo.

Setting aside the political and economic management of the fire, the consequences it has had on the functioning of hydrological processes and the erosion it has caused on soils and geomorphological elements are particularly concerning and relevant in this research. The fire is capable of destroying not only vegetation but also the dead organic matter of the soils it sweeps through, leaving it exposed to the impact of rain. Additionally, it facilitates the mobilization of sediments on the

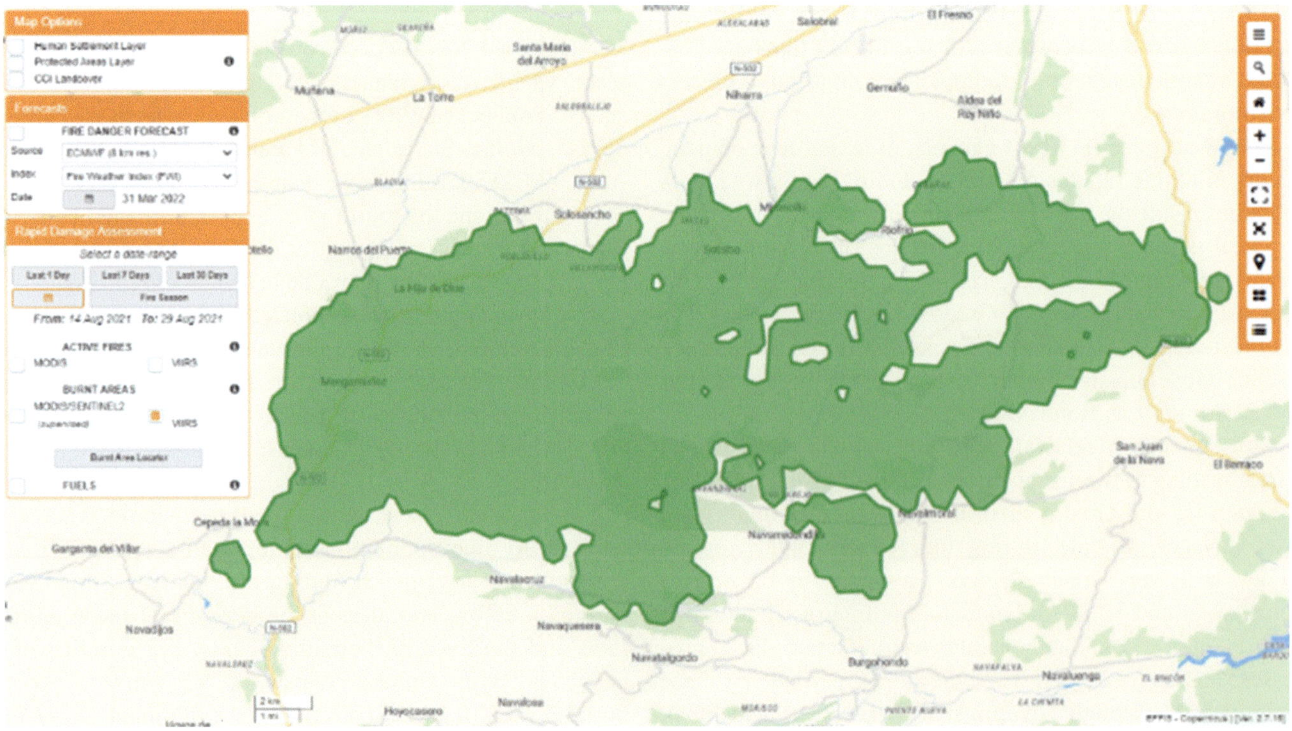


**Fig. 3.28** Burn area in Navalacruz wildfire. (Source: Copernicus Emergency Management System)

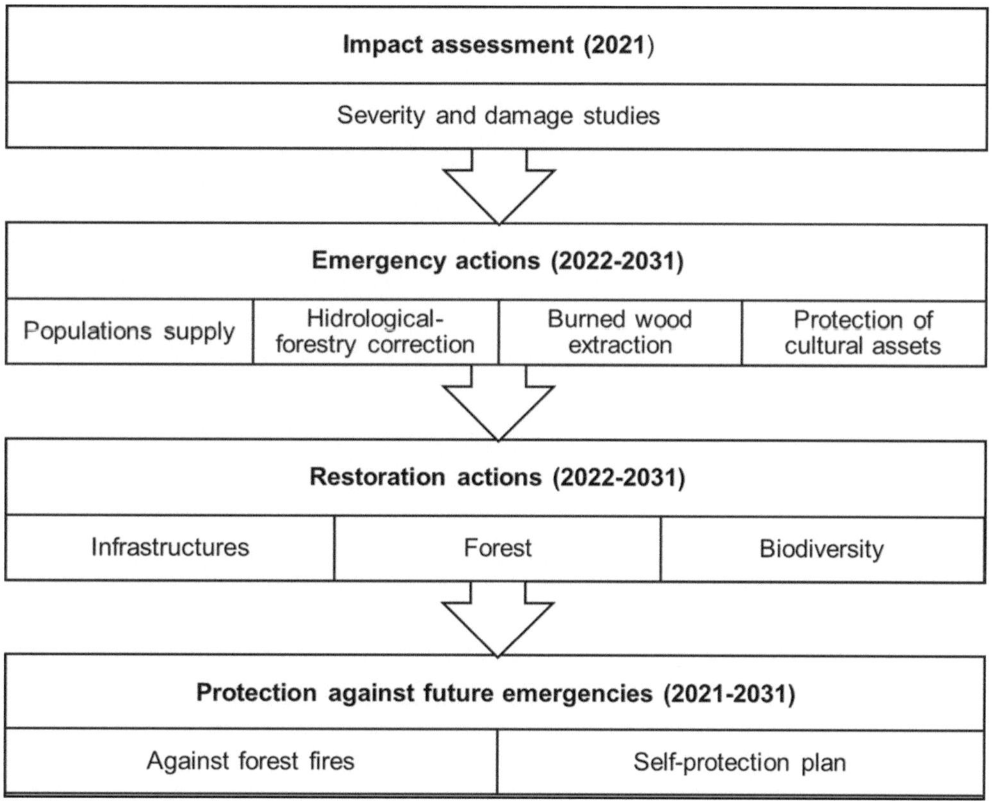


**Fig. 3.29** Time frame of the Navalacruz Fire Restoration Plan actions. Summarized from: Junta de Castilla y León (2022)

slopes, causing a strong impact on ecosystems, rivers, and lakes (Fernández & Vega, 2011). All of this leads to changes in the geomorphological and hydrological characteristics of the Sierra de la Paramera, as observed during fieldwork.

Beyond the presence of burnt vegetation and rock, evidence of scaling and fractures in granite rocks has been found as a result of the high temperatures they were exposed to (Figs. 3.30 and 3.31), with broken rock pieces and a new unburned layer being found in the field.

However, erosion is not only caused by the fire itself but also by the actions carried out afterward. Supposed recovery actions, which provide for various actions related to the extraction of burned wood in the area affected by the forest fire that occurred on August 14, 2021 in Sotalbo (Ávila), carried out by the Spanish public company TRAGSA for the cutting of damaged or partially damaged trees. Actions are causing incomprehensible erosion on slopes and roads (Fig. 3.32), as well as floods and flash floods due to increased rainfall. The residents have not been informed about the wood extraction or its use, and there is no verification of the actions carried out by Tragsa. In fact, in the last field work carried out in August 2024, it has been observed that much of the wood has not yet been extracted, being piled up for 3 years and posing a risk in the event of a new fire (Fig. 3.33).

Beyond the plans approved by public administrations, numerous initiatives have been carried out by the residents of Sotalbo and other external organizations. The work of the local organization Sotalbo Resurge is noteworthy, as they have conducted several reforestation and awareness activities in the municipality, with support from the Atlético de Madrid Foundation and a group of National Geographic Explorers. Thanks to the Atlético de Madrid Foundation, over 600 trees and aromatic plants were planted, while National Geographic has led an elaborate reforestation project led by Jordin Jon, Laia Crespo, and Sanne Derks. Accompanied by Rodrigo Gandía, a member of the Forest Management Department of the Territorial Environmental Service of Ávila, land cleaning and preparation were carried out in the summer of 2023. Subsequently, Michael O'Brien, a scientist from the Experimental Station of Arid Zones of the Spanish National Research Council (CSIC), led the reforestation process on 12 ha, where up to 30 different species were replanted with the help of over 100 volunteers. Laia Crespo, a biologist and educator, has conducted an environmental education workshop with students from the Rural Grouped School (CRA Valle Amblés) and the IESO Villa de Sotillo on the importance of forests and the consequences of wildfires.

It is now pointless to conduct an inventory of flora and fauna in the Sierras de la Paramera and la Serrota. Referring to the inventory conducted by Calvo Peña and Conde Iglesias (2022), the prefire faunal richness was based on the presence of game species (roe deer, red deer, wild boar, and fox), protected species that were present throughout the year (golden eagle, griffon vulture, sparrowhawk, royal owl). The presence

**Fig. 3.30** Granite rock affected by thermoclastics after Navalacruz wildfire

**Fig. 3.31** Broken granite in situ remains in Sierra de la Paramera

of the Iberian desman in the Adaja subbasin, with a population of about 50 individuals, was particularly relevant.

The Ministry for Ecological Transition and Demographic Challenge, along with the Duero River Basin Authority, conducted a study on the effects of the fire on habitats. The results showed that the fire affected 37.5% of the suitable habitat for the Iberian desman in the Ulaque subbasin and 67% in the La Hija de Dios basin. Numerous livestock heads perished in the fire, along with some roe deer, although there is no inventory of animals that lost their lives in the flames.

**Fig. 3.32** Slopes and paths devastated by wood extraction in Sierra de la Paramera. Date: 18.03.2021

**Fig. 3.33** Burnt logs, piled up and abandoned in the Paramera. Date: 17.08.2024

Regarding vegetation, the study area was dominated by holm oak and oak forests, distributed in the supra-mediterranean zone (between 900 and 1700 m). At Altitudes above 1700 m, corresponding to the oro-temperate zone, there was presence of Spanish broom, creeping juniper, and grasslands. It is worth noting that the holm oaks between 1200 and 1400 m in Altitude were of great importance as they were the highest population in the Peninsula, due to lower precipitation in the study area.

The supra-mediterranean floor, characterized by *Quercus pyrenaica*, was mainly found near the ports of Villatoro and Mengamuñoz, to a lesser extent in Narros del Puerto, and more scattered on the mountain slopes. This floor was accompanied by other shrub formations and wild roses. In the areas with granite outcrops typical of the region, species such as clavelinas, digitalis, and sorrel usually thrived, with a nearly relic presence of yew or rowan trees (Abad Soria & García Quiroga, 2011).

The orotemperate floor, in the high areas of Paramera and Serrota, was composed—and continues to be in La Serrota—of creeping junipers and mountain brooms, as a result of historical fire use or logging in this area. In the clearings of the broom thickets, well-drained grassland soils dominated by grasses prevail. Additionally, species associated with rocky areas such as digitalis, Gredos chamomile, and various types of carnations are present (Abad Soria & García Quiroga, 2011).

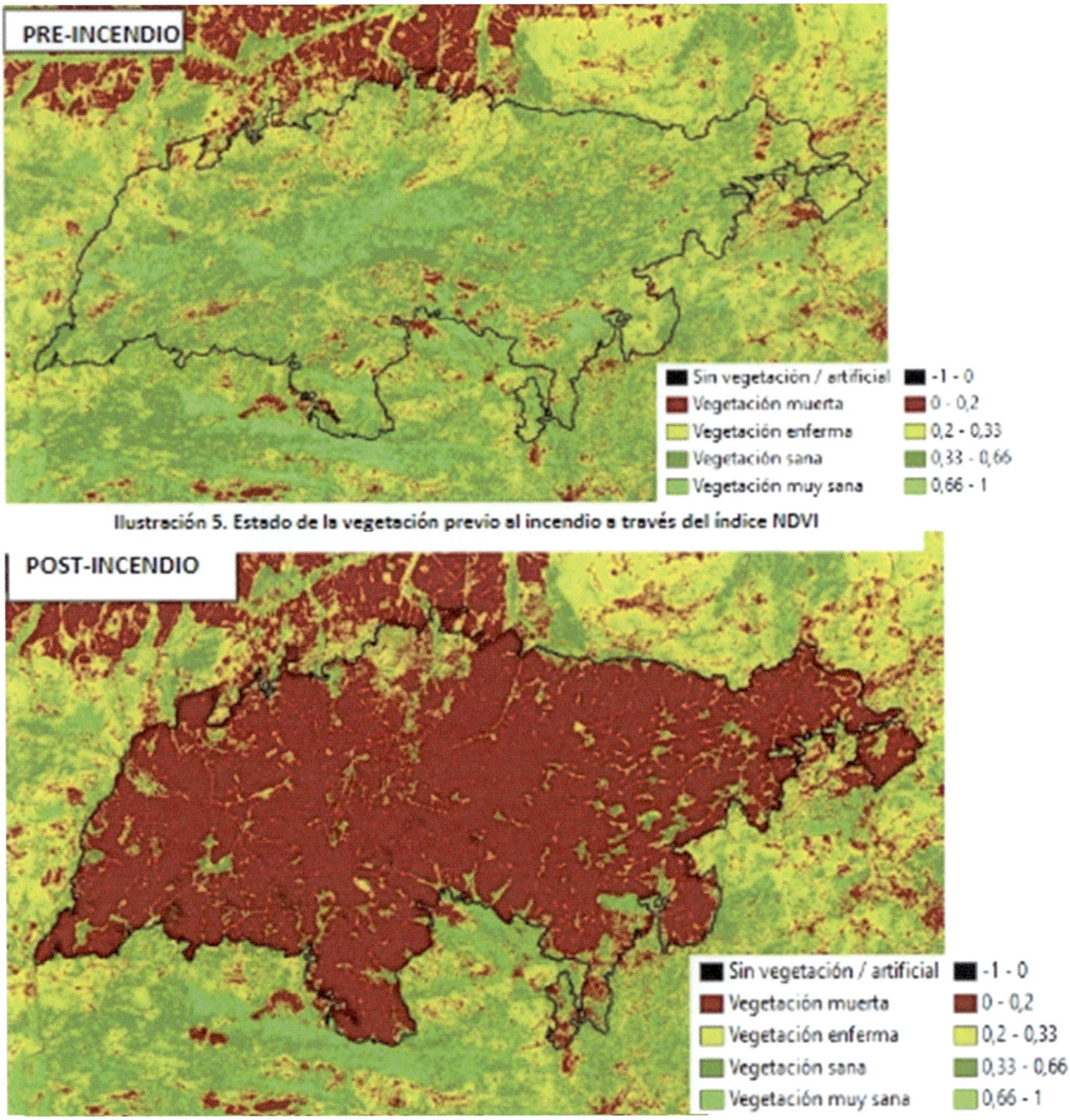


**Fig. 3.34** State of vegetation pre- and postfire in the Sierra de la Paramera. Greenish colors represent healthy vegetation before the fire; red represents dead vegetation after the fire. (Source: Junta de Castilla y León)

Reforestation of resin pine and Scots pine has been carried out for decades at the foothills and slopes of Paramera, especially in Sotalvo and Muñana, and Scots pine at the foothills and mid-slopes of La Serrota, using invasive techniques such as terracing and opening forest tracks with heavy machinery, altering the natural landscape (Abad Soria & García Quiroga, 2011). Furthermore, human activity has caused intense deforestation throughout history, due to traditional livestock use seeking new grazing lands.

The Recovery Plan developed by the Junta de Castilla y León also analyzes the damage caused by the fire on vegetation using remote sensing and telemetry techniques applied to forest areas, and following the Normalized Difference Vegetation Index (NDVI), a simple indicator of photosynthetically active biomass that allows for assessing vegetation health and distinguishing between dead, sick, healthy, and very healthy vegetation. This enables comparison of vegetation status before and after the fire (Fig. 3.34). In the Paramera area, before the fire, vegetation was mostly very healthy or healthy on the highest peaks and transition slopes, being sick at the border with Valle de Amblés. However, after the fire, the entire Paramer shows dead vegetation, except for small areas of vegetation that did not burn.

Below are some photographs taken at the same location before and after the fire by a local resident, Mario Barroso, to compare the changes in vegetation and landscape in Paramera after the fire (Figs. 3.35 and 3.36).

The damage caused by the fire is irreparable. The fear, anger, pain, and helplessness have taken a toll on the residents of La Paramera and those of us who once enjoyed its corners, witnessing an entire mountain range burning due to the lack of coordination and resources of those who are supposed to protect the environment and citizens' safety. Special mention to the men and women who fought for their villages, shovel and bucket of water in hand, protecting their homes and lives when no one else did.

La Paramera, still charred, remains a Natural Space with a natural, cultural, and human heritage that has resisted the fire. Outstanding granite formations, glacial and periglacial traces, Vetton and Visigothic castles and forts, legends and traditions, continue to make La Paramera a place of undeniable value.

### 3.1.4 Climate and Hydrology

The climate in the study area is influenced by the orography, specifically by the Sierra de Gredos, with distinctions between the Amblés Valley and the Alberche Valley, the north face and the south face. The Amblés Valley is shielded from rain by some of the storms coming from the southwest, while precipitation in the Alberche Valley ranges from 500 mm annually to almost double at the peaks of the Picos Zapateros, where it falls as snow in winter. Temperature differences of 5–7° exist between the Amblés Valley, with average annual temperatures between 10 and 12 °C, and the peaks, where cryoclastic effects are present.

Referring to AEMET data, a climatic diagram for the municipality of Sotalbo has been created (Fig. 3.37). Situated at the foot of the Picos Zapateros at 1200 m above sea level, Sotalbo has an average annual temperature of 8.8 °C and an annual accumulated precipitation of 739 mm. The maximum temperature occurs in July, with an average of 17.9 °C, and the minimum in January, with just 1.4 °C. Therefore, the temperature range is 16.5 °C. The highest precipitation is in November with 98 mm, and the lowest in July with 17 mm, with July and

**Fig. 3.35** Fuente de las Aguas Frías before and after the fire. (Author: Mario Barroso)

**Fig. 3.36** Zapateros peaks and holm oaks seen from Amblés valley. (Author: Mario Barroso)

August being the only dry months. According to the Köppen classification, it is a Csb Mediterranean climate with cool summers, prevalent in the entire Amblés Valley and the towns of Paramera and Serrota. The Csb climate is characterized by mild temperatures with cold or mild winters and dry, cool summers, typical of inland areas above 900 or 1000 m in Altitude, such as Sotalbo and the entire Amblés Valley.

At the peaks of Paramera and Serrota, the climate changes to Dsb Continental with cool summers, characterized by a continental climate with very cold, snowy winters and cool summers, typical of high-Altitude areas.

In terms of hydrology, the study area is located within two watersheds: to the north, the Duero basin, with the Adaja River as the backbone of the Amblés Valley; to the south, the Tajo basin, with the Alberche River (Fig. 3.38).

In the Natural Area, the Duero basin and the Adaja River have a significant presence. The Adaja is the second most important tributary of the Duero on its left bank after the Tormes. The Adaja River originates between the Serrota and the Sierra de Ávila, at the Fuente Berroqueña or Ortigal in Villatoro (Fig. 3.39). With a length of 163 km, it flows into the middle course of the Duero River on its left bank. The Adaja is a Mediterranean mountain river with siliceous characteristics. In its upper stretch, it forms the drainage axis of the Amblés Valley, surrounded by the Sierra de Ávila to the north and the Serrota to the south. In its lower stretch, it is mainly composed of sandy sediments, creating a floodplain 200 m wide at 1100 m above sea level.

The Hija de Dios stream and the Picuezo river are significant in the Natural Area, both being tributaries of the Adaja River. The Hija de Dios stream is a natural mountain stream that originates in the Paramera, specifically in Los Hoyuelos at an Altitude of 1810 m, where it is called the Meneagrande stream. It flows through La Hija de Dios, where it is named La Hija, through La Torre, and finally empties into the Adaja in the Valle de Amblés, in the municipality of Solosancho.

On the other hand, the Río Picuezo (Fig. 3.40), which also originates in the Paramera, in Los Acarreaderos at an Altitude of 1765 m. It flows along the shady slope of the Paramera, passing by the Picuezo peak, from which it gets its name, and passing by the old mills of Molino del Quemado, Molino de en Medium, and Molino de Abajo, which used its hydraulic power to grind wheat and barley. After passing these mills,

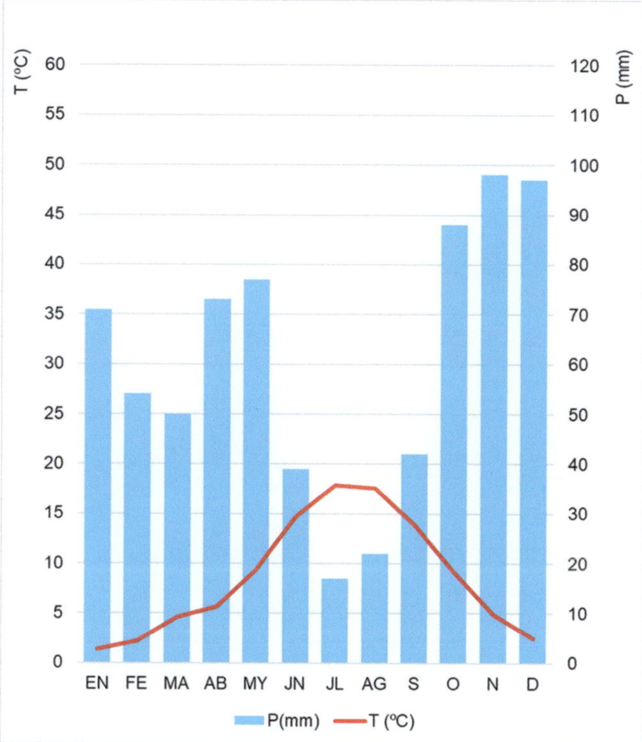


**Fig. 3.37** Sotalbo climograph. (Source: AEMET. Own elaboration)

the river flows near Palacios (Sotalbo) and continues through the Riatas valley before eventually emptying into the Adaja at Las Lagunas, at an Altitude of 1090 m. Following a fire and the lack of vegetation, the Río Picuezo has caused repeated problems of flooding and destruction as it passes through Riatas during episodes of heavy rainfall.

The rivers and streams in the study area (Ulaque River, Garganta River, Fortes River, Mayor River, Mora River, Astillero River, and Corneja River) exhibit a marked seasonal and torrential character, as they typically run dry during the summer, except for the Alberche River and streams at higher elevations. In spring, with increased precipitation and mountain snowmelt, especially from peaks above 2000 m, the highest flows occur, indicating a pluvio-nival regime (Abad Soria et al., 2022).

In La Paramera, there are natural stagnant waters in more impermeable, clayey areas, such as Navas Lagoon, which is temporary depending on rainfall, and peat bogs, particularly in the high meadows of Serrota. Additionally, there are numerous springs and sources typical of a mountainous area, such as Cabeza del Gallo Spring, Jardín Spring, Piedraluenga Spring, Cañada Spring, and Aguas Frías Spring on the slopes of the Sierra de la Paramera (Abad Soria et al., 2022).

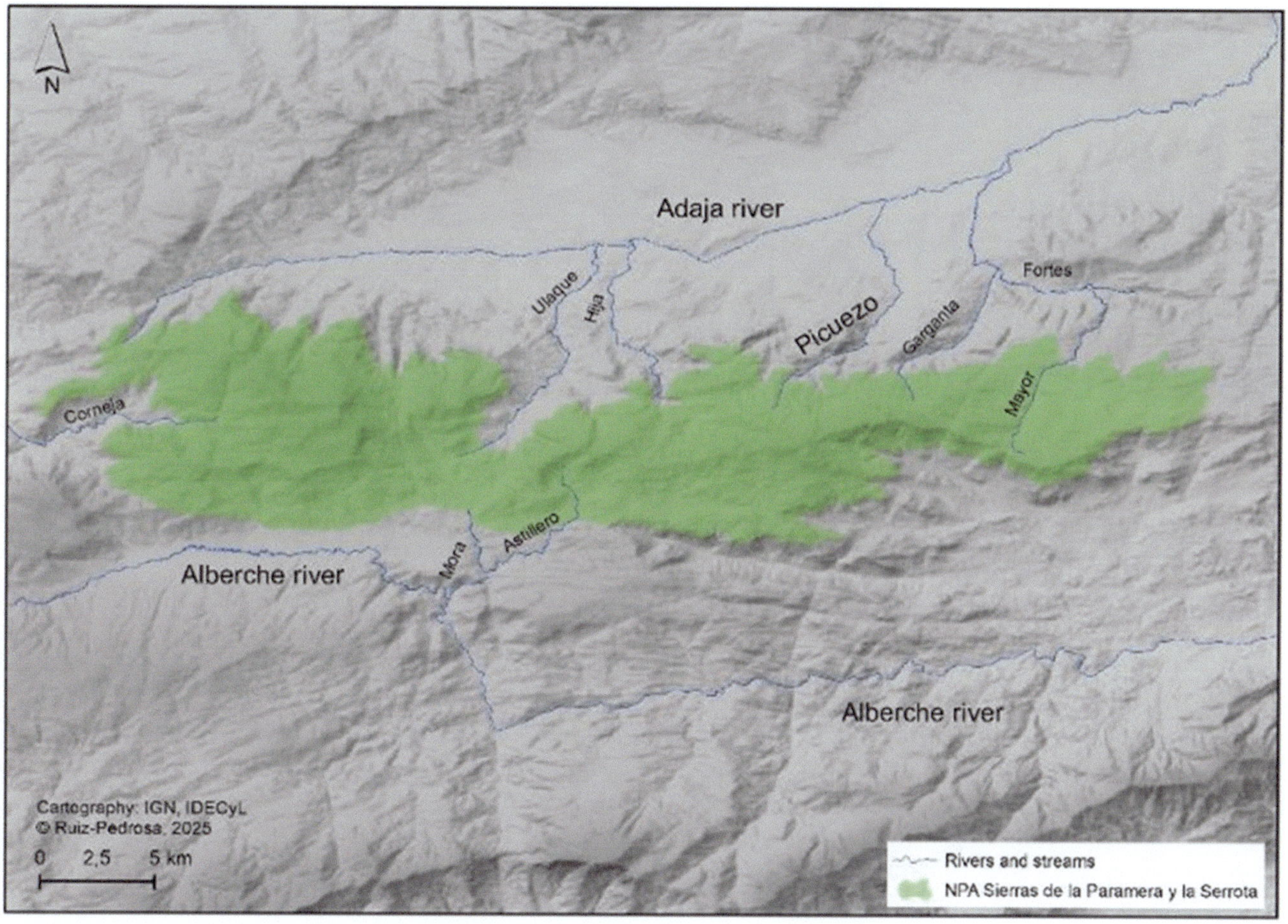


**Fig. 3.38** Main watercourses in the Paramera and Serrota mountain ranges

**Fig. 3.39** Adaja river spring

**Fig. 3.40** Picuezo river as it passes through La Hoya

## 3.2 Socioeconomic Study

### 3.2.1 Current Environmental Legislation

The Paramera and Serrota mountain ranges are currently a Special Conservation Area (ZEC in Spanish, Zone of Special Conservation) of the Natura 2000 Network, having obtained this distinction in 2015 after being a Site of Community Importance (LIC in Spanish) since 1998. In Spain, Law 42/2007, of December 13, on Natural Heritage and Biodiversity, defines Sites of Community Importance as "*those areas across the national territory that contribute significantly to the maintenance or, where appropriate, the restoration of a favourable conservation status of natural habitat types and species of community interest (...) within their natural distribution area*" (Ministry for Ecological Transition and Demographic Challenge, Date of consultation: August 2022).

The basic management and conservation plan for the Natura 2000 Protected Area—Paramera and Serrota

Mountains—establishes conservation objectives and necessary conservation measures to ensure its preservation, as well as management strategies and guidelines for the area, and the zoning of the NPA.

The plan includes a list of the Natura 2000 values that have justified its designation, highlighting those with excellent or good conservation status, both in terms of natural and seminatural habitats (Table 3.4) and essential species included in the Habitat Directive (Table 3.5).

Among the threats identified by the Management and Conservation Plan in the Sierra de la Paramera and Serrota, deforestation associated with livestock grazing and the use of fire as a management tool stands out, highlighting the threat posed by the destruction of vegetation by fires in this Natural Area. Following the Navalacruz fire in August 2021, this Plan becomes obsolete (see Sect. 3.1.3), as it resulted in the burning of most of the NPA and, therefore, of the habitats and species that justified its designation.

The NPA Sierra de la Paramera and Serrota borders three other Red Natura 2000 areas: Riverbanks of the Alberche River and its tributaries to the south, Riverbanks of the Tormes River and its tributaries to the west, and Riverbanks of the Adaja River and its tributaries to the north (Fig. 3.41).

However, on May 5, 1992, the initiation order of the Natural Resources Management Plan for the Sierra de la Paramera and Serrota NPA (1992) marked the beginning of this process. The Sierra de la Paramera and Serrota were to be incorporated into the Network of NPAs of Castilla y León under one of its protection categories: National Park, Regional Park, Natural Park, Protected Landscape, Natural

**Table 3.4** Natural and seminatural habitats present in NPA Sierras de la Paramera y la Serrota with overall excellent and good assessment

| Habitat | Excellent evaluation | Good evaluation |
|---|---|---|
| Lakes and natural dystrophic ponds | | X |
| Mountain formations of Cytisus purgans | X | |
| Grass formations with Nardus on siliceous substrates in mountainous areas | X | |
| Active high peat bogs | | X |
| Mediterranean landslides | | X |
| Siliceous rocky slopes | | X |
| Oak forest with Quercus robur and Quercus pyrenaica | | X |

**Table 3.5** Species present in NPA Sierras de la Paramera y la Serrota with overall excellent and good assessment

| Species | Excellent evaluation | Good evaluation |
|---|---|---|
| Euphydryas aurinia | | X |
| Lucanus cervus | | X |
| Lacerta schreiberi | | X |
| Canis lupus | | X |
| Iberolacerta cyreni | X | |
| Parnassius apollo | X | |
| Rana iberica | | X |

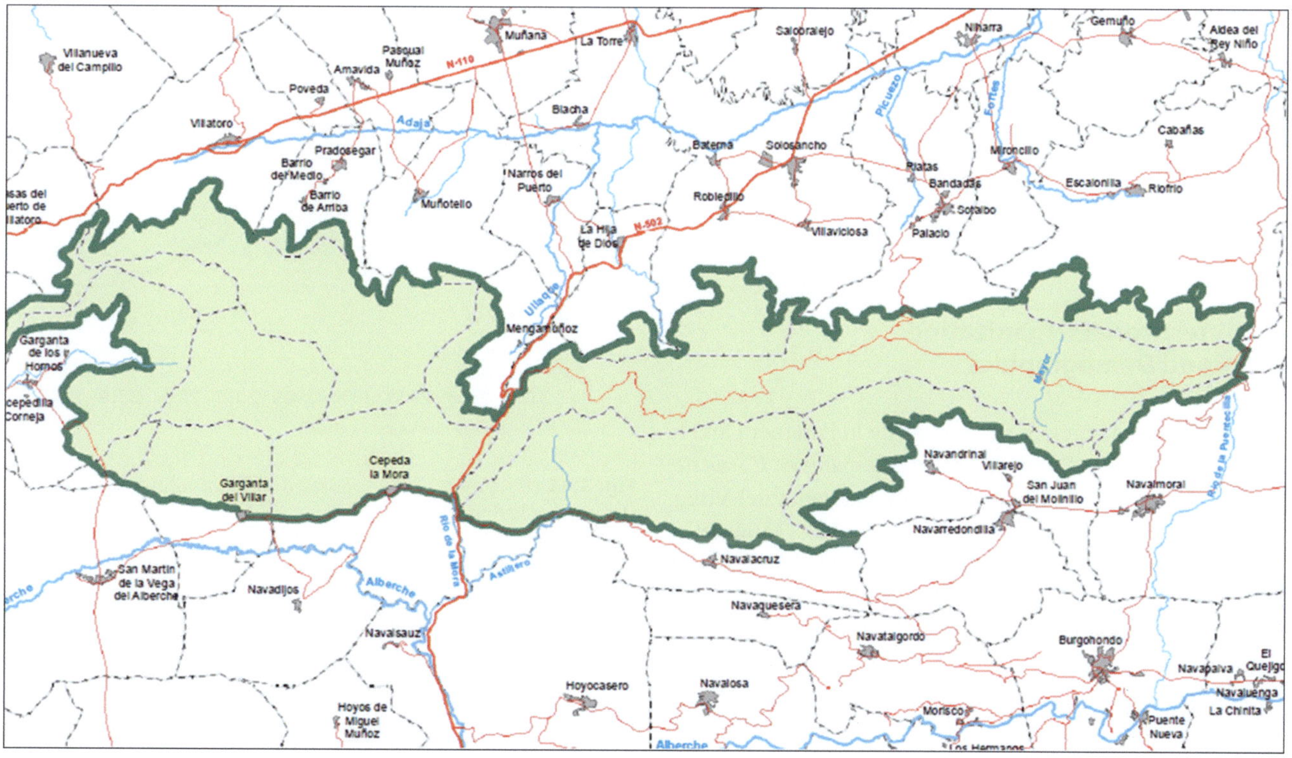


**Fig. 3.41** Perimeter of the Paramera and Serrota NPA. (Source: Junta de Castilla y León)

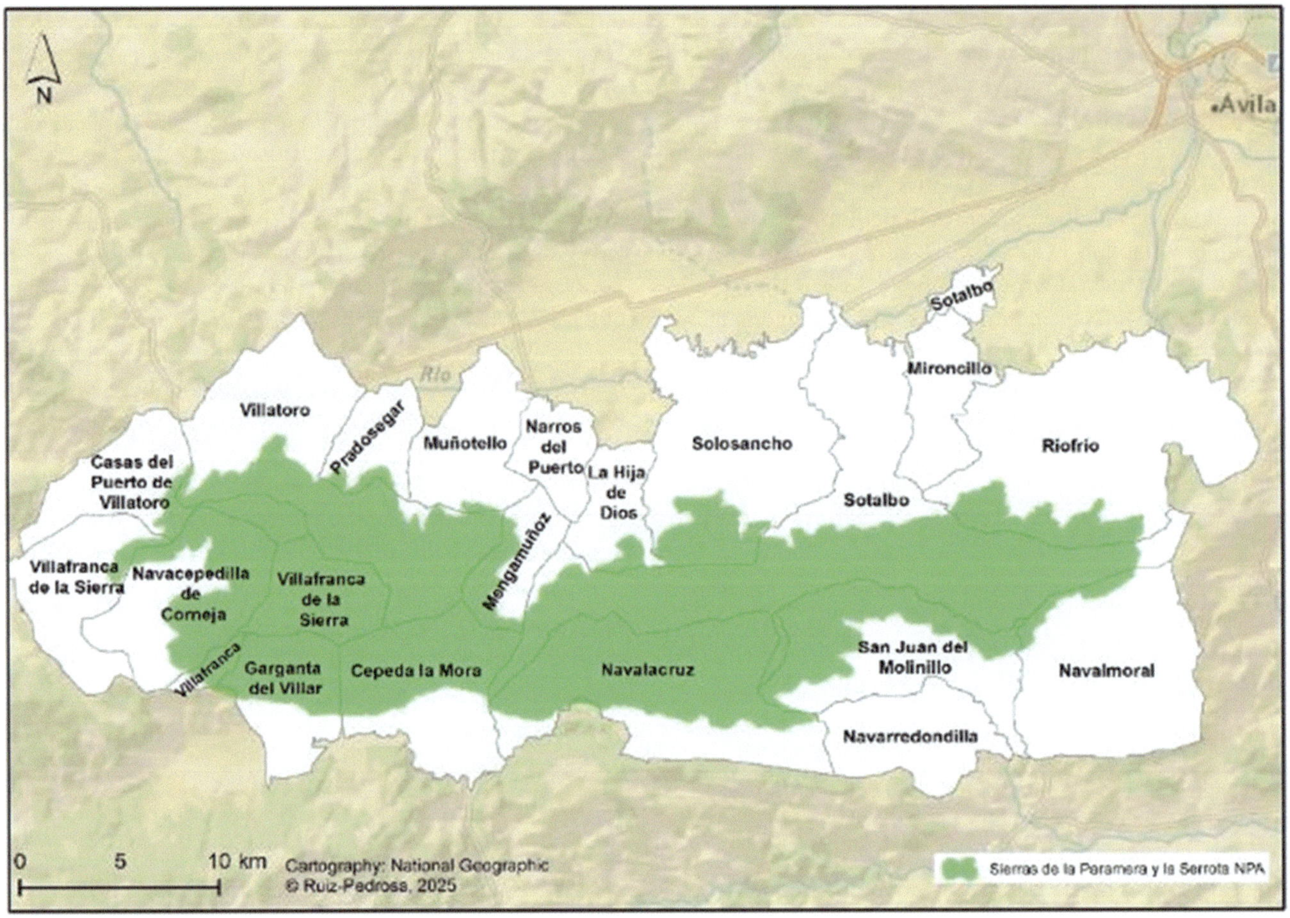


**Fig. 3.42** Municipalities that form part of the Paramera and Serrota NPA

Monument, or Nature Reserve. Despite this, no progress has been made in this process since 1992. It has been noted in some official documents that the Paramera and Serrota are designated as a Protected Landscape, indicating that this may be the intended protection category, although no official declaration has been made to date (as of March 2025).

### 3.2.2 Territorial Organization and Demographics

The socioeconomic area of the Sierras de la Paramera and la Serrota is composed of a total of 19 municipalities: Casas del Puerto de Villatoro, Cepeda de la Mora, Garganta del Villar, La Hija de Dios, Mengamuñoz, Mironcillo, Muñotello, Narros del Puerto, Navacepedilla de Corneja, Navalacruz, Navalmoral, Navarredondilla, Pradosegar, Riofrío, San Juan del Molinillo, Solosancho, Sotalbo, Villafranca de la Sierra, and Villatoro (Fig. 3.42).

The population of the municipalities under study, as in all rural areas of our autonomous community, is characterized by depopulation and aging. Since the beginning of the century, the registered population has decreased from 4500 to 3228 inhabitants in 2023 (Fig. 3.43). This represents a loss of a quarter of its population, as a result of both negative natural and migratory balances (Table 3.6). The natural balance has been negative since the 1990s, due to the decline in births due to the lack of young people of childbearing age (from 30 to just 7 in 2022) and the increase in deaths (from 60 to 69) as a result of a highly aged population.

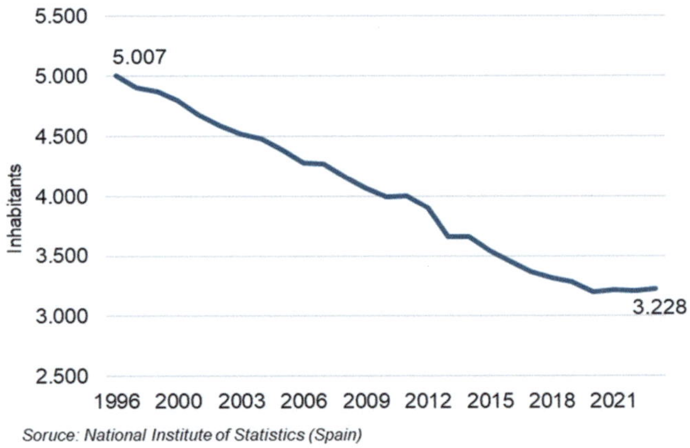


**Fig. 3.43** Evolution of the registered population in the Sierras de la Paramera y la Serrota NPA

The migration balance is negative except in 2010, when there was a significant increase in the arrival of immigrants, possibly due to the cultivation of strawberries and other agricultural works. The decline in the migration balance began in the 1990s as a result of the rural exodus that occurred during that time. Emigrations in the study area are mainly to other autonomous communities in the country, with most people from the province of Ávila moving to Madrid due to its proximity to the national capital and the close economic and social ties between Ávila and Madrid. In terms of immigrations, they have been increasing steadily over the years, although the pace has slowed since 2010. The majority of immigrants come from other autonomous communities and the same province, with a small percentage of international immigration.

**Table 3.6** Vital statistics in the study municipalities

| Vital statistics | | 1990 | 2000 | 2010 | 2022 |
|---|---|---|---|---|---|
| Births | | 30 | 24 | 11 | 7 |
| Deaths | | 60 | 83 | 71 | 69 |
| **Natural increase** | | **−30** | **−59** | **−60** | **−62** |
| **Residential variations** | | 1990 | 2000 | 2010 | 2021 |
| Outbound to | The same province | 68 | 37 | 56 | 55 |
| | Another province in the region | 1 | 7 | 4 | 7 |
| | Another region | 80 | 58 | 65 | 63 |
| | Other countries | | | 2 | 6 |
| | **Total emigrations** | **149** | **102** | **127** | **131** |
| Inbound from | The same province | 10 | 34 | 79 | 45 |
| | Another province in the region | 2 | 2 | 4 | 5 |
| | Another region | 46 | 26 | 73 | 61 |
| | Other countries | | 5 | 17 | 18 |
| | **Total immigrations** | **58** | **67** | **173** | **129** |
| **Migration balance** | | **−91** | **−35** | **46** | **−2** |

Source: Statistical Information System, Junta de Castilla y León

The population pyramid of the Paramera and Serrota mountain ranges (Fig. 3.44) shows a clearly progressive shape, with a very narrow base, as the population under 20 years old represents only 10% of the total. The majority of the population is concentrated in the 40–70 age range, with a higher proportion of male population. However, from the age of 75, women become dominant as a result of a higher life expectancy. We are facing a clearly aging population, with very low birth rates and a certain degree of masculinization that only diminishes at older ages. The future of these municipalities is therefore excessive depopulation, with no generational replacement, and the abandonment of these rural municipalities due to the low presence of permanent population, although there are numerous second homes and during holiday periods, especially in summer, the population increases significantly. However, these vacation or weekend second homes do not contribute to the development of the municipalities or their educational and health services.

If we compare the population pyramid of the NPA with that of Ávila capital (Fig. 3.45), we can see a somewhat differentiated shape, with a much wider base, which approaches

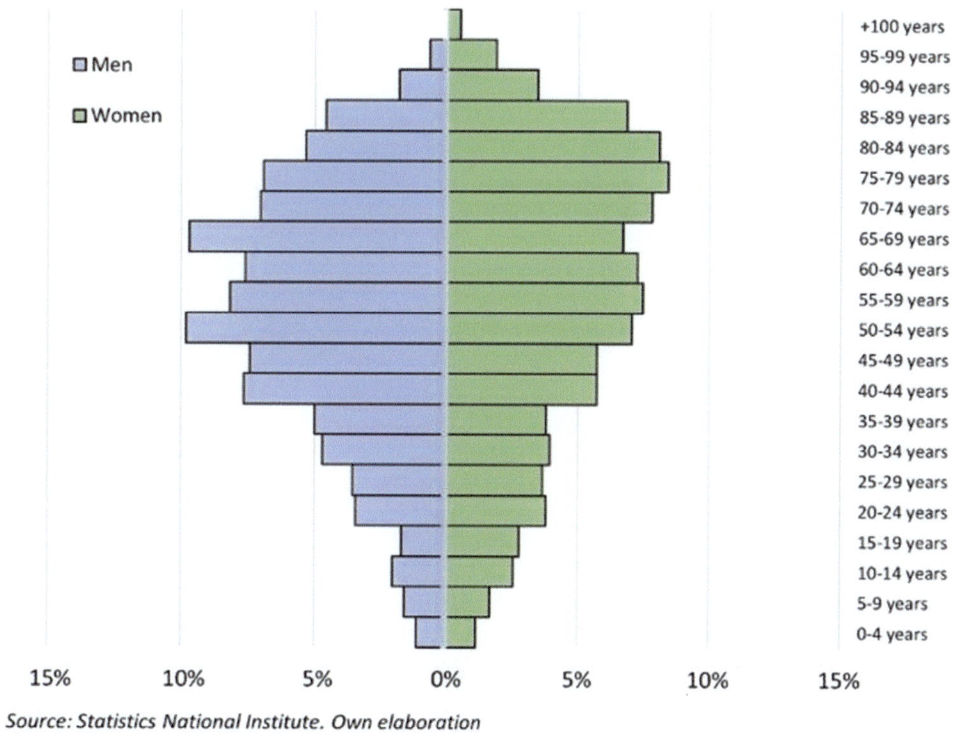


**Fig. 3.44** Population pyramid of the Sierras de la Paramera y la Serrota NPA (% of total) as of 01/01/2024

**Fig. 3.45** Population pyramid of Ávila capital (% of total) as of 01/01/2023

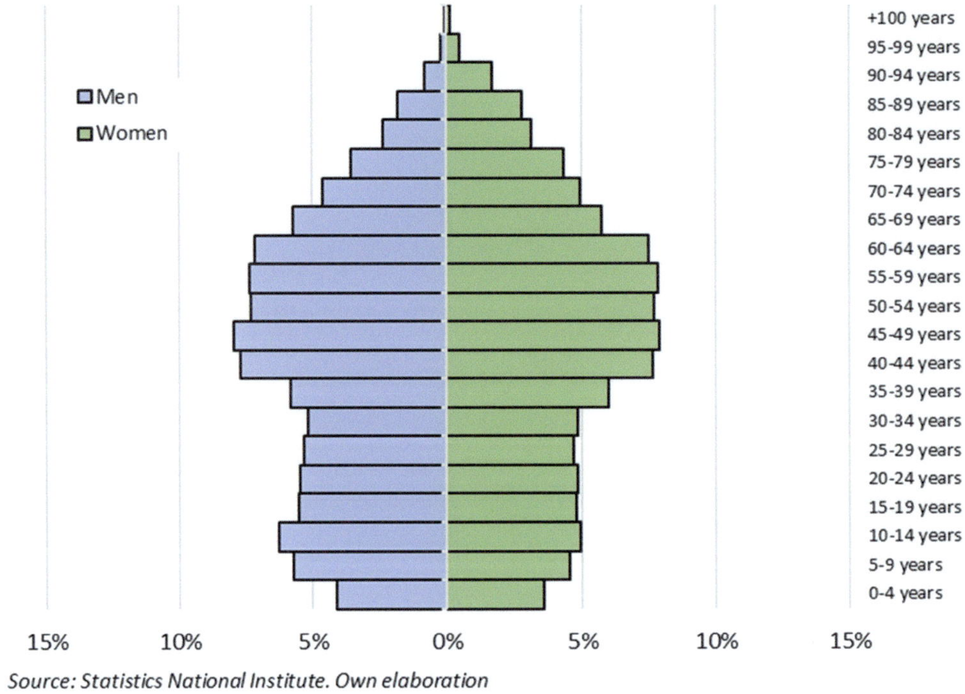


**Fig. 3.46** Distribution of the population in the municipalities of the Sierras de la Paramera and la Serrota NPA on 01/01/2023 (% of total)

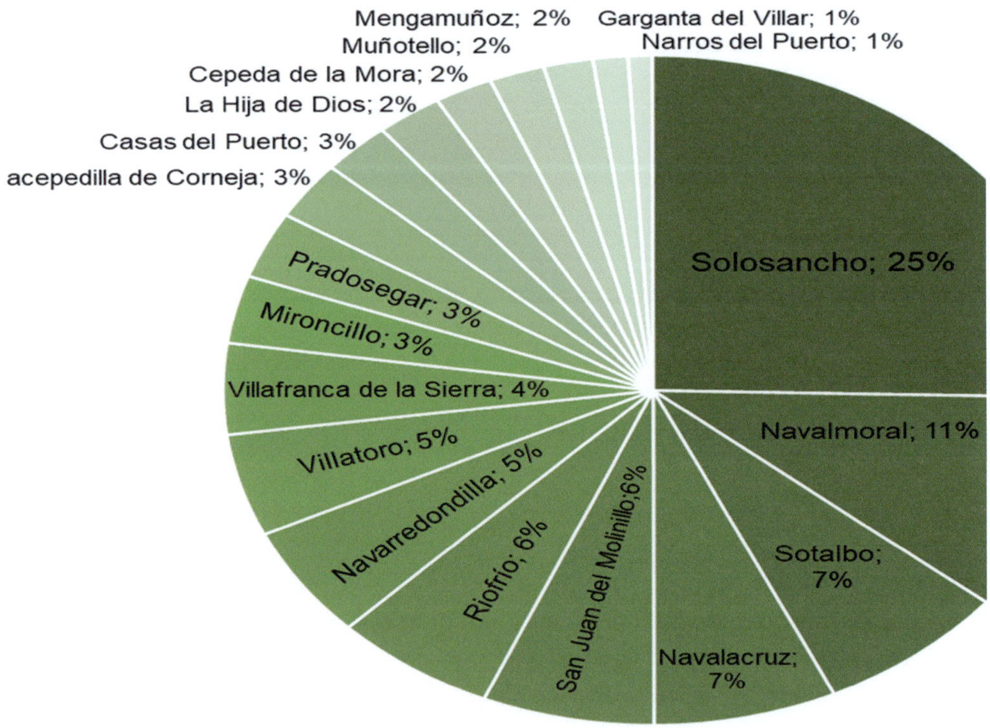


or exceeds 5% of the population in both sexes from 0 to 34 years of age (while in the Paramera and the Serrota it only occurs from 34 onward) and a much narrower peak from 70 years onward. This easily shows a greater dynamism in the capital, where most of the young population is concentrated, and the bulk of the population is in the working age population. It is evident that Ávila capital concentrates the economic, educational, and training resources of the province, although it is also an aging municipality that has been affected by depopulation due to the influence of Madrid and, to some extent, Salamanca.

Regarding the distribution of the population by municipalities, we can see that a quarter of the inhabitants live in Solosancho (Fig. 3.46), acting as the main town in the region, as we will see with other socioeconomic indicators. Solosancho is followed by Navalmoral with 11% of the total, Sotalbo and Navalacruz with 7%, and San Juan del Molinillo and Riofrío with 6%. The rest of the municipalities are sparsely populated, with Garganta del Villar and Narros del Puerto having the least population with just 1%.

### 3.2.3 Services and Economic Activity

The study area, like any rural area with a marked depopulation, has limited basic services, generally scarce, concentrated in the municipality of greater importance, Solosancho, which serves as the head of the area. In this particular case, the proximity of the most populated municipalities of Ávila capital also plays an important role, not exceeding 20 min by car, further emphasizing the dependence of the towns in the Paramera and Serrota areas on the provincial capital.

Solosancho has the highest number of doctors' offices with four, followed by San Juan del Molinillo with three. Pradosegar, Riofrío, and Villatoro have two, and the rest of the municipalities have one, except for Casas del Puerto, which does not have one. Only Solosancho, Villatoro, Navalacruz, Navalmoral, Navarredondilla, and Villafranca de la Sierra have a pharmacy (Table 3.7).

Education has been severely reduced, going from having 20 educational centers and 397 students in 1990 to just 2 centers and 53 students in 2022 (Table 3.8). These three educational centers are located in Solosancho—Ulaquillos Infant Education School, Sotalbo—Los Fresnos Rural Cluster Center, and Navalmoral—Navas del Alberche Infant and Primary Education Center. The Rural Cluster Centers of Sotalbo and Navalmoral stand out as centers attended by students from neighboring municipalities. There is no secondary education center in the area, so all students must attend a high school in the capital city of Ávila to continue their studies.

**Table 3.7** Doctors' offices and pharmacies in the municipalities of Paramera and Serrota NPA

| | Doctors' offices | Pharmacies |
|---|---|---|
| Solosancho | 4 | 1 |
| San Juan del Molinillo | 3 | X |
| Pradosegar | 2 | X |
| Riofrío | 2 | X |
| Villatoro | 2 | 1 |
| Navalacruz | 1 | 1 |
| Navalmoral | 1 | 1 |
| Navarredondilla | 1 | 1 |
| Villafranca de la sierra | 1 | 1 |
| Cepeda de la mora | 1 | X |
| Garganta del Villar | 1 | X |
| La Hija de Dios | 1 | X |
| Mengamuñoz | 1 | X |
| Mironcillo | 1 | X |
| Muñotello | 1 | X |
| Narros del Puerto | 1 | X |
| Nava cepedilla de Corneja | 1 | X |
| Sotalbo | 1 | X |
| Casas del Puerto | X | X |
| **Total** | **26** | **5** |

Source: Junta de Castilla y León. Reference date: 2022

**Table 3.8** Evolution of schools, pupils and teachers in Paramera and Serrota NPA (1990–2022)

| | 1990 | 2010 | 2022 |
|---|---|---|---|
| Educational establishments | 20 | 2 | 2 |
| Non-university students | 397 | 142 | 53 |
| Teachers | | 32 | 12 |

Source: Junta de Castilla y León

The economic activity of the municipalities of Paramera and Serrota is characterized by traditional livestock farming, with intense grazing followed by agricultural use, as well as modest development of the tertiary and tourist sectors linked to rural tourism.

The Land Use Information System (Spanish government centralized system) has been consulted and a land use map of the study area has been prepared (Fig. 3.47). The most represented land use is sclerophyllous shrubland in the perimeter of the NPA, followed by transitional shrubland and small areas of coniferous forests. However, after the Navalacruz fire, we know that the entire perimeter of the NPA lost vegetation and, currently, some species are regenerating and others are being replanted (see Sect. 3.1.3). Outside the NPA, there are some rocky areas, especially in the surroundings of Ulaca, and in the Amblés Valley dryland cultivation dominates, with small areas occupied by irrigated crops and a mosaic of crops. At the edges between the mountains and valleys, there are small extensions of deciduous forests and meadows. The Alberche Valley is also dominated by shrubland, conifers, grasslands, and small meadows and pastures.

The cultivation of strawberries in the Amblés Valley is particularly noteworthy, specifically in the municipalities of Solosancho, Niharra, and La Torre (Figs. 3.48, 3.49, and 3.50). This strawberry cultivation provides employment for thousands of seasonal workers, many of whom are of foreign nationality, between the months of April and October, the season in which strawberries are planted, harvested, and sorted. In October, 90% of the production is sent to Huelva. In the Amblés Valley, four strawberry companies stand out: Viveros Niharra, Frespalos, Viveros Carboner, and Marifrancis. An indication of the growing importance of strawberries in Ávila is the existence of new research projects related to them, such as the one being carried out by researchers Susana del Pozo and Paula de Andrés from the Escuela Politécnica Superior

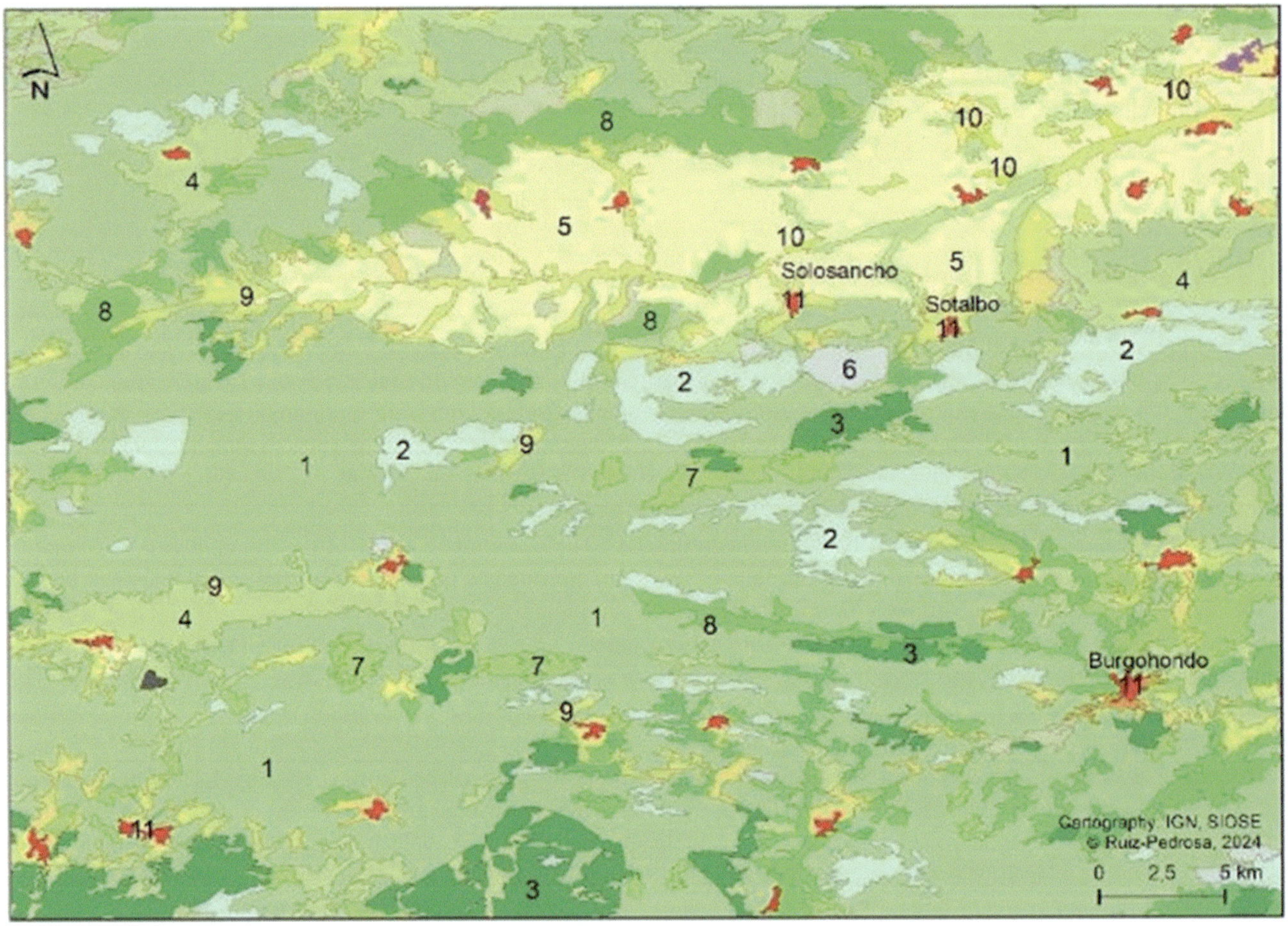


**Fig. 3.47** Land uses in the Paramera and Serrota. Legend: (1) sclerophyllous shrubs; (2) areas with sparse vegetation; (3) coniferous forest; (4) natural grasslands; (5) dryland farming; (6) rock outcrop; (7) transitional shrub forest; (8) broadleaf forest; (9) meadows and pastures; (10) irrigated farming/crop mosaic; (11) urban nucleus

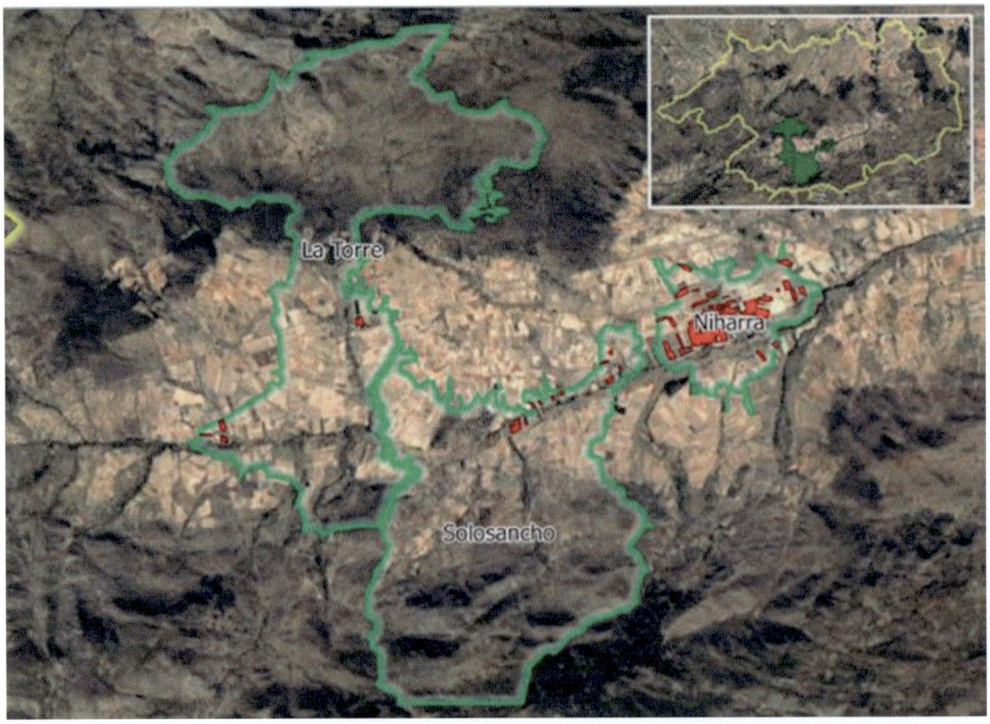


**Fig. 3.48** Location of the three municipalities dedicated to the strawberry sector within the Valle de Amblés region and details of the strawberry cultivation plots (in red). (Source: Del Pozo, S and de Andrés, P (2021))

**Fig. 3.49** Strawberry nursery in Niharra

**Fig. 3.50** Strawberry plantations in Niharra. In the background, the Zapateros Peaks

de Ávila (part of the University of Salamanca), who are developing the project titled "Sustainability in the strawberry cultivation of the Amblés Valley: Photovoltaic solar irrigation." The objective of this project is to study the economic viability of implementing photovoltaic panels to power irrigation for strawberries (Source: https://politecnicadeavila.usal.es/2021/09/13/susana-del-pozo-y-paula-de-andres-sostenibilidad-en-el-cultivo-fresero-del-valle-ambles-riego-solar-fotovoltaico/ Date: August 18, 2022).

The latest agricultural census of 2009 (Table 3.9) shows that Villatoro is the municipality with the highest number of livestock heads, followed by Riofrío and Solosancho, all three with 6000–7000 heads of cattle. Below them, with around 3000 heads, are Navalmoral and Sotalbo. As expected, in the Paramera and la Serrota areas there is a specialization in bovine livestock, with extensive farming (Fig. 3.51), highlighting the Protected Geographical Indication Carne de Ávila, with the Avileña-Negra Ibérica breed. There is also a relatively important presence of goat farming (Fig. 3.52).

Regarding the data provided by the State Public Employment Service (SEPE, Spanish government) on contracts by municipalities and sectors (Fig. 3.53), the negligible weight of the industrial sector is evident, with the construction sector slightly higher. However, the services sector always stands out over agriculture and livestock, with a significant increase in contracts in both sectors in May–September, linked to the harvest season in agriculture and summer and increased services in the tertiary sector.

**Table 3.9** Heads of livestock by municipality in Paramera and Serrota NPA

| | Cattle | Goat | Rabbits | Horse | Sheep | Pig | Other cattle | Dairy cows | Total |
|---|---|---|---|---|---|---|---|---|---|
| Villatoro | 2647 | 11 | | 2 | 1129 | 455 | 2629 | 18 | 6891 |
| Riofrío | 2895 | 45 | 1 | 74 | 841 | 2 | 2892 | 3 | 6753 |
| Solosancho | 1772 | 28 | 2 | 71 | 1831 | 1164 | 1683 | 89 | 6640 |
| Navalmoral | 1467 | 74 | 2 | 275 | 666 | 3 | 1446 | 21 | 3954 |
| Sotalbo | 998 | 132 | 1 | 3 | 1375 | 6 | 831 | 167 | 3513 |
| Garganta del Villar | 1195 | | | 1 | 225 | | 1195 | | 2616 |
| Navalacruz | 814 | 6 | | 129 | 635 | | 814 | | 2398 |
| San Juan del Molinillo | 758 | 13 | | 3 | 779 | 3 | 738 | 2 | 2296 |
| Casas del Puerto | 643 | 29 | | 15 | 451 | | 643 | | 1781 |
| Cepeda de la mora | 186 | | | 21 | 427 | | 996 | 9 | 1639 |
| Mengamuñoz | 622 | | | | 11 | | 622 | | 1255 |
| Muñotello | 13 | | | 8 | 78 | | 887 | 143 | 1129 |
| Villafranca de la sierra | 441 | 52 | | 45 | 85 | 6 | 441 | | 1070 |
| Mironcillo | 448 | | 4 | 5 | 4 | 154 | 445 | 3 | 1063 |
| Navarredondilla | 419 | | | 14 | | 2 | 419 | | 854 |
| La Hija de Dios | 264 | | | 3 | 58 | | 253 | 11 | 589 |
| Pradosegar | 153 | | | 12 | | | 153 | | 318 |
| Navacepedilla de Corneja | 98 | 32 | | 45 | | | 98 | | 273 |
| Narros del Puerto | 47 | | | 3 | 157 | | 47 | | 254 |
| Total | 15,880 | 422 | 10 | 729 | 8752 | 1795 | 17,232 | 466 | |

Source: Agricultural Census 2009, Junta de Castilla y León

**Fig. 3.51** Extensive livestock farming at Peña Cabrera

In recent years, there has been a certain dynamism in the field of rural tourism in the NPA. Thus, it can be seen that the area has a total of 60 rural accommodation options, 16 restaurants, and 3 hotels (Table 3.10).

In the face of the inaction of the Junta de Castilla y León in the protection and dissemination of the Sierras de la Paramera and la Serrota, it is the Diputación de Ávila, with the support of the municipalities, that is taking the initiative in promoting the natural and cultural heritage. Thus, in 2023, the municipalities of Sotalbo, Solosancho, Mironcillo, and Riofrío, at the initiative of the Provincial Council of Ávila, have developed a collaborative project for the design of the first signposted and georeferenced routes in the NPA. A total of four routes (Fig. 3.54) that aim to highlight the natural environment, environmental activities, and sustainable tourism.

Similarly, initiatives to promote cultural heritage are also being carried out. In Villaviciosa, a visitor center for the Vettonian Castro of Ulaca is being built, funded by the

**Fig. 3.52** Sheep and goat grazing. Provincial road 409 from Riatas to Solosancho

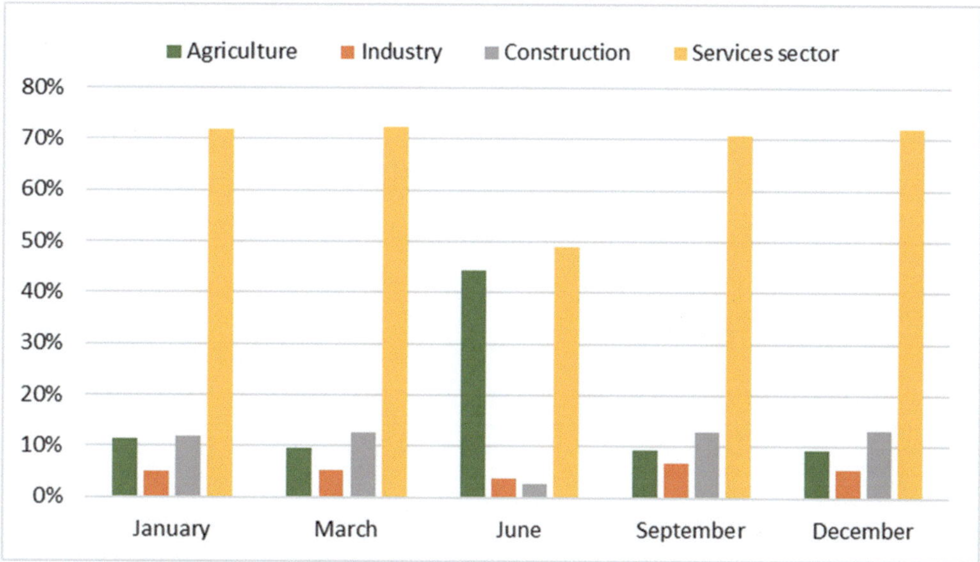


**Fig. 3.53** Contracts by sector in the study area (year 2023). (Source: SEPE)

Ministry of Culture of the Junta de Castilla y León due to the initiative of the Solosancho Town Hall. Additionally, in Riatas, the Ethnographic Museum of Riatas was opened in October 2021.

### 3.2.4 Cultural Heritage

The towns of Paramera and Serrota and their people possess a tangible and intangible cultural heritage of undeniable value and richness. In this mountainous area of Ávila, a series of traditions, celebrations, and arts linked to the territory, its resources, and the Sierra where they have grown have been woven throughout history.

There are four Cultural Interest Sites in the area, two of them of great importance and relevance in the area of study and for its residents: the Castle of Aunqueospese—Manqueospese for the locals—and the Abandoned Village of Ulaca.

The Castle of Manqueospese or Aunqueospese is located on a hill in the Sierra de la Paramera, in the municipality of Mironcillo, with an imposing view of the entire Amblés Valley. Built between the fourteenth and fifteenth centuries, and declared a Cultural Interest Site on June 3, 1931, it is constructed using the topography and the granite relief, with large granite blocks acting as walls (Figs. 3.55, 3.56, and 3.57).

Neighbors say that the castle (known to them as Manqueospese, although the official name is Aunqueospese)

**Table 3.10** Tourism establishments in the NPA Sierras de la Paramera y la Serrota (as of 2022)

| | Hotels | Rural tourism housing | Restaurants |
|---|---|---|---|
| Solosancho | 1 | 21 | 4 |
| Pradosegar | | 6 | |
| Villatoro | | 6 | 1 |
| Casas del Puerto | | 4 | 1 |
| Navacepedilla de Corneja | | 3 | 1 |
| Riofrío | | 3 | |
| Cepeda de la mora | | 2 | |
| Mironcillo | | 2 | |
| Navalacruz | | 2 | 1 |
| Navalmoral | 2 | 2 | 6 |
| Navarredondilla | | 2 | |
| Sotalbo | | 2 | |
| Mengamuñoz | | 1 | 2 |
| Muñotello | | 1 | |
| Narros del Puerto | | 1 | |
| San Juan del Molinillo | | 1 | |
| Villafranca de la sierra | | 1 | |
| Garganta del Villar | | | |
| Total | 3 | 60 | 16 |

Source: Junta de Castilla y León. Reference date: 2021

gets its name from a love story between a lady and a knight, with her parents opposing their romance and the castle being their meeting place. Thus, they would consummate their love in the castle, "Manqueospese" to their families. Currently, the castle is privately owned, and it is in a deplorable state of abandonment and dirtiness inside and in its surroundings, with its main accesses walled up due to criminal activities. Given its poor state of conservation, access to the interior posed a risk to visitors. Therefore, one can only enjoy this imposing castle by visiting its surroundings and the views it offers of the entire Amblés Valley and the Picos Zapateros.

The Despoblado de Ulaca or Castro de Ulaca, located in the municipality of Villaviciosa, has a rich historical and archaeological heritage, around which a series of traditions have been created, especially among the inhabitants of Solosancho and Villaviciosa. Located on the top of Cerro del Castillo on one of the foothills of the Sierra de la Paramera, almost in contact with the Amblés Valley, it is a fortified Vetton settlement from the late Iron Age. Its enormous interest lies in the fact that with over 60 ha and more than 3000 m of walls, it is one of the largest in Hispanic Celtica. It houses some exceptional Celtic structures in excellent condition,

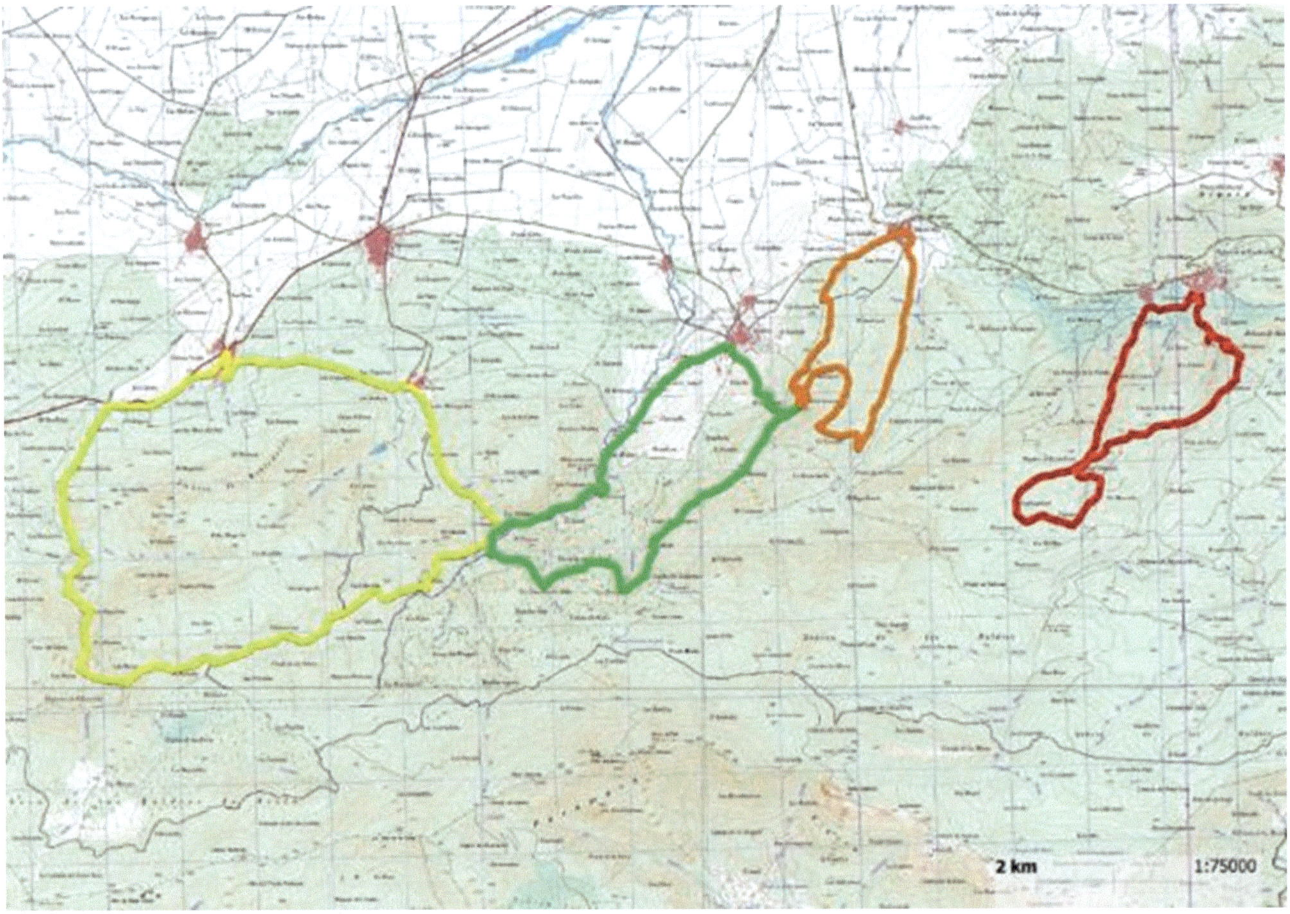


**Fig. 3.54** Routes in the Sierra de la Paramera and Serrota. (Source: Solosancho Town Council)

**Fig. 3.55** Main facade of Manqueospese Castle

**Fig. 3.56** Side of the Manqueospese Castle, integrated into the large granite blocks

including a 3-km perimeter wall, a sanctuary, ritual sauna, the remains of over 250 houses, the central tower of the settlement, and spectacular granite quarries that showcase how the Vettones utilized the geological and geomorphological resources offered by the Sierra. The Ulaca community may have had around 1500 inhabitants, according to Ruiz Zapatero (2005), who disappeared after the Roman conquest and likely settled in what is now the capital city of Ávila. The Castro was not extensively studied until the late 1990s when a team of Spanish and British archaeologists from the Complutense University of Madrid and the University of Sheffield, led by Ruiz Zapatero, Álvarez Sanchís, and J. Collis, began excavation and archaeological reconnaissance work. Since 2016, the Complutense team has been conducting topographic, photogrammetric, and geophysical work at various points of the oppidum, as well as several cleaning and excavation campaigns. In recent years, a team of geophysicists led by Miguel Ángel Maté Gonzalez from the University of Salamanca has been working on the digitalization of the Castro and applying remote sensing techniques to

**Fig. 3.57** Southern face of Manqueospese Castle

**Fig. 3.58** Walls of access to the Ulaca hillfort

characterize the structures that are still unexcavated (Figs. 3.58, 3.59, and 3.60).

From the vicinity of Ulaca come two verracos, sculptures carved in granite representing bulls and pigs, typical of the territory occupied by the Vettones, and are related to the protective function of livestock or as boundary markers for livestock areas (Ruiz Zapatero, 2005). The most spectacular of the verracos, over 2 m in length, is currently located next to the church of Solosancho (Fig. 3.61), while the second one is preserved next to the castle of Villaviciosa (Fig. 3.62).

The connection of the population with the hillfort and its Celtic roots is such that every August the Celtic Moon is celebrated in Solosancho and the population centers that make up the municipality (Baterna, Robledillo, and Villaviciosa). It is a festival set in the Celtic aesthetic and culture that combines tradition, history, and culture, past and present. Guided tours and theatrical performances in the hillfort, falconry displays, archery, craft markets, and archaeological conferences are some of the activities that have taken place in the more than 15 editions that have been held, with the 18th edition scheduled for 2024 (Fig. 3.63). One of the nights of the Celtic Moon, the Solosancho theater group performs a play at the Higher of sacrifices of the Ulaca hillfort (Fig. 3.64), usually related to some mythological story of the

**Fig. 3.59** Higher of sacrifices in the Ulaca hillfort

**Fig. 3.60** Granite quarries in the Ulaca hillfort

**Fig. 3.61** Solosancho verraco

**Fig. 3.62** Villaviciosa verraco

Celtic peoples. A magical night where hundreds of residents (and visitors from Ávila and the surrounding areas) gather at the hillfort (Fig. 3.65).

Also linked to the Celtic world and a cultural use of granite modeling, we can highlight the myth around the so-called "Canto de los Responsos" (Fig. 3.66), a large granite block located southwest of the hillfort, on top of which there is a pile of small stones. Some authors argue that it preserves a folkloric ritual related to the Celtic funerary world. The ritual consisted of throwing stones onto the granite block, which is over 3 m high, in order to ward off spirits (Almagro Gorbea, 2006). However, among the

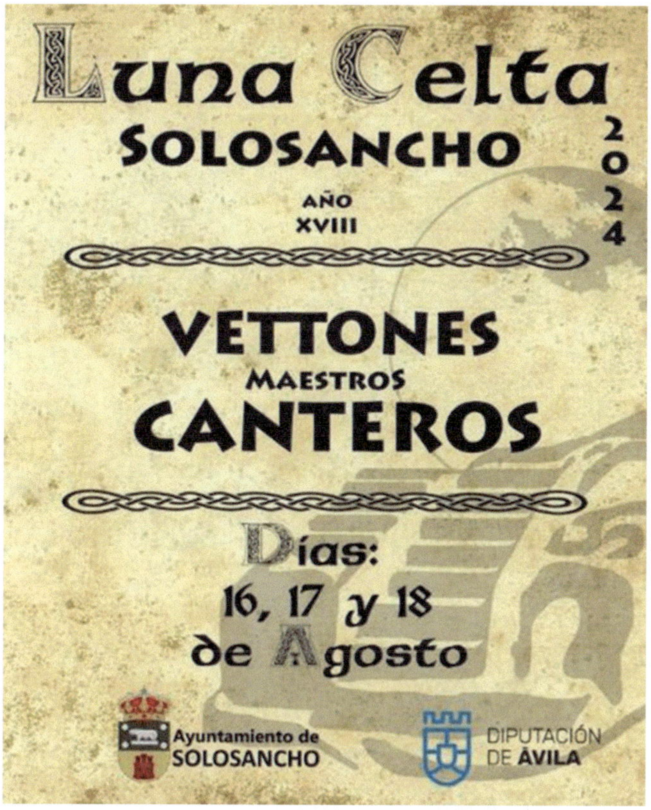


**Fig. 3.63** Poster of the XVIII Celtic Moon. (Source: Solosancho Town Hall)

locals, this practice of throwing a "responsory" has at least two meanings. The first is linked to the traditional practice of grazing, as the rock is located on a path used by carts from the village of Villaviciosa to the pastures of La Paramera. Thus, the shepherds would throw a stone at the canto, and if it stayed on top and did not fall, it would ensure a good journey. Nowadays, a more generic use is to throw the stone and make a wish, and similarly, if it stays on top and does not fall, the wish will come true.

The residents of La Paramera have always been linked to their territory, especially to the mountain that surrounds them. Traditionally, recreational activities and enjoyment of nature are carried out, combining culture and natural heritage. A good example is the tradition of families gathering at different points in the Sierra on Easter Sunday to eat the typical "hornazo" from Ávila. The Town Councils, especially those of Sotalbo and Solosancho, promote sports activities in La Paramera, such as the famous Celtibike (Fig. 3.67), a bike ride that starts in Solosancho and goes through the Amblés Valley and La Paramera—with a different route each year—or different mountain routes, such as the ascent to Peña Cabrera promoted by the Town Council of Sotalbo in 2022.

**Fig. 3.64** Theatrical performance at the sacrificial Higher of Ulaca. Celtic Moon of 2022

**Fig. 3.65** Celtic Moon, August 16, 2024

**Fig. 3.66** Canto de los Responsos in Villaviciosa

**Fig. 3.67** Celtibike, 2022 poster. (Source: Solosancho Town Hall)

## 3.3 Paramera and la Serrota Geomorphosites Inventory

The geomorphosites inventoried in the Sierras de la Paramera and la Serrota reflect their fractured and elevated morphostructure and their granite and glacial landforms. With a total of 12 geomorphosites, two of them are of structural attribution, two are fluvial, one geomorphosite is of fluvio-granitic attribution, and the rest are of granitic attribution (Table 3.11).

There are three exceptional geomorphosites: geomorphosite N°4 Granite rock thermal weathering, as a result of the large forest fire in Navalacruz in 2022; N°7 Serrota glacial cirques, being the only area in the NPA where we find glaciation; and N°12 Picuezo river giant's kettle, as this riverbed is where the most spectacular kettles are formed. Out of 12 geomorphosites, only two have high accessibility: N°8 Adaja river spring, and N°11 Puerto de Mengamuñoz fault, the first one less than 3 km from a parking, and the Mengamuñoz mountain pass fault accessible by car via the N-502 road. In the rest, medium accessibility is characterized by the absence of marked trails, as there are hardly any signposted geotourism routes. Additionally, the paths are rocky, more than 3 km long, and have relatively steep slopes. In terms of interest, eight geomorphosites have high interest, and four have medium interest. The latter are characterized by a greater geomorphological specialization and lower landscape and cultural value.

Geomorphosites distribution is not homogeneous with respect to the Natura 2000 Network zone, with five of them located outside of it (Fig. 3.68). These sites have been selected based on their geoheritage value, as the delimitation of Red Natura 2000 is based on biogeographical criteria and does not consider abiotic heritage. It is worth noting the concentration of geomorphosites in the surroundings of the Ulaca semihorst and the Picuezo river valley, where granite modeling and river incision create the most spectacular landscapes.

### 3.3.1 Geomorphosite N°1 Ulaca Semihorst

The Ulaca semihorst, known as the Castle Hill, is located on the northern slope of the Sierra de la Paramera horst, on the border with the Amblés Valley graben to the north and the Picuezo River to the south (Table 3.12, Fig. 3.69). In Ulaca, outstanding examples of granite landforms can be appreciated, with the most characteristic forms being nubbin and rock platforms, followed by tors, domes, half-domes, and pedestal rocks among the largest, as well as gnammas, tafone, and polygonal cracking among the microforms.

Although it is outside the NPA, the Despoblado de Ulaca has been a Cultural Interest Site since 1986, and it has one of the most important Vettonian hillforts in the Iberian Peninsula, where the granite forms are distributed according to their location, which in turn has conditioned the uses and modes of exploitation by the Vettones (see Sect. 5.3.1.2).

The hillfort has informational panels and a recently marked trail, although it is rocky and has a significant slope. The landscape value is high, with a panoramic view of the Amblés Valley, the Paramera, and the Zapateros Peaks.

**Table 3.11** Inventoried geomorphosites in Sierras de la Paramera y la Serrota

| N° | Geomorphological attribution | Type | | Accessibility | Interest |
|---|---|---|---|---|---|
| 1 | Granite | Place | Representative | Medium | High |
| 2 | Granite | Place | Representative | Medium | High |
| 3 | Fluvial | Place | Representative | Medium | High |
| 4 | Granite | Element | Exceptional | Medium | Medium |
| 5 | Granite | Element | Representative | Medium | Medium |
| 6 | Granite | Place | Representative | Medium | High |
| 7 | Glacial | Place | Exceptional | Medium | High |
| 8 | Fluvial | Place | Representative | High | High |
| 9 | Structural | Place | Representative | Medium | Medium |
| 10 | Granite | Place | Representative | Medium | High |
| 11 | Structural | Place | Representative | High | Medium |
| 12 | Fluvio-granite | Place | Exceptional | Medium | High |

Numbers are geomorphosites: 1. Ulaca Semihorst, 2. Manqueospese Granite castle, 3. Picuezo river alluvial fan, 4. Granite rock thermal weathering, 5. Responsos Granite isolated rock, 6. Picos Zapateros granite mountain ranges, 7. Serrota glacial cirques, 8. Adaja river spring, 9. Garganta Honda fault-line valley, 10. Gavilanes and Chinitas granite system, 11. Puerto de Mengamuñoz fault, 12. Picuezo river giant's kettle

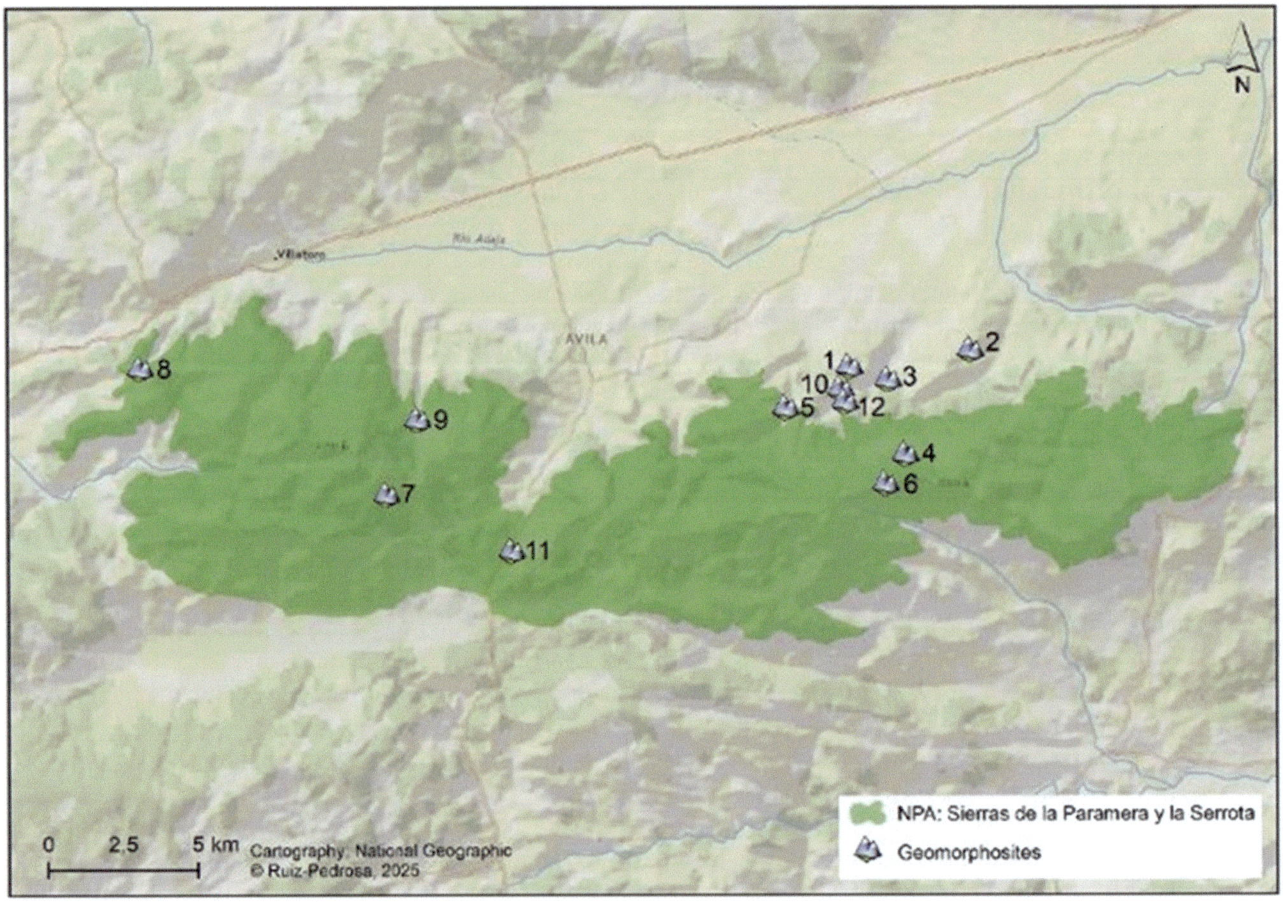


**Fig. 3.68** Sierras de la Paramera y la Serrota inventoried geomorphosites location. Numbers are geomorphosites described in Table 3.11

**Table 3.12** Semihorst of Ulaca geomorphosite

| Geomorphosite identification sheet | | | |
|---|---|---|---|
| Identification | **Name: Ulaca semihorst** | **Site: Ulaca archaeological site** | **N°1**<br>Altitude: 1508 m |
| Situation | Municipality: Villaviciosa, Ávila | Coordinates: 40.525181; −4.883174 | |
| Geomorphological values | Type | Site | |
| | | Representative | |
| | Genesis | Granite weathering | |
| | Morphology: Description, morphostructures, erosion | Flat-topped hill, on the northern slope of the sierra de la Paramera horst, at the boundary with the Amblés Valley graben to the north and the Picuezo River to the south. Outstanding examples of granite landforms can be seen, including domes, tors, nubbins, pedestal rock, and gnammas. It also offers a good view of the granite ridges of the Zapateros Peaks to the south-southwest | |
| | Dynamics | Granite weathering | |
| | Chronology | Miocene—present | |
| | Main interest | Granite landforms | |
| | Secondary interest | Structural | |
| | Representativeness | High | |
| | Geomorphosite attribution | Granite | |
| Additional values | Landscape | Panoramic view of the valley, the Sierra, and the granite modeling itself | |
| | Cultural elements | Castro de Ulaca | |
| | Economic | Touristic | |
| | Social | Recreation and cultural. Celebration of the Celtic Moon | |
| | State of conservation | Good | |
| | Actual uses | Cultural, touristic, and archaeological | |
| | Infrastructure | Interpretative panels about the hillfort, path (poorly signposted) | |
| | Impacts | Paths eroded by fire (August 2021), human influx | |
| | Legal situation | Ulaca Cultural Site | |
| Accessibility | Walking distance | 2 km from the parking lot (45 min) | |
| | Slope | 300 m of positive slope | |
| | Security | Medium (marked path but with loose stones) | |

### 3.3.2 Geomorphosite N°2 Manqueospese Granite Castle

Manqueospese granite castle (Table 3.13, Fig. 3.70) is located south of Mironcillo, west of the Garganta River, and its interest lies in its adaptation to the topography and granite landforms, using large granite blocks as part of its walls, with the construction completely embedded in the granite. Like the rest of the Paramera region, other granite landforms can be seen in its surroundings.

It offers views of the Amblés graben, Ávila capital, and the rest of the Paramera and Zapateros peaks. The castle was declared a Cultural Interest Property in 1991, although it is currently abandoned, and has a history related to its name (see Sect. 3.2.4). The access road starts from the village of Mironcillo, with a 3-km route and a positive slope of 250 m.

### 3.3.3 Geomorphosite N°3 Picuezo River Alluvial Fan

The Picuezo River alluvial fan is located in the Molino del Quemado Recreational Area, in the municipality of Sotalbo (Table 3.14, Fig. 3.71). It is an excellent example of active fluvial transport and sedimentation, as it has been recently formed after the Navalacuz wildfire, due to the accumulation of sandstones and ashes. The high erosive force has caused damage in the recreational area.

The presence of hydraulic flour mills with a granite wheel and a structure that included a room for the miller is interesting, although they are in poor condition (Fig. 3.72). Access to the recreational area is easy, with a 1.8-km route from Sotalbo and a 50-m slope.

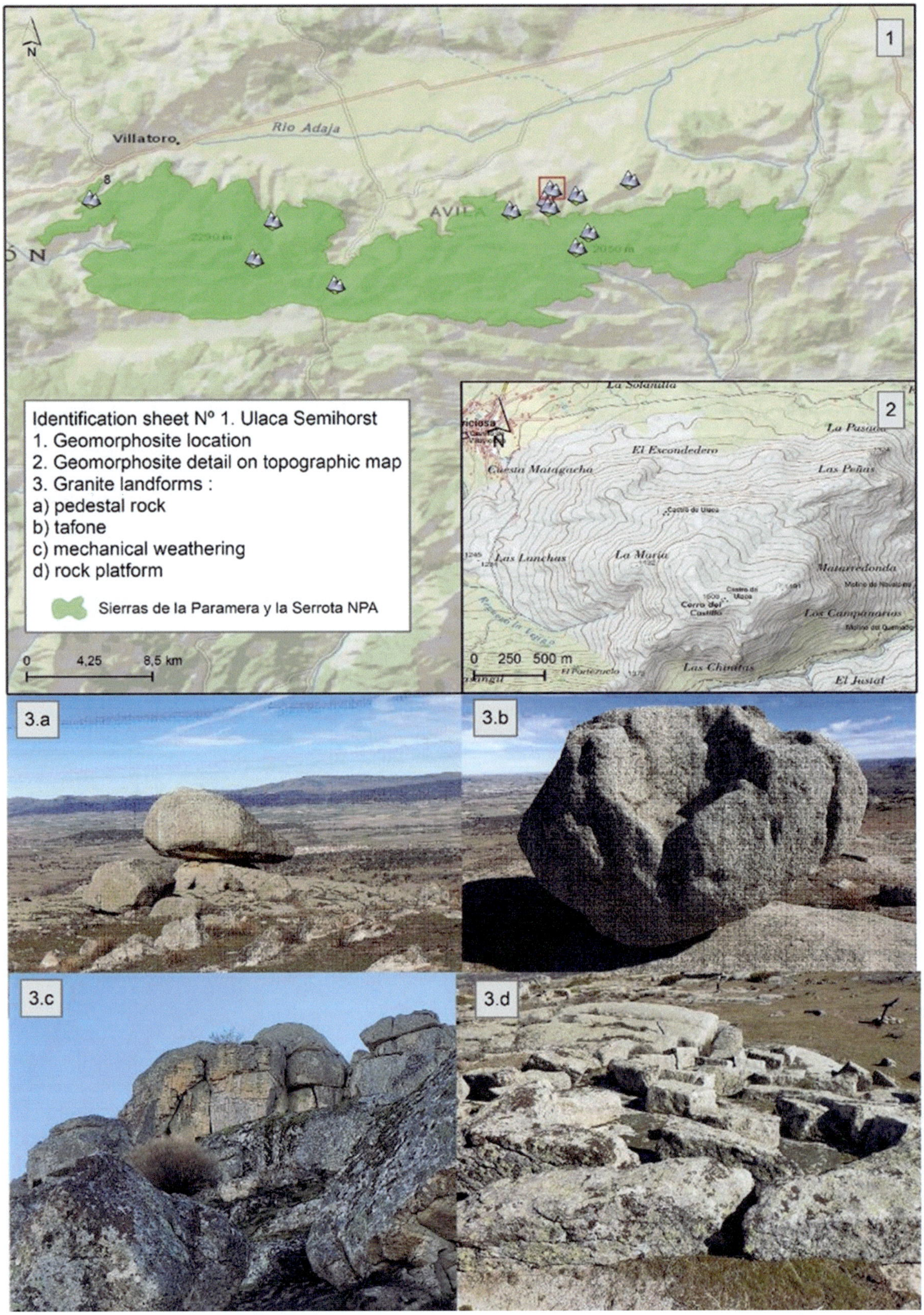


**Fig. 3.69** Ulaca semihorst identification sheet

**Table 3.13** Manqueospese granite castle geomorphosite

| Geomorphosite identification sheet | | | |
|---|---|---|---|
| Identification | **Name: Manqueospese granite castle** | **Site: Manqueospese castle** | **N°2**<br>Altitude: 1372 |
| Situation | Municipality: Mironcillo, Ávila | Coordinates: 40.34306; −4.838682 | |
| Valores geomorfológicos | Type | Site | |
| | | Representative | |
| | Genesis | Granite weathering | |
| | Morphology: description, morphostructures, erosion | Castle located in a semi-hill to the south of Mironcillo, to the west of the Garganta River. Adapted to the orography and the granite landforms, using large granite blocks as part of its walls, the construction is completely embedded in the granite. It offers views of the Amblés graben and the capital of Ávila | |
| | Dynamics | Granite weathering | |
| | Chronology | Miocene—present | |
| | Main interest | Granite landforms | |
| | Secondary interest | Germanic relief (horst-graben) | |
| | Representativeness | High | |
| | Geomorphosite attribution | Granite | |
| Additional values | Landscape | Panoramic view of the Amblés Valley and the capital of Ávila | |
| | Cultural elements | Aunqueospese castle | |
| | Economic | Touristic | |
| | Social | Cultural, recreation | |
| | State of conservation | Medium (garbage, construction materials) | |
| | Actual uses | Cultural, touristic | |
| | Infrastructure | Path | |
| | Impacts | Navalacruz wildfire (August 2022) | |
| | Legal situation | Aunqueospese Castle | |
| Accessibility | Walking distance | 3 km from Mironcillo (40 min) | |
| | Slope | 250 m of positive slope | |
| | Security | High (marked path) | |

### 3.3.4 Geomorphosite N°4 Granite Rock Thermal Weathering

This Geomorphosite is an exceptional example of granite erosion due to thermal weathering (Table 3.15, Fig. 3.73), caused by thermoclasty processes associated with the Navalacruz wildfire. The detached rock fragments due to high temperatures and marks on the granite are easily recognizable in situ.

The rock is located on the path to the summit of Zapatero peak from Fuente de las Aguas Frías, 3 km and 200 m slope, at a common access point to the summit. The landscape value is moderate, with a limited view of the Amblés Valley.

### 3.3.5 Geomorphosites N°5 Responsos Granite Isolated Rock

Responsos Granite Rock is a large granite block isolated by the chemical weathering action of granite, of special interest because its location on the side of the road that gives access to the Paramera from Villaviciosa makes it stand out in the environment. Additionally, there is a folkloric ritual related to the Celtic funerary world linked to it (see Sect. 3.2.4). Access on foot involves a 3.6-km walk from Villaviciosa with a 260-m slope, although it is halfway along the Veredas route and the access to the Navasangil hillfort, making it a passageway location. It is also possible to access by car via a forest track (Table 3.16 and Fig. 3.74 ).

### 3.3.6 Geomorphosite N°6 Picos Zapateros Granite Mountain Ranges

Picos Zapateros granite mountain ranges (Table 3.17, Fig. 3.75) are the quintessential image of the Sierra de la Paramera, visible and recognizable from several kilometers away. They are irregular pyramidal shapes, formed by frost shattering or cryoclastic in the vertical joints of the granite. The succession of ridges forms the ridge of the Zapateros

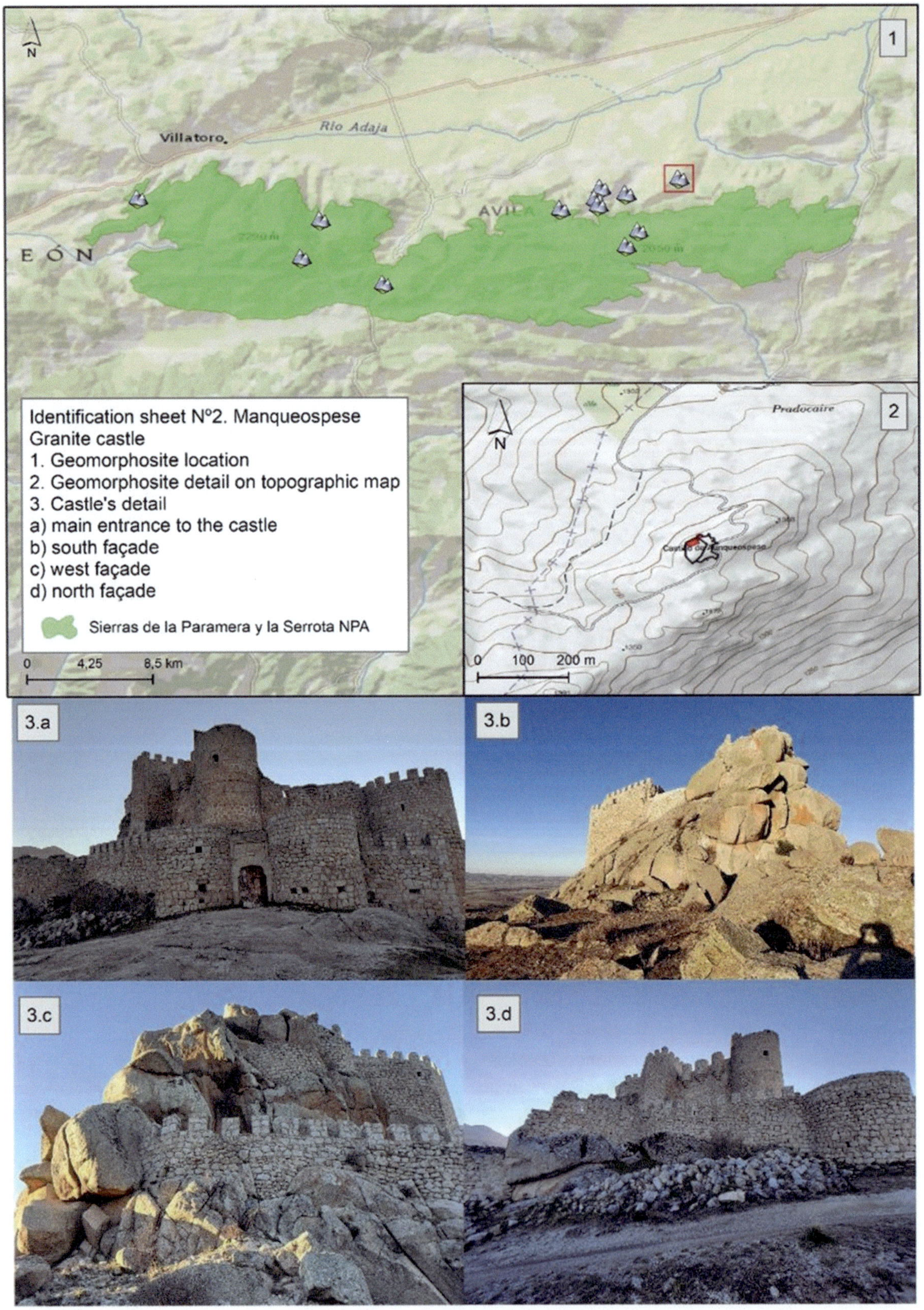


**Fig. 3.70** Manqueospese granite castle identification sheet

**Table 3.14** Picuezo river alluvial fan identification sheet

| Geomorphosite identification sheet | | | |
|---|---|---|---|
| Identification | **Name: Picuezo river alluvial fan** | **Site: Molino del Quemado Recreational Area** | **N°3**<br>Altitude: 1200 |
| Situation | Municipality: Sotalbo, Ávila | Coordinates: 40.523363; −4.870172 | |
| Geomorphological values | Type | Site<br>Representative | |
| | Genesis | Fluvial | |
| | Morphology: description, morphostructures, erosion | Picuezo River alluvial fan generated by the accumulation of sandstones and ashes from the Navalacruz wildfire due to erosion at the broken outlet of the Molino del Quemado. An example of traditional water use, with local hydrological alteration, and deterioration due to abandonment and fire | |
| | Dynamics | Fluvial transport and sedimentation | |
| | Chronology | Actual | |
| | Main interest | Fluvial | |
| | Secondary interest | – | |
| | Representativeness | Medium | |
| | Geomorphosite attribution | Fluvial | |
| Additional values | Landscape | – | |
| | Cultural elements | Remains of old mills | |
| | Economic | – | |
| | Social | Recreation | |
| | State of conservation | Poor (infrastructure damage from the fire) | |
| | Actual uses | Recreation | |
| | Infrastructures | Molino del Quemado recreational area (damaged) | |
| | Impacts | Wildfire, human use | |
| | Legal situation | – | |
| Accessibility | Walking distance | 1.8 km from Palacios (20 min) | |
| | Slope | 50 m of positive slope | |
| | Security | High (marked path) | |

peaks: Risco del Sol (2113 m), Cancha Morena (2122 m), Pico Zapatero (2158 m), and Risco Redondo (2054 m). They are an excellent example of weathering and sanding of granite, and the landscape value is high, being able to see Ávila capital, the Valle del Amblés, and the Alberche Valley on a clear day.

The accessibility is medium, typical of a high mountain peak, and has two access points from both valleys: from Sotalbo, at the Fuente de las Aguas Frías or from the town itself, and from Navandrinal. In any case, a slope of over 700 m must be overcome.

### 3.3.7 Geomorphosites N°7 Serrota Glacial Cirques

Serrota glacial cirques (Table 3.18, Fig. 3.76) are an exceptional example of glaciation in the Sierras de la Paramera and the Serrota, as it is the only site where glacial features are present. There are three intensely degraded Pleistocene glacial cirques located on the eastern slope of the Serrota: the Media Luna glacier, La Honda, and Las Serradillas, with the latter two being the largest. The cirques themselves and the frontal and lateral moraine deposits in the southeast are clearly visible, although they are of limited magnitude and development, and heavily eroded by the hydrological network.

The landscape value of the geomorphosite is high, at an Altitude of 2292 m, providing a complete view of Ávila, the Paramera, and the Valle de Amblés. Accessibility is moderate, due to a steep slope (950 m) and a rocky, unmarked path. The glaciers of the Serrota have attracted the interest of hikers and researchers since the first half of the twentieth century, indicating a process of heritage recognition of the Serrota.

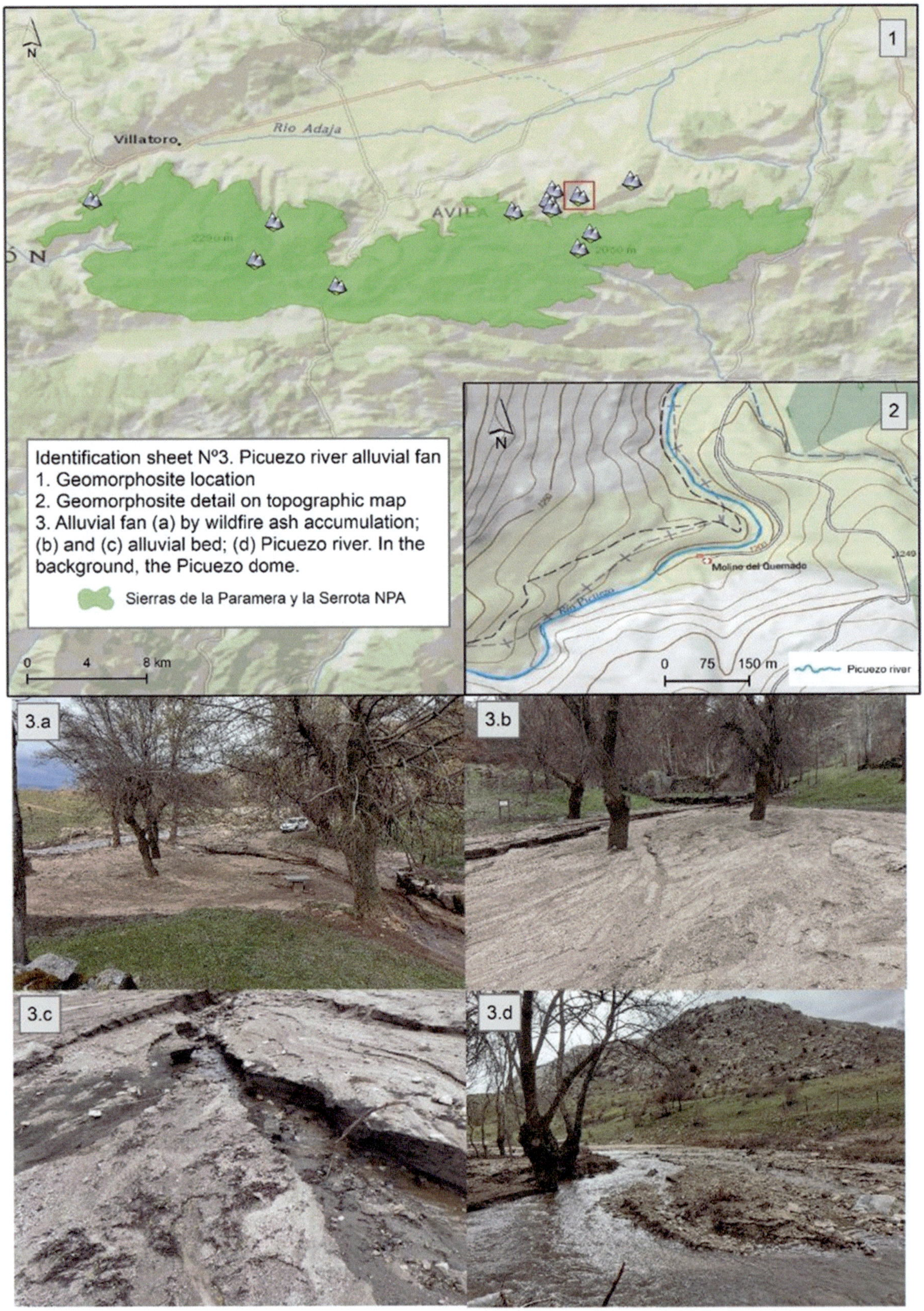


**Fig. 3.71** Picuezo river alluvial fan identification sheet

**Fig. 3.72** Molino del Quemado ruins (old mill), in Sotalbo

**Table 3.15** Granite rock thermal weathering geomorphosite

| Geomorphosite identification sheet | | | |
|---|---|---|---|
| Identification | **Name: Granite rock thermal weathering** | **Site: Granitic boulder in the Sierra de la Paramera (ascent to Zapatero peak from Sotalbo)** | **N°4**<br>Altitude: 1700 |
| Situation | Municipality: Sotalbo, Ávila | Coordinates: 40.500851; −4.862873 | |
| Geomorphological values | Type | Element | |
| | | Exceptional | |
| | Genesis | Thermal weathering | |
| | Morphology: description, morphostructures, erosion | Granitic boulder clearly affected by thermal weathering processes associated with the Navalacruz wildfire. At the base of the boulder, detached fragments can be seen, as well as scaling marks on the granite itself | |
| | Dynamics | Granite thermal weathering | |
| | Chronology | Present | |
| | Main interest | Thermal weathering | |
| | Secondary interest | Granite | |
| | Representativeness | Medium | |
| | Geomorphosite attribution | Granite | |
| Additional values | Landscape | Reduced view of the Amblés Valley and the Paramera | |
| | Cultural elements | – | |
| | Economic | – | |
| | Social | – | |
| | State of conservation | Good | |
| | Actual uses | Recreation, hiking | |
| | Infrastructure | Track, trail | |
| | Impacts | Navalacruz wildfire | |
| | Legal situation | Sierras de la Paramera and la Serrota NPA | |
| Accessibility | Walking distance | 3 km from Fuente de las Aguas Frías | |
| | Slope | 200 m of slope | |
| | Security | Medium, unmarked path, rocky terrain | |

**Fig. 3.73** Granite rock thermal weathering identification sheet

**Table 3.16** Responsos granite isolated rock geomorphosite

| Geomorphosite identification sheet | | | |
|---|---|---|---|
| Identification | **Name: Responsos Granite isolated rock** | **Site: Canto de los Responsos** | **N°5** |
| | | | Altitude: 1440 |
| Situation | Municipality: Villaviciosa, Ávila | Coordinates: 40.514522; −4.909844 | |
| Geomorphological values | Type | Element | |
| | | Representative | |
| | Genesis | Granite weathering | |
| | Morphology: description, morphostructures, erosion | Granitic boulder isolated by the chemical weathering of granite | |
| | Dynamics | Granite weathering | |
| | Chronology | Miocene—present | |
| | Main interest | Granite | |
| | Secondary interest | – | |
| | Representativeness | Medium | |
| | Geomorphosite attribution | Granite | |
| Additional values | Landscape | – | |
| | Cultural elements | Folkloric ritual related to the Celtic funeral world | |
| | Economic | – | |
| | Social | Cultural and ritual | |
| | State of conservation | High | |
| | Actual uses | Cultural, touristic | |
| | Infrastructures | Interpretative panels and path | |
| | Impacts | Pile of stones at the top (human) | |
| | Legal situation | – | |
| Accessibility | Walking distance | 3.6 km from Villaviciosa | |
| | Slope | 260 m of positive slope | |
| | Security | High (marked path) | |

### 3.3.8 Geomorphosite N°8 Adaja River Spring

The Adaja River springs in the northern slope of La Serrota receives the waters from this Sierra at the Fuente del Ortigal (Table 3.19, Fig. 3.77). The waters of the Adaja emerge between porphyritic granites and flow toward the Amblés Valley, forming a V-shaped valley. In this upper stretch, the riverbed is formed by blocks, pebbles, and gravel, while in the floodplain of the Amblés tectonic basin, it forms deposits of sand and silt, creating a flat valley floor. It originates in a valley along a fault line with a SW–NE direction, a fracture that channels the waters from the ridges of La Serrota to the north and south of El Colladillo, in a typical location for water outcrops (springs, seeps) in granite environments. The spring is considered not for its flow but as the point that marks the furthest extent of the Adaja River.

The importance of this geomorphosite lies in the relevance of the Adaja River itself in the study area, which articulates the entire Amblés Valley (see Sect. 3.1.4). The landscape value is moderate, with views of the Paramera, and it is easily accessible, with a distance of 2.2 km from Puerto de Villatoro and a slope of 150 m.

### 3.3.9 Geomorphosite N°9 Garganta Honda Fault-Line Valley

Garganta Honda fault-line valley is located in the area that gives it its name, in the municipality of Muñotello (Table 3.20, Fig. 3.78). It is a straight valley through which the Arroyo de Canto Moreno flows in a V shape, following a north–south fracture. It acted as a proglacial drainage for the Media Luna and La Honda glaciers, and granite outcrops can be seen on the slopes.

This geomorphosite reflects the morphostructure of the study area, conditioned by the structural and faulted relief, as well as the granite landforms and the functioning of the glacial features of the Serrota. It is used for livestock grazing, and the accessibility is 3 km from Pradosegar, with a slope of 400 m along an unmarked trail.

### 3.3.10 Geomorphosite N°10 Gavilanes and Chinitas Granite System

The Gavilanes and Chinitas geomorphosite constitute one of the most spectacular granite landscapes in the study area (Table 3.21, Fig. 3.79). Located in the municipality of

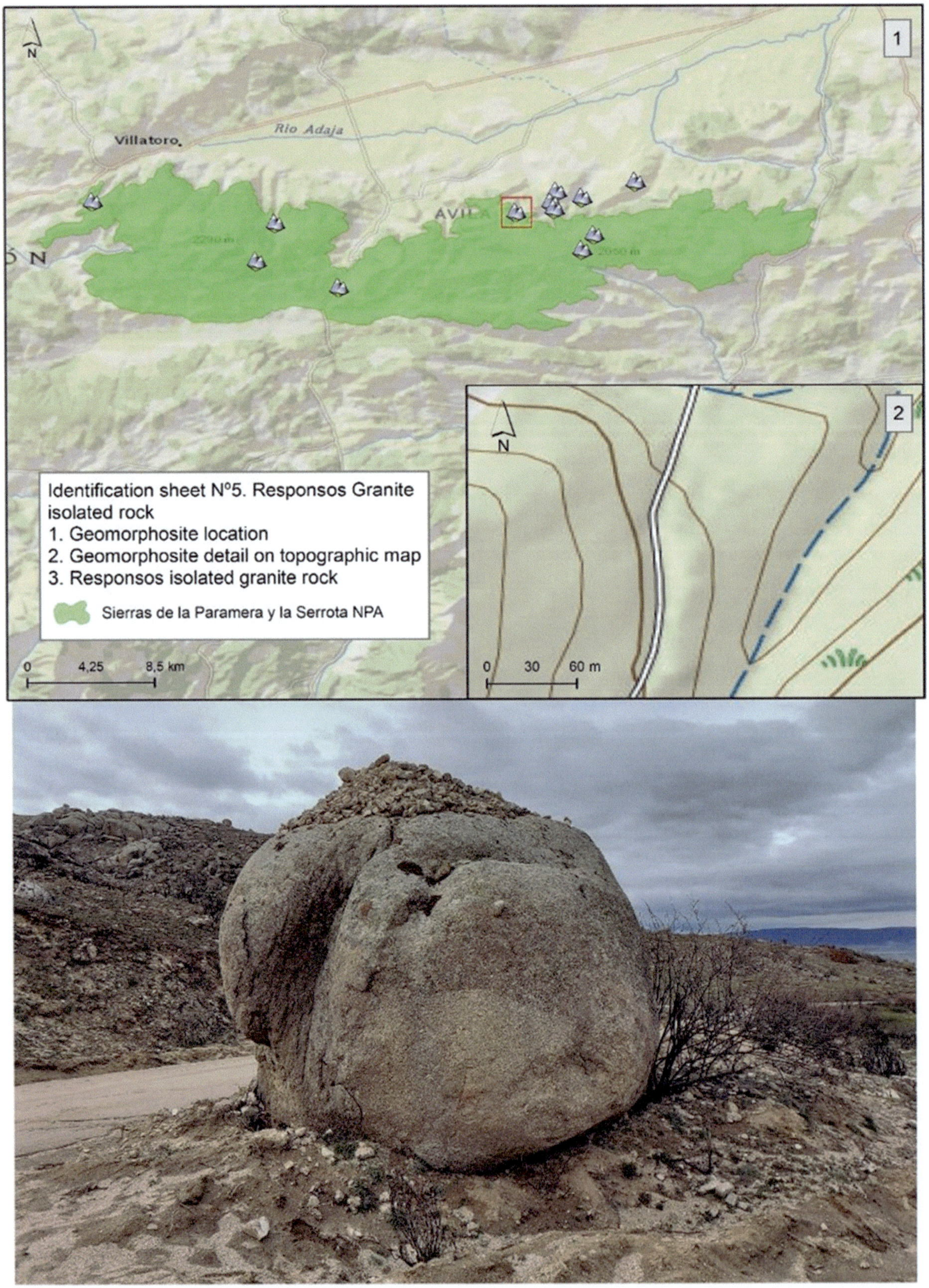


**Fig. 3.74** Responsos granite isolated rock identification sheet

**Table 3.17** Picos Zapateros granite mountain ranges geomorphosite

| Geomorphosite identification sheet | | | |
|---|---|---|---|
| Identification | **Name: Picos Zapateros granite mountain ranges** | **Site: Pico Zapatero–Cancha Morena–Risco del Sol** | **N°6**<br>Altitude: 2158 |
| Situation | Municipality: Sotalbo, Ávila | Coordinates: 40.492145; −4.870210 | |
| Geomorphological values | Type | Site | |
| | | Representative | |
| | Genesis | Granite | |
| | Morphology: description, morphostructures, erosion | Irregular shapes, generally with a prismatic or pyramidal tendency, whose origin is due to the action of ice and water through frost weathering or cryoclastics, taking advantage of the vertical joints present in the granite block. It is called a granite ridge when it appears individually, or a ridge if several associated and continuous ridges are present, as is the case in the so-called Zapateros peaks of the Sierra de la Paramera, which from east to west are: Risco del Sol (2113 m), Pico Zapatero (2158 m), and Risco Redondo (2054 m) | |
| | Dynamics | Mechanical weathering | |
| | Chronology | Miocene—present | |
| | Main interest | Mechanical weathering | |
| | Secondary interest | Granite | |
| | Representativeness | High | |
| | Geomorphosite attribution | Granite | |
| Additional values | Landscape | Views of Ávila capital and the Amblés and Alberche valleys | |
| | Cultural elements | – | |
| | Economic | Touristic | |
| | Social | Touristic, recreation | |
| | State of conservation | Good | |
| | Actual uses | Tourism, sports | |
| | Infrastructures | Track, path | |
| | Impacts | Erosion from the 2021 wildfire | |
| | Legal situation | Sierras de la Paramera and la Serrota NPA | |
| Accessibility | Walking distance | 8.7 km from Sotalbo | |
| | Slope | 990 m of positive slope | |
| | Security | Medium (marked path, rocky areas) | |

Villaviciosa, in the extension of the Ulaca semihorst, it consists of a highly degraded and exhumed elevation due to the sanding processes of the granite, giving rise to all kinds of granite landforms, both large and small, highly visible and recognizable. Particularly noteworthy are the two sites that give the geomorphosites their name: the Gavilanes Peaks, two sharp-edged granite ridges, and the blocks of Las Chinitas, two huge isolated granite blocks.

The landscape value of the geomorphosites is high, with a complete view of the Picuezo valley and the rest of the Paramera with the Zapateros Peaks, as well as cultural elements such as the shepherd's enclosures made using the blocks of Las Chinitas (Fig. 3.80). Accessibility is moderate, due to the rocky state of the path and the 300 m of positive slope that separate Villaviciosa from the Gavilanes peaks.

### 3.3.11 Geomorphosite N°11 Puerto de Mengamuñoz Fault

The Puerto de Mengamuñoz (mountain pass) is the dividing line between the Sierra de la Paramera and the Sierra de la Serrota. It is a subvertical fracture running in a NNE–SSO direction through which the national road N-502 passes, making it an important communication axis. This is a straight regional fault (Table 3.22, Fig. 3.81) that has caused the division of the tectonic pillar into two smaller units: on one side, the Paramera to the east, which is more extensive but does not exceed 2000 m in Altitude, and on the other side, the Serrota to the west, a more uplifted block than the Paramera reaching 2292 m at its main summit.

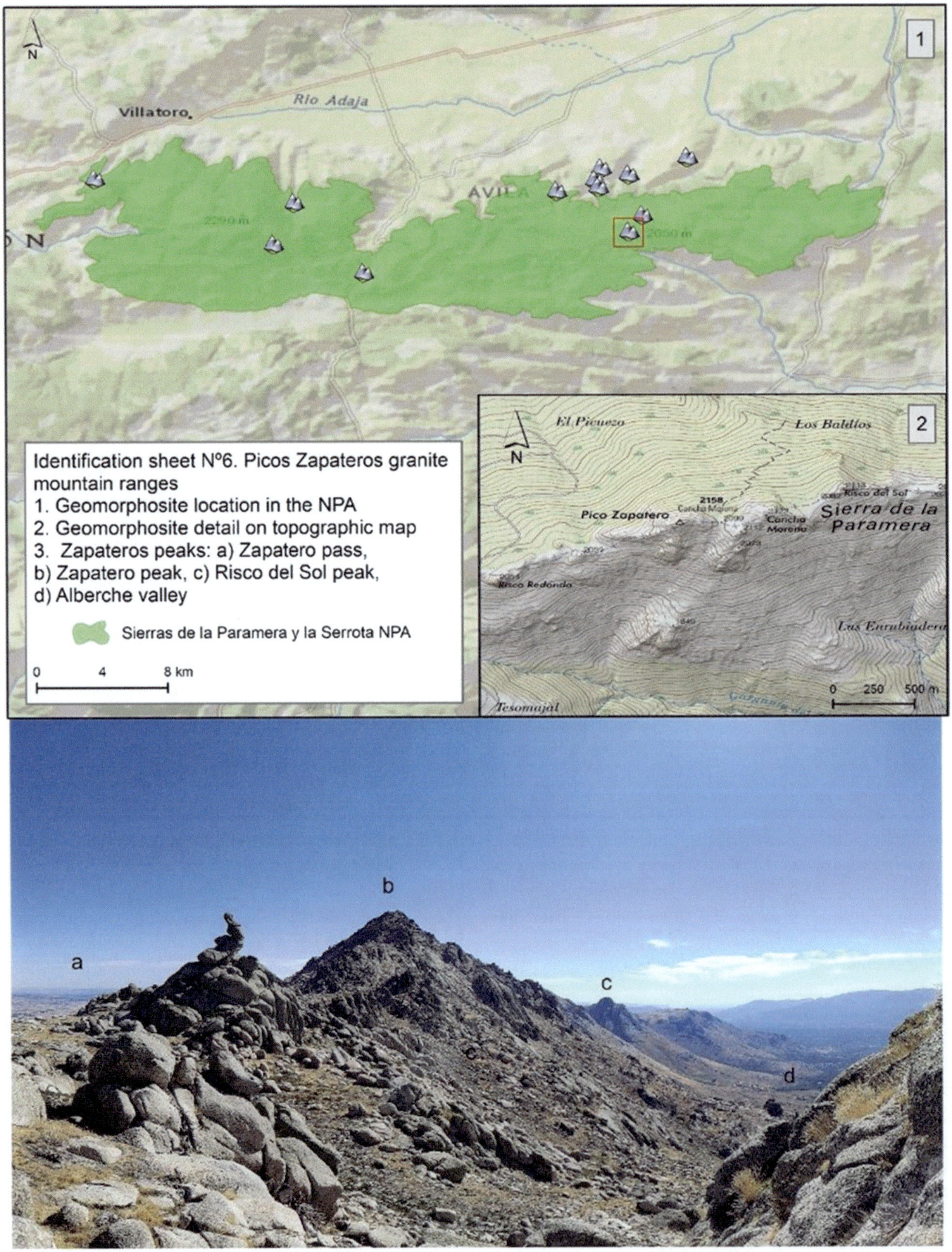


**Fig. 3.75** Picos Zapateros granite mountain ranges identification sheet

**Table 3.18** Serrota glacial cirques geomorphosite

| Geomorphosite identification sheet | | | |
|---|---|---|---|
| Identification | **Name: Serrota glacial cirques** | **Site: La Serrota** | **N°7** |
| | | | Altitude: 2292 |
| Situation | Municipality: Pradosegar | Coordinates: 40.499450; −5.078918 | |
| Geomorphological values | Type | Site | |
| | | Exceptional | |
| | Genesis | Glacial | |
| | Morphology: description, morphostructures, erosion | Five Pleistocene glacial cirques located on the eastern slope of the Sierra de la Serrota: Las Cerradillas, La Honda, Media Luna, Belesar, and Los Hornillos. The frontal and lateral moraine deposits are visible in the southeast, eroded by the hydrological network | |
| | Dynamics | Glaciation | |
| | Chronology | Pleistocene—present | |
| | Main interest | Glacial | |
| | Secondary interest | Granite | |
| | Representativeness | High | |
| | Geomorphosite attribution | Glacial | |
| Additional values | Landscape | Panoramic view of Ávila, La Paramera, and Valle de Amblés | |
| | Cultural elements | Shepherd's huts and pastoral infrastructures | |
| | Economic | Touristic | |
| | Social | Touristic, recreation | |
| | State of conservation | High | |
| | Actual uses | Hiking | |
| | Infrastructures | Path | |
| | Impacts | Hiking affluence | |
| | Legal situation | Sierras de la Paramera y la Serrota NPA | |
| Accessibility | Walking distance | 7 km from Pradosegar | |
| | Slope | 950 m of positive slope | |
| | Security | Medium, unmarked rocky path | |

### 3.3.12 Geomorphosite N°12 Picuezo River Giant's Kettle

Picuezo River giant's Kettles are located in the area of La Hoya (Table 3.23, Fig. 3.82) and are a magnificent example of this granite landform. They are concavities generated during flood events on the porphyritic granite of the Picuezo riverbed, developed on a gentle slope in a corridor-like manner, forming a linear succession of spherical pools.

The landscape value is high, with views of the Picuezo valley and the surrounding landforms (Chinitas, Gavilanes, Picos Zapateros), and a medium accessibility, with a rocky, unmarked path 2.5 km from Villaviciosa.

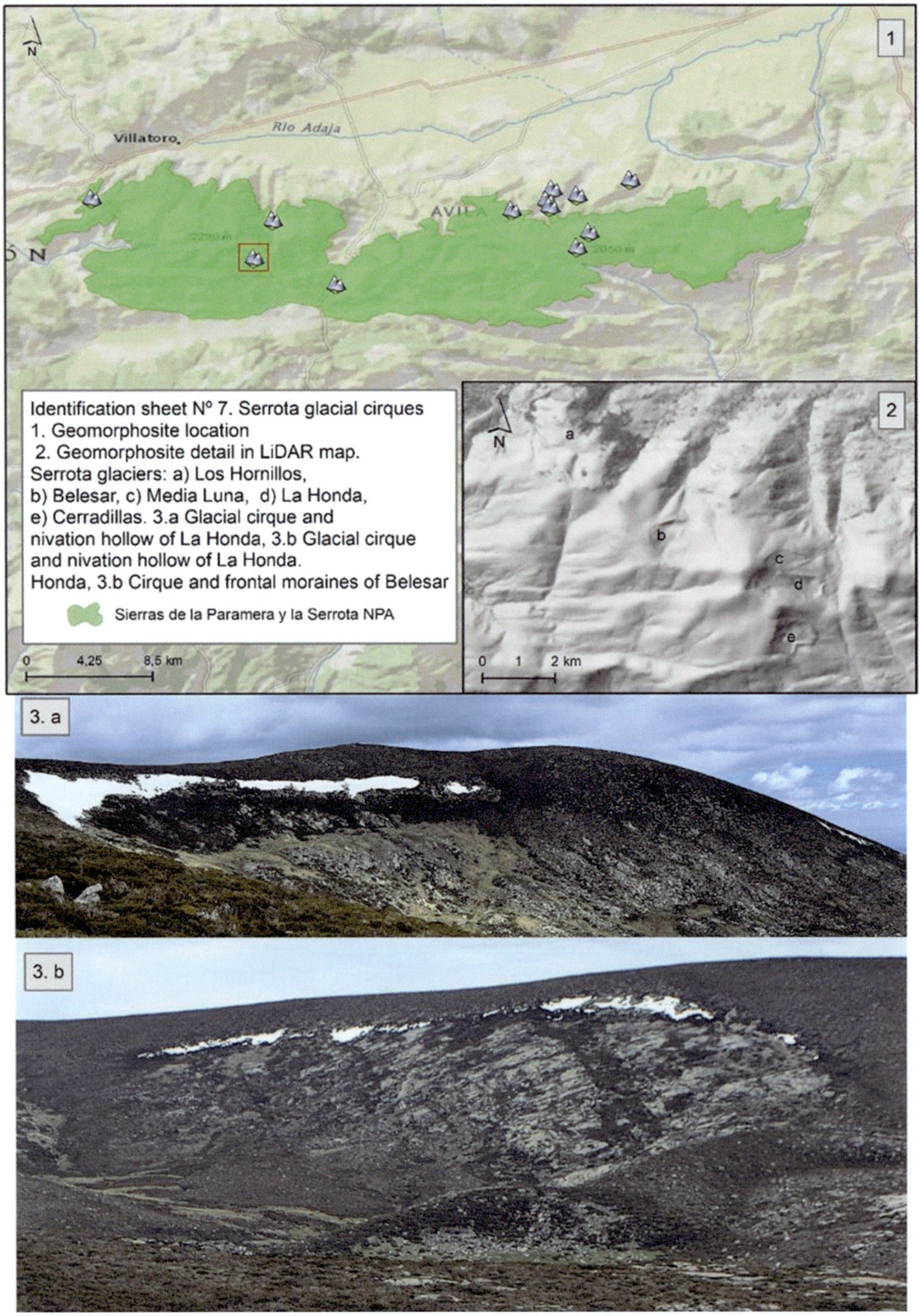


**Fig. 3.76** Serrota glacial cirques identification sheet

**Table 3.19** Adaja river spring geomorphosite

| Geomorphosite identification sheet | | | |
|---|---|---|---|
| Identification | **Name: Adaja river spring** | **Site: Fuente del Ortigal** | **N°8**<br>Altitude: 1500 |
| Situation | Municipality: Villatoro | Coordinates: 40.522447; −5.163114 | |
| Geomorphological values | Type | Site<br>Exceptional | |
| | Genesis | Fluvial | |
| | Morphology: description, morphostructures, erosion | Adaja river spring at the Ortigal Spring, receiving the waters from the northern slope of La Serrota. An example of water emergence between porphyritic granites, forming a V-shaped valley with a bed made up of blocks, pebbles, and gravel in its upper stretch. As it flows through the Amblés Valley, it forms the floodplain of the tectonic trench, with sandy deposits | |
| | Dynamics | Active (fluvial) | |
| | Chronology | Paleozoic—Holocene | |
| | Main interest | Fluvial | |
| | Secondary interest | Hydrological, structural | |
| | Representativeness | High | |
| | Geomorphosite attribution | Fluvial | |
| Additional values | Landscape | Medium | |
| | Cultural elements | – | |
| | Economic | Livestock farming | |
| | Social | Recreation, hiking | |
| | State of conservation | Good | |
| | Actual uses | Recreation | |
| | Infrastructures | Spring | |
| | Impacts | – | |
| | Legal situation | Sierras de la Paramera and la Serrota NPA | |
| Accessibility | Walking distance | 2.2 km from Puerto de Villatoro | |
| | Slope | 150 m of positive slope | |
| | Security | High, path | |

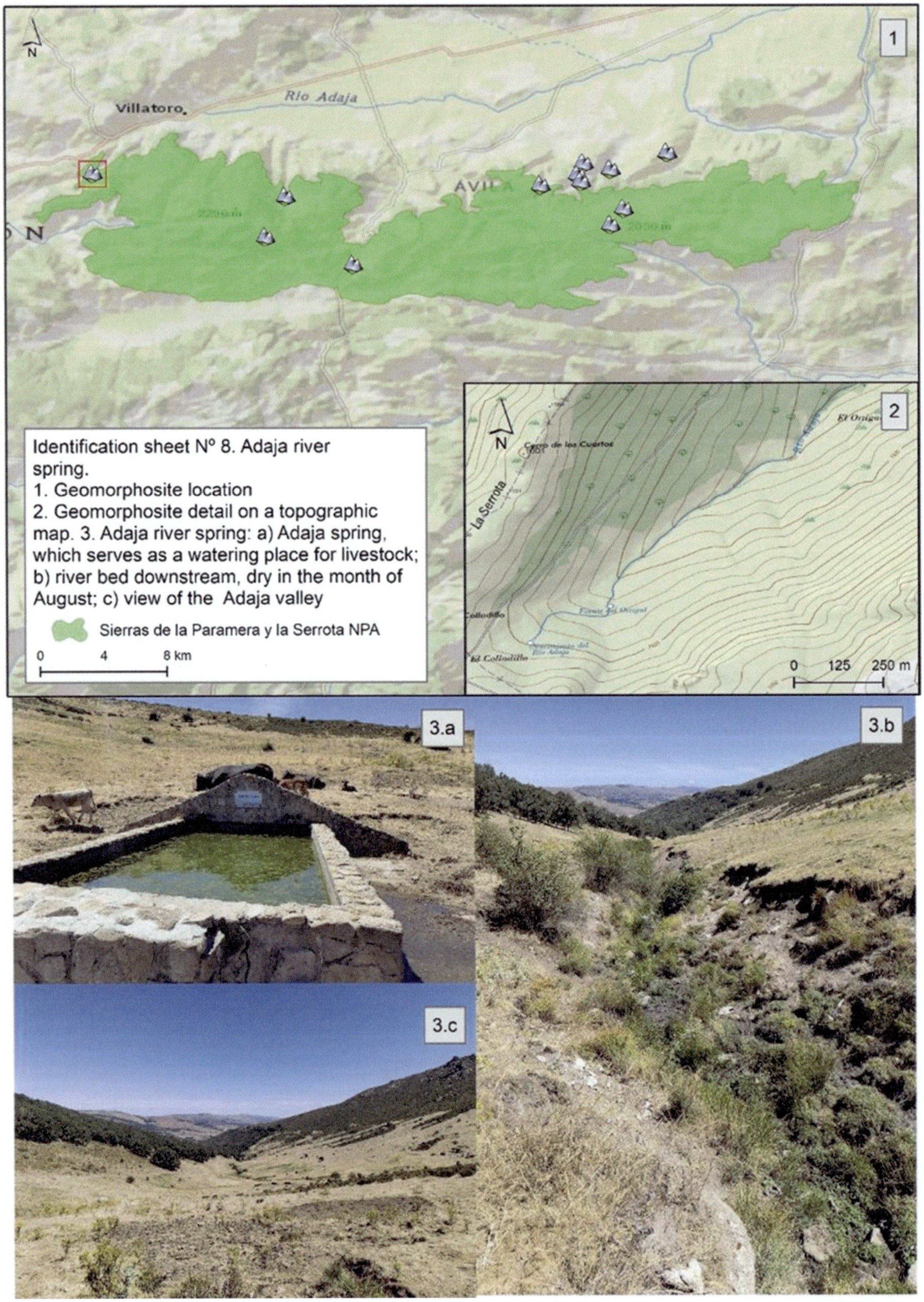


**Fig. 3.77** Adaja river spring identification sheet

**Table 3.20** Garganta Honda fault-line valley geomorphosite

| Geomorphosite identification sheet | | | |
|---|---|---|---|
| Identification | **Name: Garganta Honda fault-line valley** | **Site: Garganta Honda** | **N°9**<br>Altitude: 1630 |
| Situation | Municipality: Muñotello | Coordinates: 40.509153, −5.054864 | |
| Geomorphological values | Type | Site | |
| | | Representative | |
| | Genesis | Structural | |
| | Morphology: description, morphostructures, erosion | Valley in a straight line, in a V shape, following a north-south fracture. It was the proglacial drainage of the Media Luna and la Honda glaciers. The slopes have granite outcrops | |
| | Dynamics | Structural and fluvial-torrential | |
| | Chronology | Miocene—Pleistocene | |
| | Main interest | Structural | |
| | Secondary interest | Fluvial-torrential and granite | |
| | Representativeness | High | |
| | Geomorphosite attribution | Structural | |
| Additional values | Landscape | High | |
| | Cultural elements | – | |
| | Economic | Grassland | |
| | Social | Recreation | |
| | State of conservation | High | |
| | Actual uses | Livestock farming and hiking | |
| | Infrastructures | Trails | |
| | Impacts | – | |
| | Legal situation | Sierras de la Paramera and la Serrota NPA | |
| Accessibility | Walking distance | 3 km from Pradosegar | |
| | Slope | 400 m of positive slope | |
| | Security | Medium, unmarked trail | |

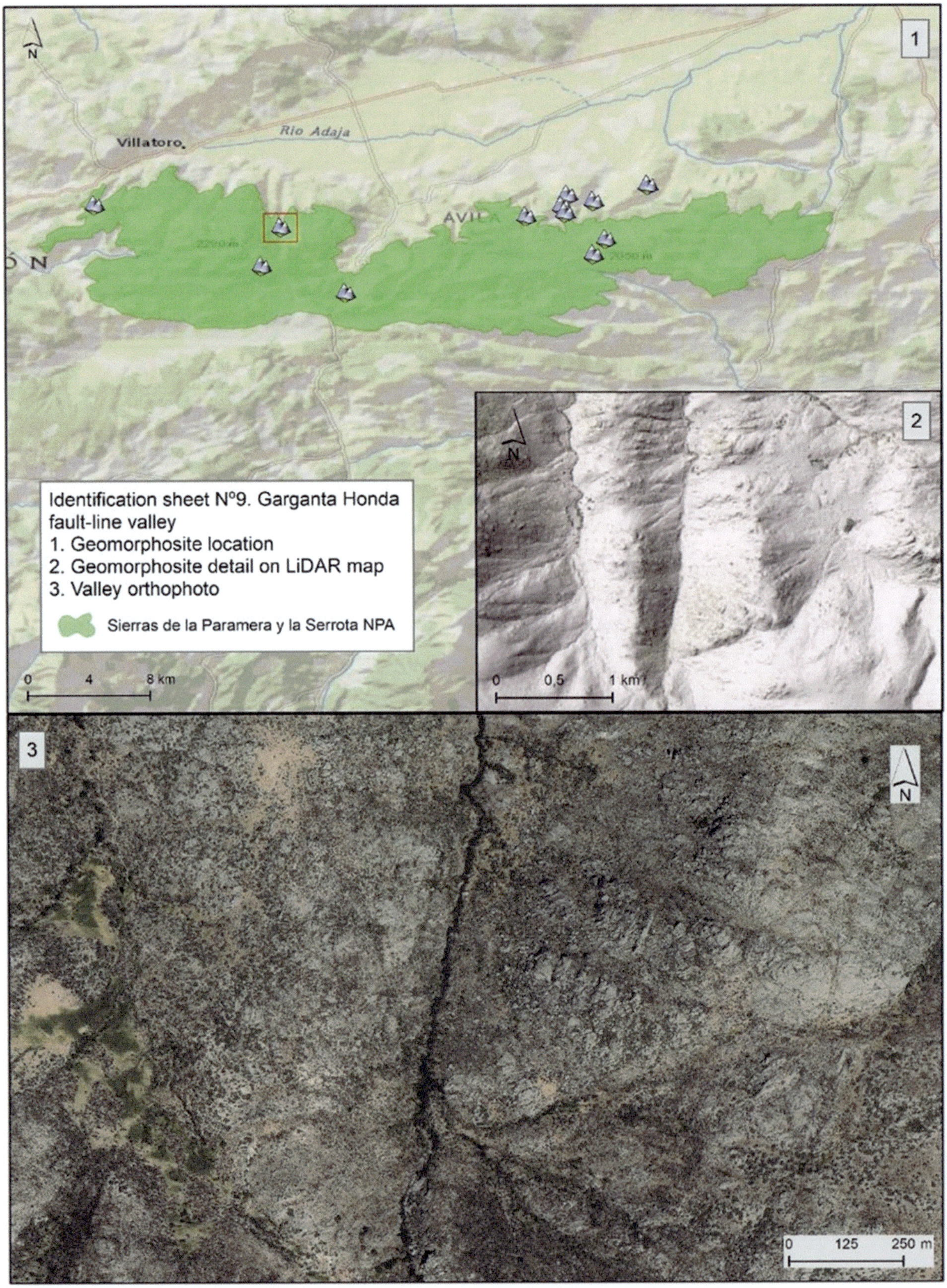


**Fig. 3.78** Garganta Honda fault-line valley identification sheet

**Table 3.21** Gavilanes and Chinitas granite system geomorphosite

| Geomorphosite identification sheet | | | |
|---|---|---|---|
| Identification | **Name: Gavilanes and Chinitas granite system** | **Site: El Gavilan-Las Chinitas** | **N°10**<br>Altitude: 1297 |
| Situation | Municipality: Villaviciosa | Coordinates: 40.519967, −4.884095 | |
| Geomorphological values | Type | Site | |
| | | Representative | |
| | Genesis | Granite | |
| | Morphology: description, morphostructures, erosion | Granite landscape resulting from the exhumation of granite due to the weathering of the rock, where a variety of granite landforms can be observed, both large and small, highlighting the blocks of Las Chinitas and the ridges of Gavilán | |
| | Dynamics | Granite | |
| | Chronology | Miocene—present | |
| | Main interest | Granite | |
| | Secondary interest | Structural | |
| | Representativeness | High | |
| | Geomorphosite attribution | Granite | |
| Additional values | Landscape | High, panoramic view of the Picuezo valley and the Zapateros Peaks | |
| | Cultural elements | Pasture enclosures in artificial caves | |
| | Economic | Livestock farming | |
| | Social | Recreation, hiking | |
| | State of conservation | Good | |
| | Actual uses | Hiking, livestock farming | |
| | Infrastructures | Path | |
| | Impacts | – | |
| | Legal situation | – | |
| Accessibility | Walking distance | 4 km from Villaviciosa | |
| | Slope | 200 m to Las Chinitas–300 m to Gavilan | |
| | Security | Medium, path with loose stones | |

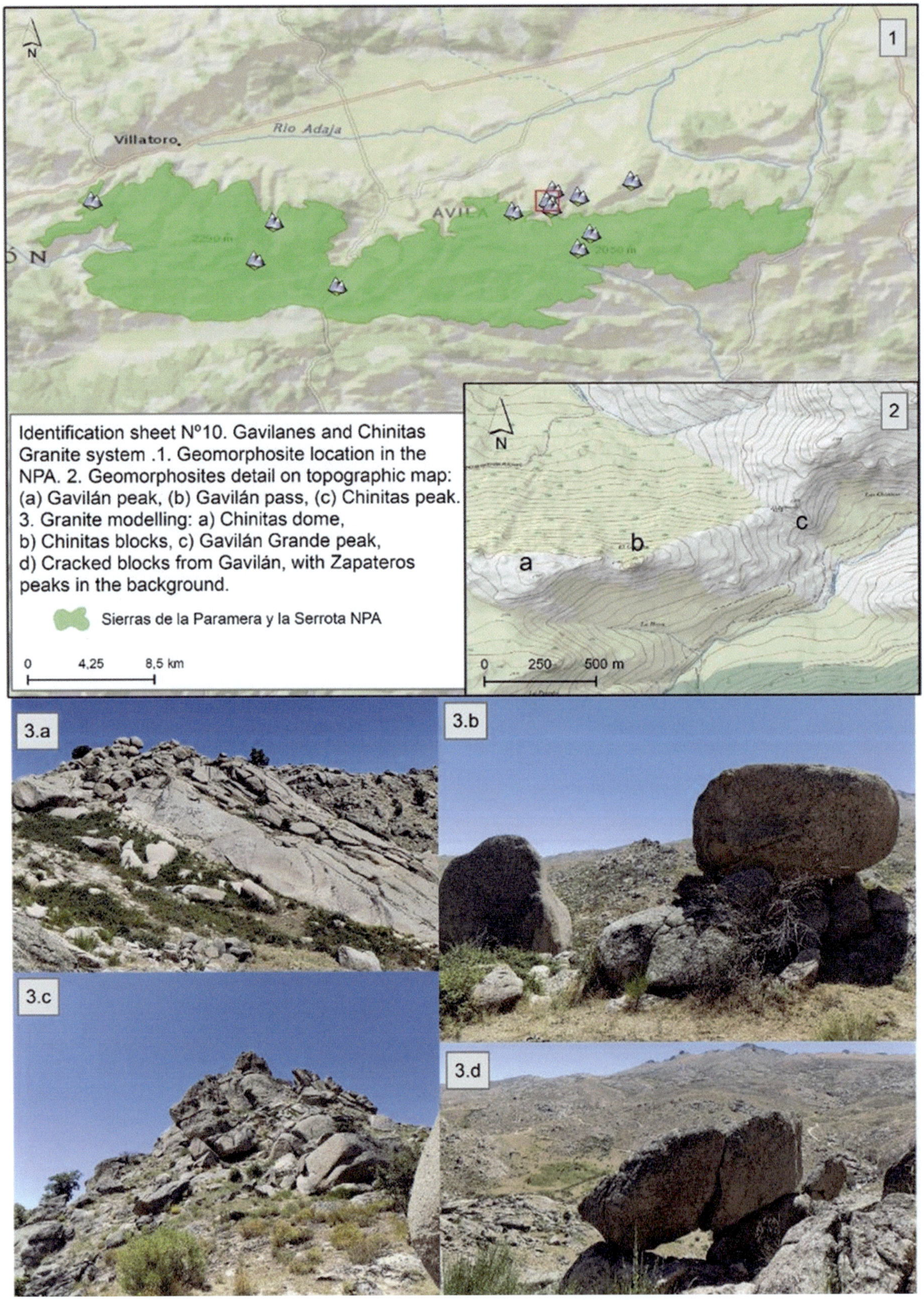


**Fig. 3.79** Gavilanes and Chinitas granite system identification sheet

**Fig. 3.80** Shepherd's enclosures at the base of Las Chinitas

**Table 3.22** Puerto de Mengamuñoz fault geomorphosite

| Geomorphosite identification sheet | | | |
|---|---|---|---|
| Identification | **Name: Puerto de Mengamuñoz fault** | **Site: Puerto de Mengamuñoz** | **N°11**<br>Altitude: |
| Situation | Municipality: Mengamuñoz | Coordinates: 40.497135; −5.002432 | |
| Geomorphological values | Type | Site | |
| | | Representative | |
| | Genesis | Structural | |
| | Morphology: description, morphostructures, erosion | Subvertical fracture with NNE–SSW orientation forming the Mengamuñoz Pass. Its relevance lies in serving as a boundary between the two mountain ranges: La Paramera (to the east) and La Serrota (to the west) | |
| | Dynamics | Structural | |
| | Chronology | Paleozoic | |
| | Main interest | Structural | |
| | Secondary interest | – | |
| | Representativeness | High | |
| | Geomorphosite attribution | Structural | |
| Additional values | Landscape | Medium, partial view of the mountain ranges | |
| | Cultural elements | – | |
| | Economic | – | |
| | Social | – | |
| | State of conservation | Medium, impact of road construction | |
| | Actual uses | Mountain pass | |
| | Infrastructures | National road N-502 | |
| | Impacts | National road | |
| | Legal situation | Sierras de la Paramera and la Serrota NPA | |
| Accessibility | Walking distance | Accessible by car | |
| | Slope | – | |
| | Security | High | |

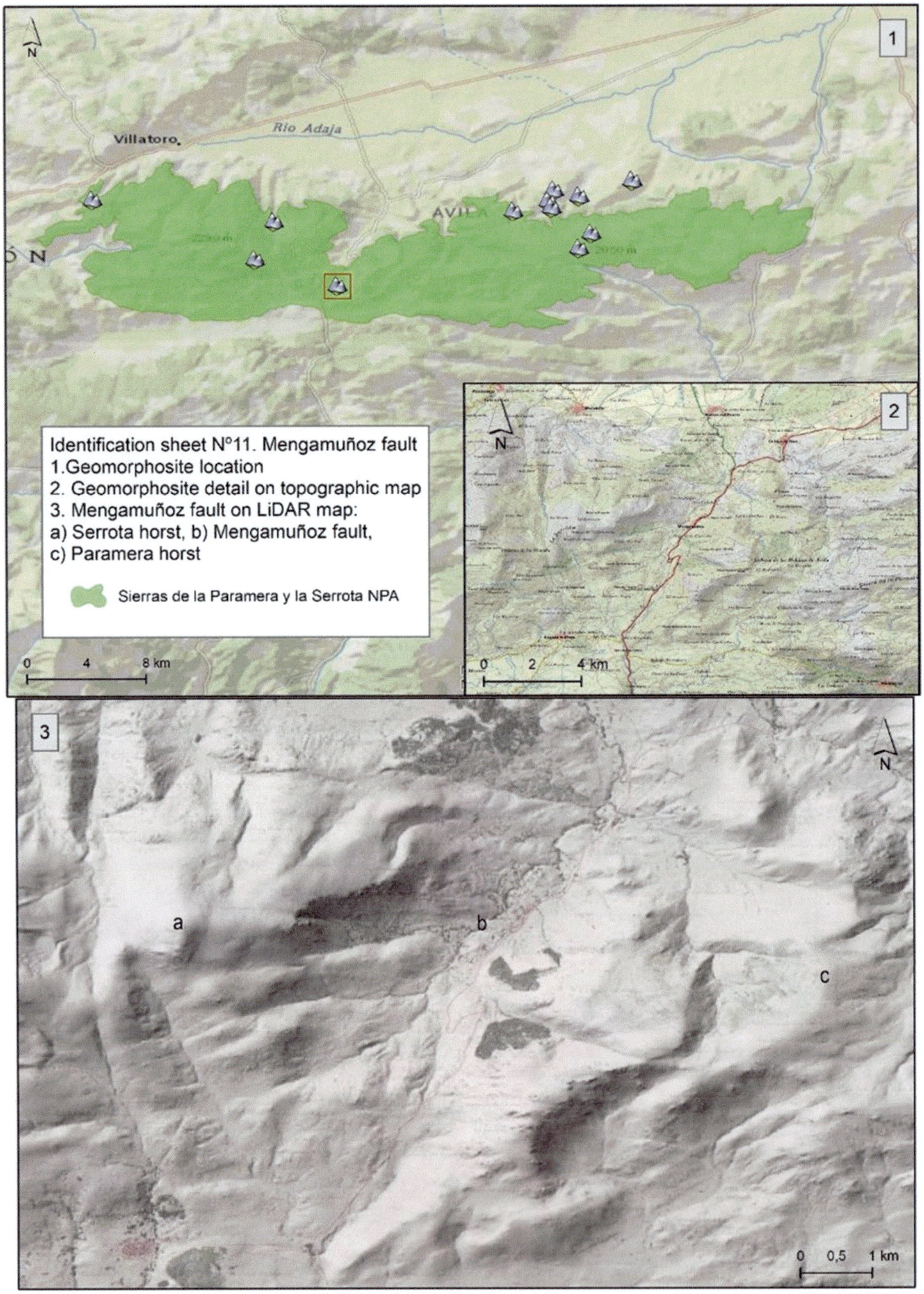


**Fig. 3.81** Puerto de Mengamuñoz fault identification sheet

**Table 3.23** Picuezo river giant's kettle geomorphosite

| Geomorphosite identification sheet | | | |
|---|---|---|---|
| Identification | **Name: Picuezo river giant's kettle** | **Site: Río Picuezo en La Hoya** | **N°12**<br>Altitude: 1313 |
| Situation | Municipality: Villaviciosa | Coordinates: 40.517192; −4.886917 | |
| Geomorphological values | Type | Site | |
| | | Exceptional | |
| | Genesis | Fluvio-granite | |
| | Morphology: description, morphostructures, erosion | Concavities generated on porphyritic granite in the riverbed of the Picuezo River in the La Hoya area. Developed on a gentle slope, they appear in a corridor-like manner as a result of erosion caused by the Picuezo during times of high water flow, which carries quartzite pebbles, forming spherical pools | |
| | Dynamics | Fluvial | |
| | Chronology | Carboniferous—present | |
| | Main interest | Fluvial | |
| | Secondary interest | Granite | |
| | Representativeness | High | |
| | Geomorphosite attribution | Fluvio-granite | |
| Additional values | Landscape | High, view of the Picuezo valley and the surrounding landforms (Chinitas, Gavilanes, Picos Zapateros) | |
| | Cultural elements | – | |
| | Economic | – | |
| | Social | Recreation, swimming | |
| | State of conservation | Good | |
| | Actual uses | Recreation, swimming | |
| | Infrastructures | Fenced for livestock | |
| | Impacts | – | |
| | Legal situation | – | |
| Accessibility | Walking distance | 2.5 km from Villaviciosa | |
| | Slope | 130 m of positive slop | |
| | Security | Medium, undefined path | |

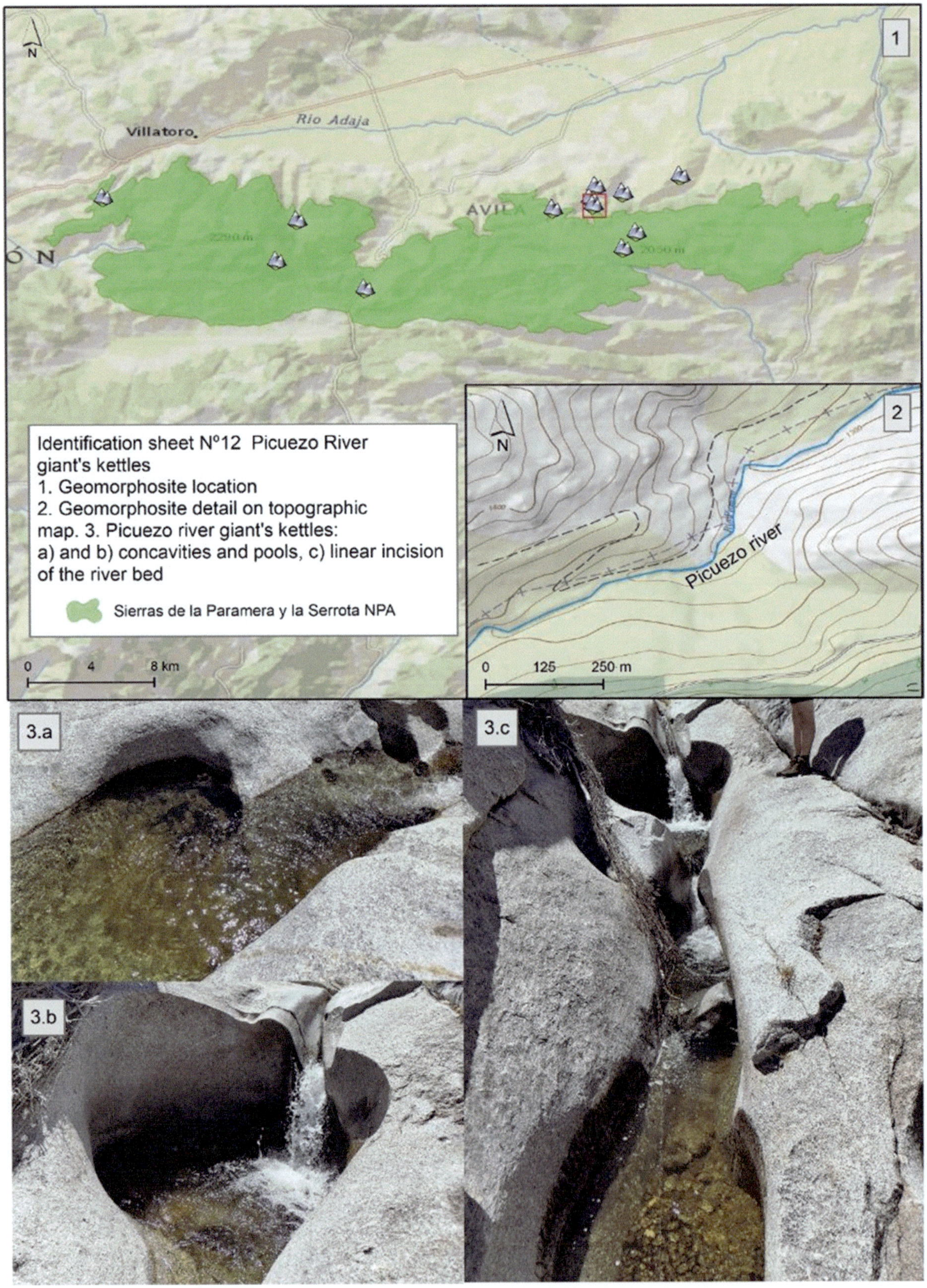


**Fig. 3.82** Picuezo river giant's kettle identification

# 4 Cañón del Río Lobos Natural Park

## 4.1 Physical Environment Study

Cañón de Río Lobos Natural Park (hereinafter CRLNP) is located in the provinces of Burgos and Soria and covers an area of 12,244 ha between 951 and 1352 m above sea level. Located in the southern foothills of the Iberian Mountain Range, it is in the transition between the Duero plain to the south and the Urbión and Neila mountain ranges to the north (Fig. 4.1).

### 4.1.1 Geological Context

For the geological contextualization, the analysis previously conducted and published in Serrano et al. (2020) will be used as a reference.

The Natural Park is located on the southern edge of the Cameros Basin, in a platform environment that generated the alternation of limestones, marls, sandstones, and conglomerates, characteristic of the North Castilian carbonate platform, with materials from the Upper Cretaceous (Floquet, 1991; Salas et al., 2001; Sopeña et al., 2004).

In the Natural Park of Cañón del Río Lobos, materials from the Cretaceous, from Aptian to Santonian, predominantly outcrop, with lesser amounts of Paleogene and Neogene. The emplacement of surface structures that have implications on the relief begins in the early stages of the Alpine Orogeny, responsible for the structures of the Mesozoic cover of the southern edge of the Iberian Range. By the end of the Mesozoic and the beginning of the Paleocene, the emergence of Paleozoic materials from the chain, the La Demanda and Mansilla mountain ranges to the north, deforms the limestone depressions of the ancient depression into a succession of broad anticlines and synclines. This uplift involved erosive processes and widespread leveling that led to the export of sediments and the filling of the southern depression, the Duero. It also generates grooves and valleys that will be immediately filled by Tertiary materials, Paleocene-Eocene, remnants of which remain in the eastern portion of the Park.

During this period, the anticlinal and synclinal folds of the west of the Park with a NE–SW direction are located, as well as the wide synclinal of the central portion. An immediate phase of extension allows for the continued erosion of these mountain ranges, and a general flattening occurs from the Medium Eocene to the upper Eocene. Although these structures and erosions will be decisive in shaping the relief, the characteristic relief structures and landforms of the current landscape have not yet formed.

During the transition from Eocene to Oligocene, in the compressive tectonic phase (Pyrenean Phase) of the Alpine orogeny, the deformations of the Mesozoic cover end. Folding continues, with overturned structures resulting from stresses from the north and northeast, and fractures and thrust faults with a NE–SW direction that compartmentalize the folds are generated. At the same time, there is a differential uplift, greater to the northeast and lesser to the southwest, which creates a succession of NE–SW fractures, very dense to the northeast of the Park and more spaced out to the southwest. All of these will be fundamental in shaping the relief later on. During a second phase of deformation, compressive at the end of the Oligocene and beginning of the Miocene (Savica Phase), the Iberian foreland is uplifted, structures are emplaced, and there is a widespread erosion of the geological structures.

Thus, during the Oligocene, the surface geological structures that characterize the Natural Park are emplaced, guided by the wide synclinal fold of the southwest, with its axis raised toward the northwest, where Cretaceous and Paleocene materials are deformed and uplifted to the northeast of the Lobos River. All of this is compartmentalized by NW–SE fractures and divided by the significant NW–SE Arganza fracture. To the south, it continues with the overturned and fractured anticline that is limited by immersion and the peri-


R. M. Ruiz Pedrosa, *Geomorphological Heritage as Didactic and Geotouristic Resources*, Geoheritage, Geoparks and Geotourism, https://doi.org/10.1007/978-3-032-06205-5_4

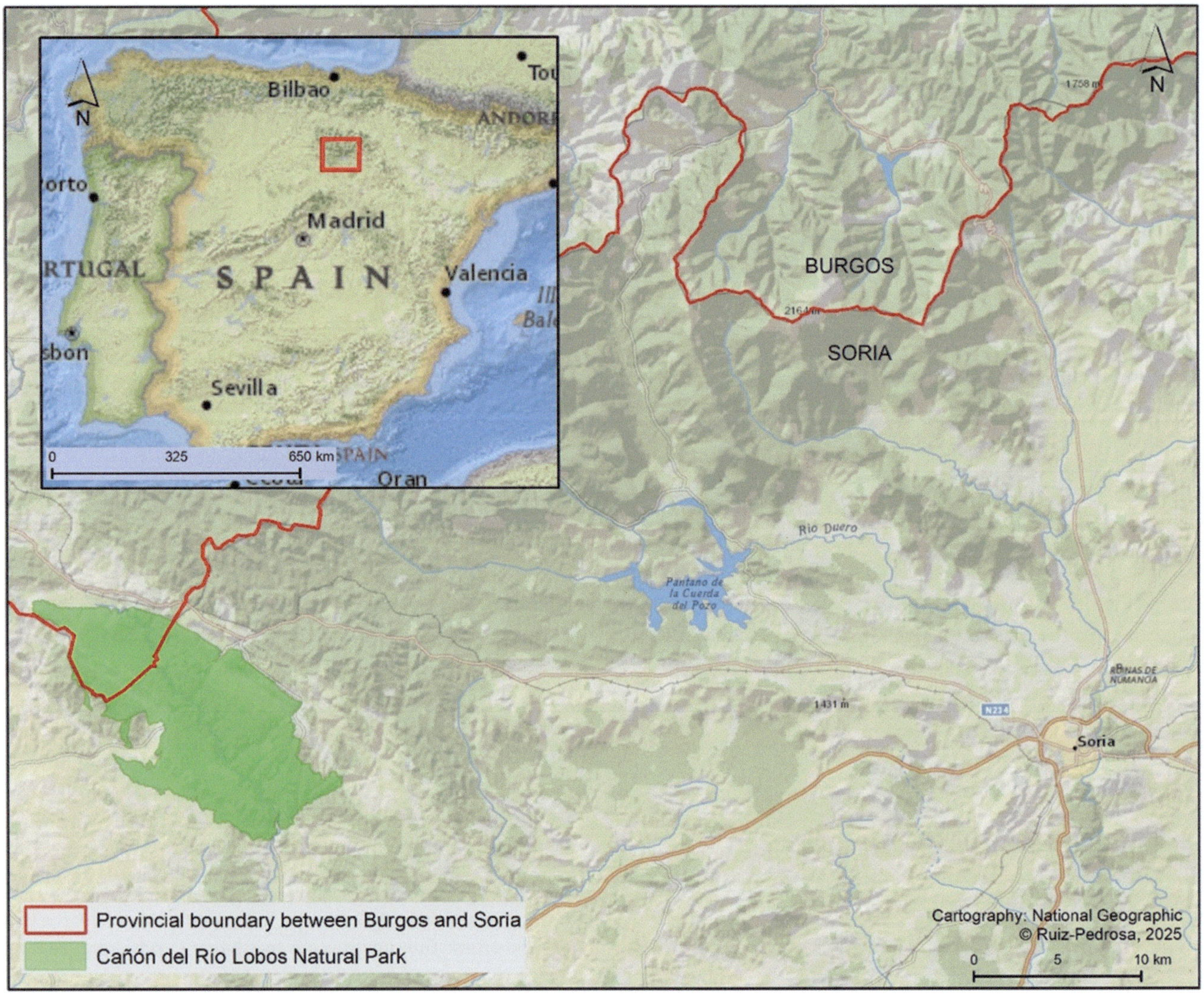


**Fig. 4.1** Location of the Cañón del Río Lobos Natural Park, 55 km west of the capital city of Soria

anticline closure with the southeast synclinal fold, and to the northwest with the elevated synclinal in the portion of the perisyncline closure, strongly compartmentalized by NE–SW fractures.

### 4.1.2 Geomorphology: Structural and Karstic Landforms

The Park landscape is formed by a plateau located between 1100 and 1200 m above sea level, with cliffs of 100 and 200 m on its edges and crossed in its central part by the valley and canyon of the Lobos River. The edges form very asymmetrical mountain ranges and wide valleys sandwiched between the ridges, while the 26 km long limestone canyon has steep walls, creating strong contrasts in the landscape. The dominance of limestone and the associated karst processes explain most of the karstic forms, both surface (exokarstic, such as dolines and karren) and underground (endokarstic, caves, and pits), that dominate the landscape and determine the distribution of ecosystems or human uses that adapt to the geomorphological conditions. It is a complex structural relief defined by the presence of perched synclines, present to the east and west of the Natural Park, structural lineaments, and hogs back. These forms have been modified by Tertiary erosion surfaces on which karst processes have developed. Six morphostructural units have been defined in the Lobos River Canyon (for a detailed description, see Serrano et al., 2020): Lobos River syncline depression, Nafría orthocline depression, Arganza fault-line valley, Hontoria dissymmetric perched syncline, Santa María combe, and Navas perched syncline (Fig. 4.2).

The fluvial-karst canyon articulates the entire Natural Park, from drainage to karstic action and it dissected the plateaus. It acts as a base level and in it and its surroundings, the main active geomorphological processes take place. The

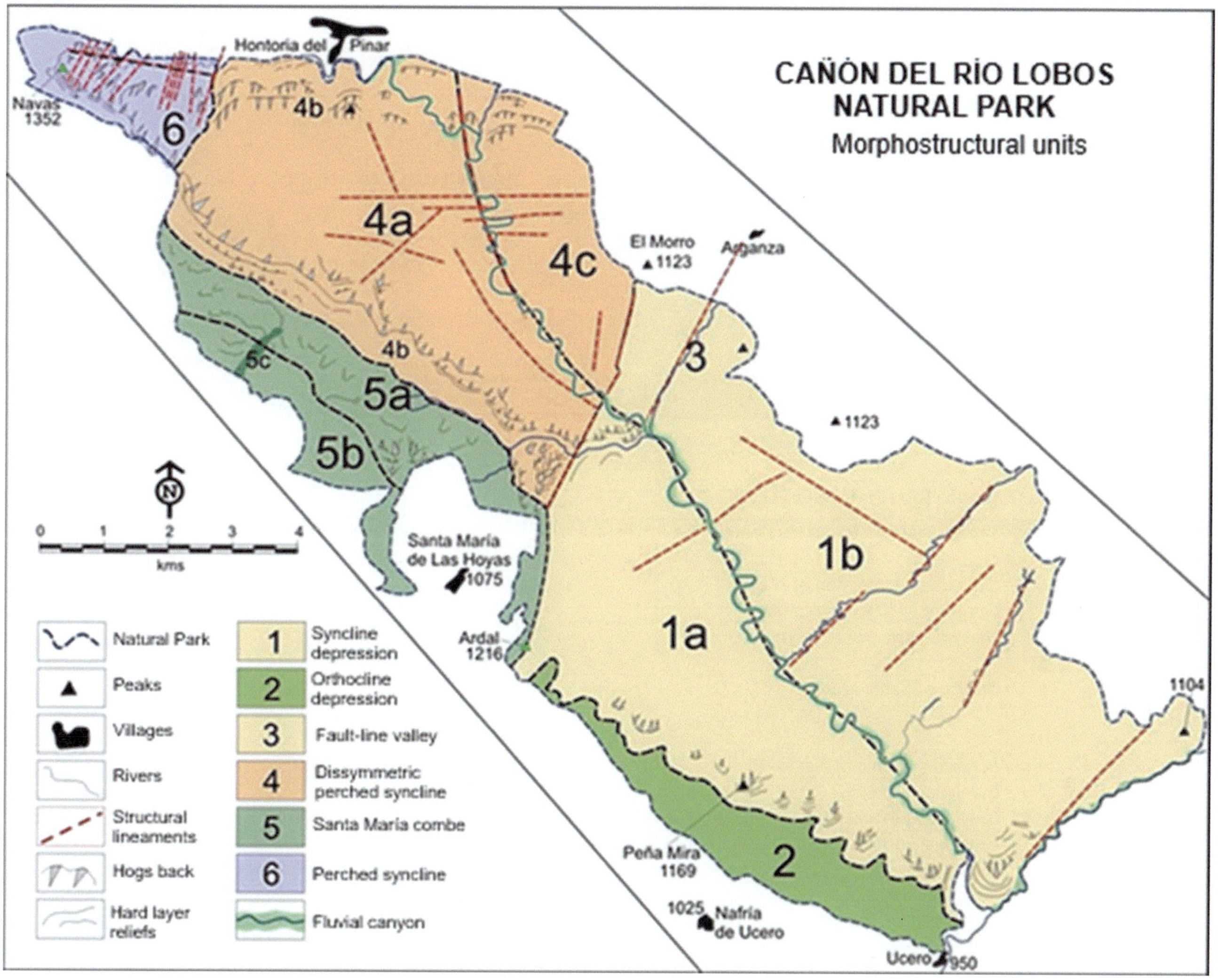


**Fig. 4.2** Morphostructural units in the Cañón o Lobos Natural. (Source: Serrano et al., 2020)

Natural Park presents two differentiated environments, with clear contrasts to the east and west of the Lobos River. If the eastern portion shows a greater capacity for incision and development of river valleys, in the western portion karstic processes dominate. Throughout the Natural Park, karstic forms and processes dominate, active and responsible for both major forms, the holokarstic canyon or closed depressions, as well as minor forms, dolines, and karrens. The juxtaposition of structural, karstic, and incised fluvial relief forms over a wide erosion surface shapes the unique relief of the sierra's edge.

The canyon cuts through the limestone from north to south, forming a narrow valley between vertical walls that dominate the relief and construct the archetypal image of the Natural Protected Area (NPA) (Fig. 4.3). The canyon is bordered to the east and west by wide and high plateaus where ravines, valleys, and karstic depressions have developed, breaking the plain. The margins of the Natural Park are formed by asymmetrical ridges, gentle slopes, and rocky walls that form a high plateau separated by surrounding valleys.

Table 4.1 shows the main landforms of the Rio Lobos Natural Park. There are 33 different landforms of a structural, karstic, fluvial, and slope nature that, along with the geomorphological map, show the diversity and complexity of landforms and their spatial distribution. Many of them are of geotouristic interest due to their geomorphological nature (1), or as a landscape component (2), or related to cultural elements (3).

### 4.1.3 Climate and Hydrology

The climate of the CRLNP is continental-Mediterranean (Fig. 4.4), with an average annual precipitation of 678 mm, highest in spring and autumn-winter, and a moderate summer drought. The average annual temperature is 9.4 °C with highs in July (18.8 °C) and lows in January (1.8 °C). In the

**Fig. 4.3** Río Lobos canyon and San Bartolomé hermitage, representative image of the NPA

**Table 4.1** Main landforms of the Rio Lobos Natural Park

| | | | | |
|---|---|---|---|---|
| Structural | Syncline depression of río Lobos | | Monocline crests (3) | |
| | | | Fault-line valley | |
| | Orthocline depression of Nafría | | Erosive orthocline depression | |
| | Fault-line valley of Arganza | | Fault-line valley | |
| | Dissymmetric perched syncline (2, 3) | | Monocline crests (3) | |
| | | | Fault-line valley | |
| | Combe of Santa María | | Combe (1) | |
| | | | Mont derivé (1) | |
| | | | Cluse (2) | |
| | Perched syncline of Navas | | Perched syncline (2) | |
| Erosional surfaces | Erosional surfaces 1 | Great high plateaus, ~1150 to 1250 m (2) | | |
| | Erosional surfaces 2 | High plateaus, 1090–1120 m | | |
| | Erosional surfaces 3 | Small labeling, 1050–1110 m | | |
| Karst | Exokarst | Fluviokarstic canyon (2, 3) | | |
| | | Karren or lapiaz (1) | Structural karren | |
| | | | Covered karren | |
| | | | Semicovered karren | |
| | | Skinholes or dolines (1) | Solution sinkholes | Funnel sinkhole |
| | | | | Basin sinkholes |
| | | | Collapse sinkholes | |
| | | | Subsidence sinkholes (alluvial sediments) | |
| | | Karstic closed depressions (2) | | |
| | Endokarst | Caves | Chasms (1) | |
| | | | Pits (3) | |
| | | | Galleries and halls (1, 3) | |
| | | | Speleothem (1) | |
| | | | Breccia (1) | |
| Fluvial | Erosional | River cuts (1) | | |
| | | Erosional heads (1) | | |
| | | V-shaped valley (1) | | |
| | Accumulation | Fluvial terraces (1) | T1, +70 m | |
| | | | T2, +45 to 50 m | |
| | | | T3, +25 to 30 m | |
| | | | T4, +8 to 10 m | |
| | | Alluvial fan (1) | | |
| | | Fluvio-lacustrine plane (2, 3) | | |
| Slopes | Landslide | Calving, toppless (1) | | |
| | | Masivos | Techos | |
| | | | Abrigos de roca | |
| | Slopeslide | Rotational slopeslide (2) | | |
| | | Translational slopeslide | | |
| | Debris | Debris talus (1) | | |
| | | Debris cones (1) | | |
| | Richter denudation slope (1) | | | |

Geotouristic potential values: (1) geomorphic, (2) landscape, (3) linked to cultural elements plateaus above 1000–1100 m of altitude, summers do not exceed 20 °C and cold winters have 6 months with average temperatures <10 °C, 15–20 days of snow per year, and more than 100.

Regarding hydrology, the main watercourse in the Park is the Lobos river, belonging to the Duero basin (Fig. 4.5). The Lobos river originates in the area of Mata Blanca, in Mamolar (Burgos). It flows for 16 km in a southeast direction, and once it enters the canyon it forms along its path, it enters the province of Soria. It has a total length of 38.3 km, after which it merges with the Chico river, at which point it loses its name when it meets the source of the Ucero river.

The Lobos River flows through the park in a narrow and meandering course, entering with a moderate annual flow rate (38 $m^3/s$) and flowing over fluvial deposits where individual channels and stable fluvial bars alternate (Serrano et al., 2020). Within the Natural Park, the riverbed of the Lobos river has two well-differentiated sections: the upper stretch, between Hontoria del Pinar and the Puente de los Siete Ojos (Fig. 4.6), has the steepest slope and where the flow is channeled through underground conduits (Segovia, 2008); the lower portion, between the Puente de los Siete Ojos and the Ucero river spring, has a more moderate slope and is characterized by karstic outlets and springs, with continuous fluvial accumulations in the valley bottom (Serrano et al., 2020).

On the other hand, the Ucero river originates from the La Galiana spring. This is the most important spring in the park, where the filtered waters from the karst system of the Cañón del Río Lobos emerge. In the eastern part of the park, there are valleys enclosed by vertical limestone walls, through which rivers such as the Navaleno and the Valderrueda stream flows. The Navaleno River, which runs through the fault-line valley of Arganza, also serves as a sink for the waters of the Lobos river. On the other hand, the Valderrueda stream is characterized by a meandering course with high sinuosity. In both cases, these are watercourses that flow on the surface of erosion, with low slope, and adapt to fracture lines and structural weaknesses, impacting as the cover rises (Serrano et al., 2020).

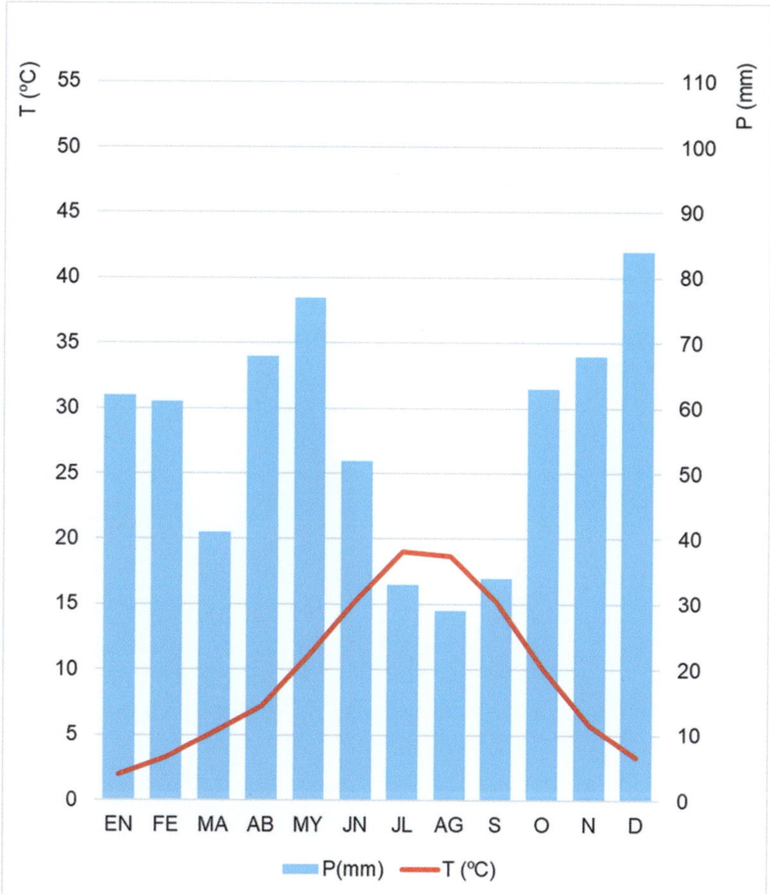


**Fig. 4.4** Cañón del Río Lobos climograph. (Source: AEMET. Own elaboration)

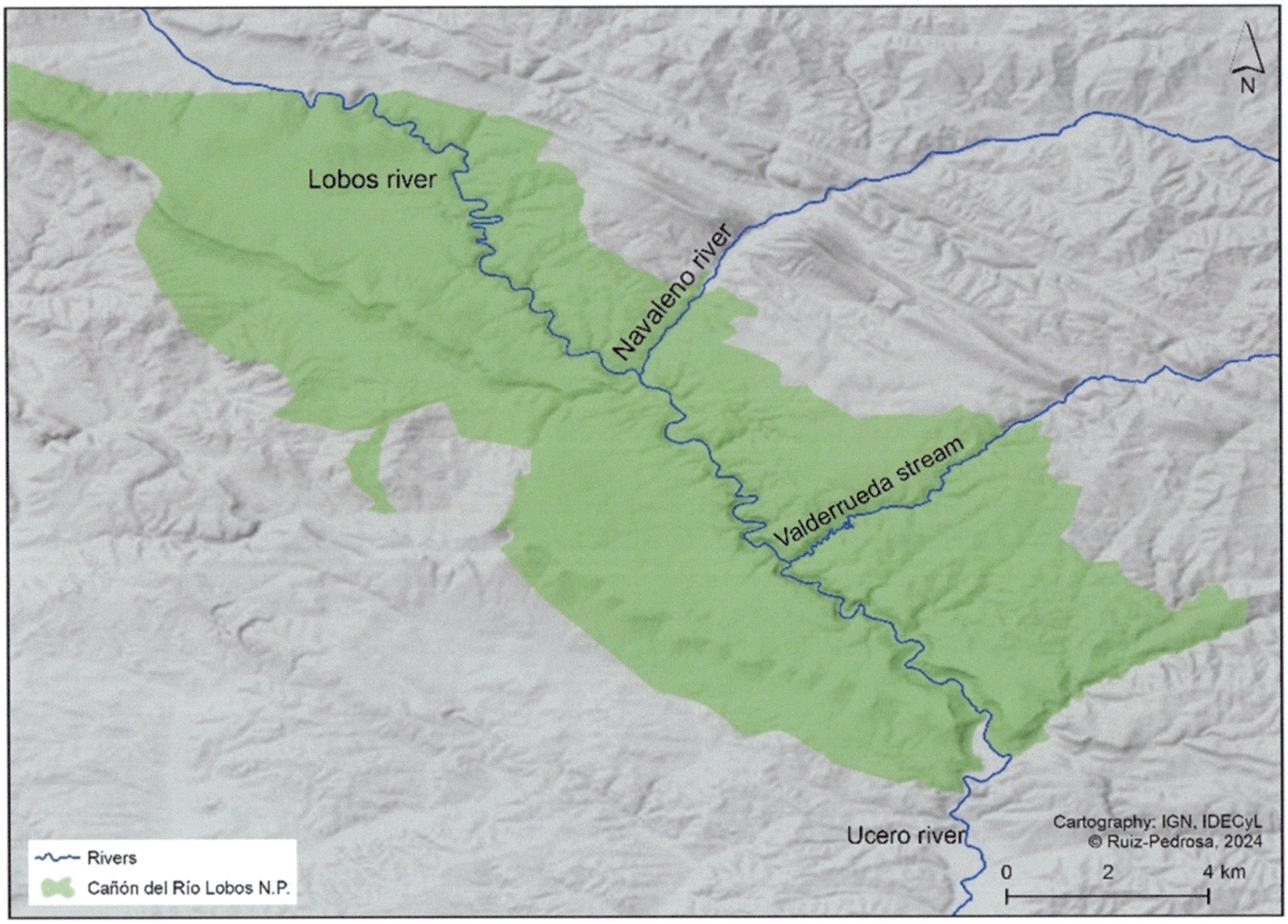


**Fig. 4.5** Main watercourses of the CRLNP

**Fig. 4.6** Bridge of the Seven Eyes, with the Lobos River completely dry (date 15/08/2024)

## 4.2 Socioeconomic Study

### 4.2.1 Current Environmental Legislation

The Natural Park Cañón del Río Lobos was one of the first NPA in Castilla y León, declared in 1985 in the provinces of Soria and Burgos. It had previously been included in the National Inventory of Outstanding Landscapes (1975) and in the Open Inventory of Special Natural Protected Areas (1980), and its declaration as a Park was considered necessary to advance in its protection and management.

Despite its early declaration, and nearing 40 years, the Park continues to be immersed in a long legal recognition process that has not yet been completed. The Declaration Decree, despite justifying the declaration due to its spectacular karstic modeling, does not include an inventory of the abiotic or biotic elements to be protected, not even a description. It establishes compatible uses (traditional activities and orderly use of its productions) and prohibited actions (use of pesticides or poisons, littering, and use of fire), and does not consider any urban planning limitations. A very poor decree that became obsolete in 1991 with the approval of Law 8/1991 on Natural Areas of the Community of Castilla y León and Law 42/2007 on Natural Heritage and Biodiversity. This law stipulates that all Natural Parks must have a Natural Resources Management Plan (NRMP) as a planning instrument. In 2008, the drafting process of the NRMP for the Cañón del Río Lobos began in the provinces of Burgos and Soria. The next and last news on this matter is that on June 17, 2009, the period of public information, hearing, and consultation of the Park's NRMP was opened, without it being approved to this day.

This NRMP of 2009 becomes even more obsolete when in 2015 Law 4/2015 on the Natural Heritage of Castilla y León is approved, at which point the processing NRMP should have been reactivated. Having been able to consult the Summary Document that was made public in June 2009, we can see that the NRMP proposal does include an inventory and a dispositive part. The inventory includes 12 points of geological interest, for which a map or more details could never be consulted. The dispositive part establishes the zoning proposal into Reserve Zones, Limited Use Zones, and Compatible Use Zones (for more details, see Ruiz-Pedrosa, 2019).

Therefore, by July 2024, the first Natural Park declared in Castilla y León, one of the most visited, about to turn 40 years, and having experienced two regional Natural Heritage laws, does not have an effective planning instrument beyond the Declaration Decree, which is absolutely scant in all respects.

On the other hand, since 1998, the Cañón del Río Lobos has been part of the Natura 2000 Network as a Special Conservation Area and Special Conservation Area for Birds (Fig. 4.7). The Special Conservation covers an area of 12,238 ha, 83% of the Natural Park's area, while the Special Conservation Area for Birds covers 10,201 ha. Both areas share 83.36% of the total area (see Fig. 4.7). In the basic management and conservation plan of the Red Natura network, the Cañón del Río Lobos is defined as one of the most spectacular landscapes in the center of the peninsula, with strong landscape contrasts between the river and the cliffs. It also mentions the historical forest exploitation in the area, specifically of the black pine for resin use, with tourism currently driving the area. All habitat codes and species are described by state of conservation and overall assessment, many of them in unfavorable and inadequate state of conservation.

The main threats to the CRLNP are related to public use, poplar forest crops, the decrease in sheep farming, overgrazing, and invasive species. The large influx of people, as well as the caving activities carried out in the caves and chasms of the Park, are particularly concerning (Fig. 4.8).

### 4.2.2 Territorial Organization and Demographics

The surface of the Natural Park is located in two different regions: the Pinares region and the Burgo de Osma region. There are eight municipalities that are part of the socioeconomic area of the Park (previously seven, but with the proposal of the NRMP, Espeja de San Marcelino is included): Hontoria del Pinar, the only one in Burgos; San Leonardo de Yagüe, Casarejos, Herrera de Soria, Ucero, Nafría de Ucero, Santa María de las Hoyas, and Espeja de San Marcelino, in Soria. The municipalities that contribute the most surface area to the Park are Hontoria del Pinar and Santa María de las Hoyas, with 21.9% and 21.2%, respectively.

These municipalities are home to a total of 3132 inhabitants (as of January 1, 2023). Like in La Paramera and in most rural areas of Spain, the population of the CRLNP has decreased significantly since the 1990s (Fig. 4.9). If there were 4053 registered residents in 1996, the number of inhabitants has decreased by almost a thousand, as a result of both negative natural and migratory balances (Table 4.2), with very low birth rates, an aging population, and high levels of mortality.

The natural balance of the Park has been negative for the last 30 years, with a stagnant number of births in the 2020s, and with a mortality rate that, although peaked in 2010, has significantly decreased in 2022 after the pandemic deaths. Regarding the migratory balance, it was positive in 2022, especially due to the influx of residents from the province of Soria and other regions, with little immigration from other countries. However, in previous years, it was negative, characterized by emigration within

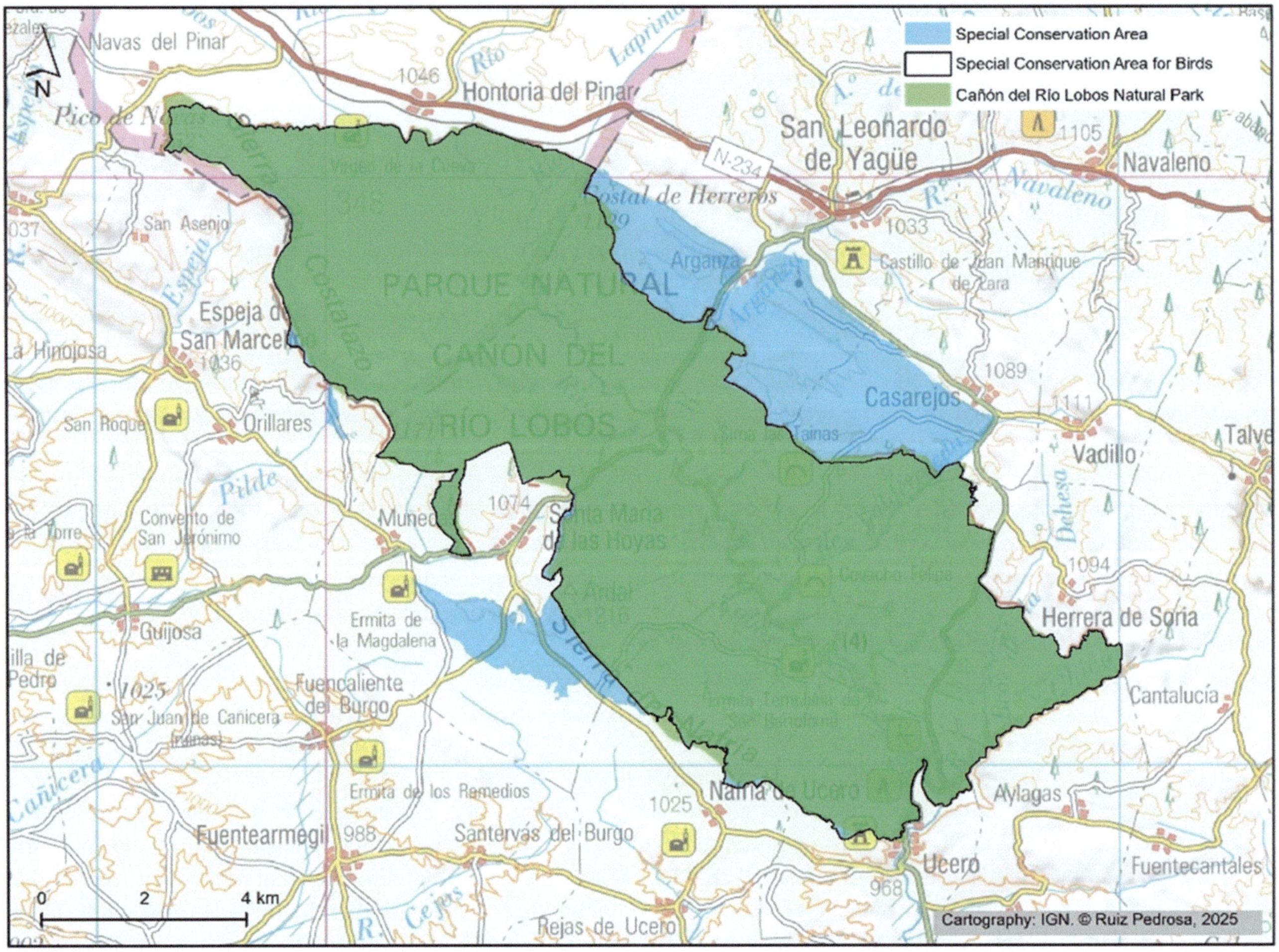


**Fig. 4.7** Environmental protection figures in CRLNP

the same province and to other autonomous communities. This peak in residential influx in recent years may be due to the "urban exodus," i.e., the return to rural and natural areas of new residents, generally younger people, fleeing the city. A phenomenon that has intensified after the pandemic and is bringing some dynamism back to certain rural areas.

The socioeconomic area of the Park is characterized by a clearly aging and regressive population, with the majority of the population between 40 and 64 years old, very low birth rates and very few births, and a higher proportion of women than men, especially from the age of 80, due to greater female longevity (Fig. 4.10).

The demographic profile of the provincial capital is not very different, also regressive, although it has a higher proportion of people under 15 years old, and the majority of the population is younger, between 35 and 59 years old (Fig. 4.11).

Regarding the distribution of the population by municipalities, San Leonardo de Yagüe is home to 63% of the population, followed by Hontoria del Pinar with 20% (as of 01/01/2023, Fig. 4.12). Thus, we know that San Leonardo de Yagüe serves as the regional capital of the Park, as other socioeconomic indicators will also reflect, as it is where the majority of services and economic activities are concentrated. The least populated municipalities are Herrera de Soria (with 0.03% of the population), Nafría de Ucero with 1%, and Ucero with 3%.

### 4.2.3 Services and Economic Activity

In the CRLNP, San Leonardo de Yagüe is the most dynamic, most populous, and economically powerful municipality, so it is also where the majority of services are concentrated. It is the only municipality with a health center (Table 4.3), with the rest of the municipalities having clinics. As for pharmacies, there are only in Hontoria del Pinar and San Leonardo de Yagüe.

In the CRLNP, the number of educational centers has decreased from 6 to 4 from 1990 to 2022, and the number of students has been reduced by almost half (Table 4.4). These educational establishments are located in Casarejos, with the

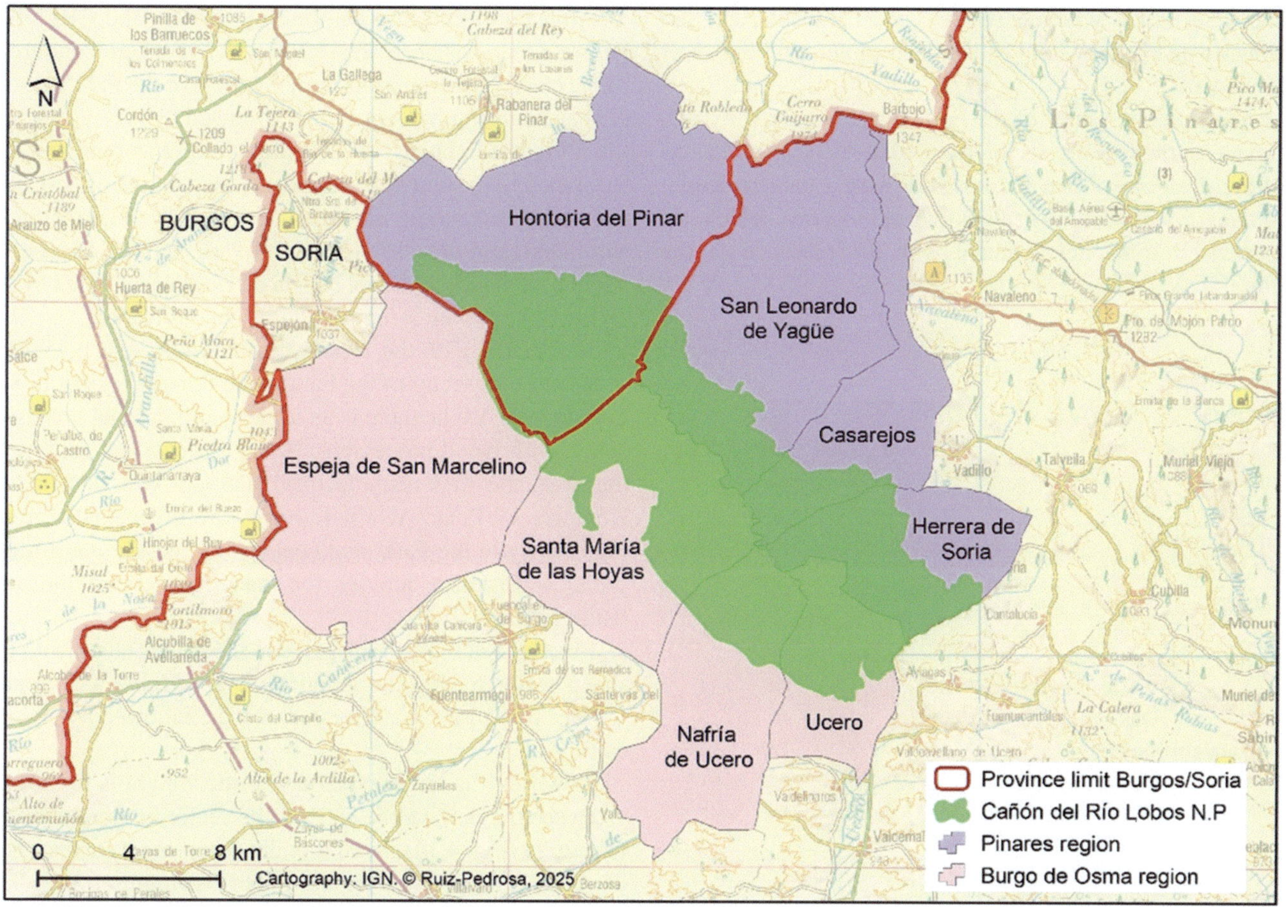


**Fig. 4.8** Municipalities and districts in the CRLNP

**Fig. 4.9** Population evolution in CRLNP (1990–2022)

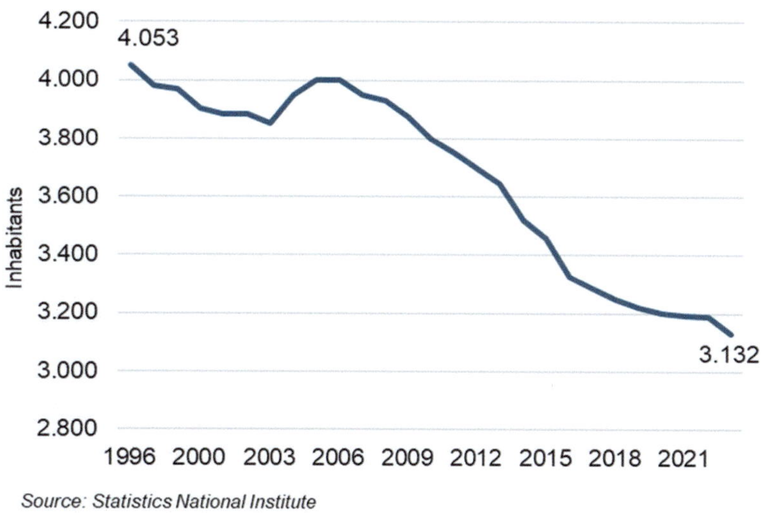

Pinares Sur Primary School (CEIP) and 21 students, and San Leonardo de Yagüe, with 3 educational centers and a total of 288 students.

The economic sector of the Park is characterized by two main areas: the forestry sector, focused on wood exploitation and the presence of a wood industry with a long tradition; and the services sector, specifically rural tourism, which has experienced some development with the presence of the Natural Park.

**Table 4.2** Vital statistics in CRLNP (1990–2022)

| Vital statistics | | 1990 | 2000 | 2010 | 2022 |
|---|---|---|---|---|---|
| Births | | 26 | 22 | 16 | 25 |
| Deaths | | 44 | 41 | 51 | 36 |
| **Natural increase** | | **−18** | **−19** | **−35** | **−11** |
| **Residential variations** | | 1990 | 2000 | 2010 | 2021 |
| Outbound to | The same province | 39 | 22 | 64 | 39 |
| | Another province in the region | 14 | 20 | 23 | 31 |
| | Another region | 31 | 39 | 38 | 54 |
| | Other countries | | | 10 | 10 |
| **Total emigrations** | | **84** | **81** | **135** | **134** |
| Inbound from | The same province | 30 | 13 | 37 | 61 |
| | Another province in the region | 17 | 14 | 18 | 28 |
| | Another region | 33 | 30 | 31 | 57 |
| | Other countries | 1 | 11 | 17 | 22 |
| **Total immigrations** | | **81** | **68** | **103** | **168** |
| **Migration balance** | | **−3** | **−13** | **−32** | **34** |

Source: Statistical Information System, Junta de Castilla y León

After consulting the land use system for the CRLNP, a land use map has been created (Fig. 4.13), where coniferous forest clearly dominates, occupying most of the Park's extension and surrounding areas. It is followed by mixed forest in the northeast boundary of the Park and transitional wooded shrubland in the southern boundary. Also, in the eastern sector of the Park, there are areas occupied by sclerophyllous shrubland, and outside the Natural Park, rainfed farming in Espejas de San Marcelino, Nafría de Ucero, and Santa María de las Hoyas, mainly focused on wheat cultivation.

Associated with the presence of conifers, and belonging to the western sector of the Park in the Pinares region, in San Leonardo de Yagüe there is an important industrial hub associated with wood processing. Being a traditional activity in the area, the first resin factory in Spain was installed in Hontoria del Pinar. Also noteworthy are mycology, a growing activity in the Park, and beekeeping, which was a traditional subsistence activity for many farmers in the area (Ruiz-Pedrosa, 2019).

Livestock activity is relegated to the background, with the sheep sector having the greatest weight (Table 4.5), representing 70% of the total livestock heads, followed by pig and cattle.

However, the largest socioeconomic weight is in the services sector, as it can be seen in the sector contracts for the year 2023 (Fig. 4.14). In all months, it represents at least half of the contracts, even in December assuming 100%. The industry also has a significant weight throughout the year, while agriculture only in the harvest months.

**Fig. 4.10** Population pyramid of Cañón del Río Lobos as of 1/01/2023 (% of total)

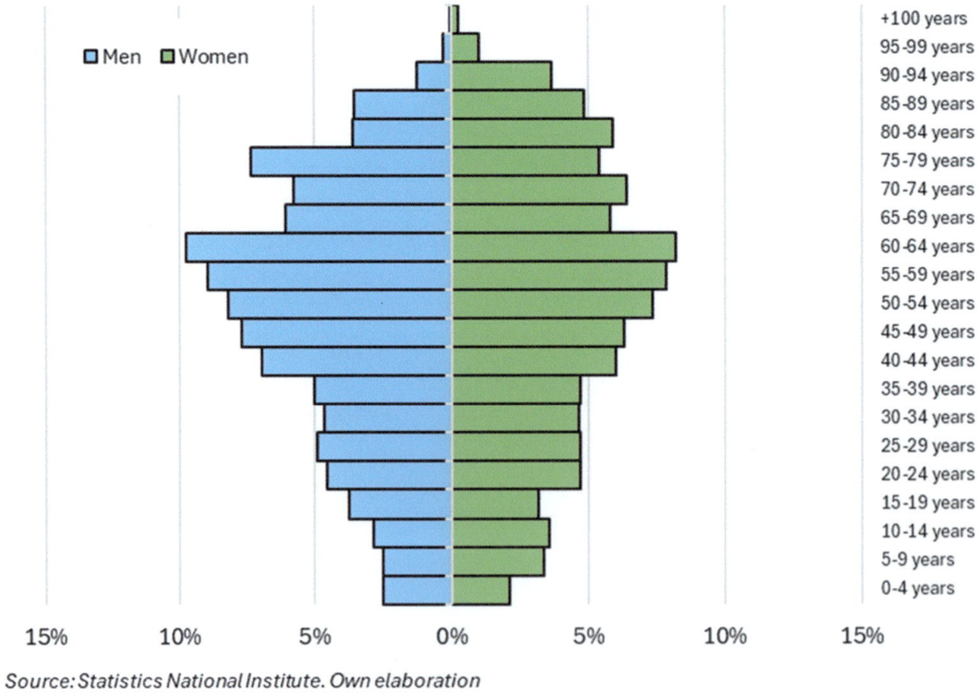

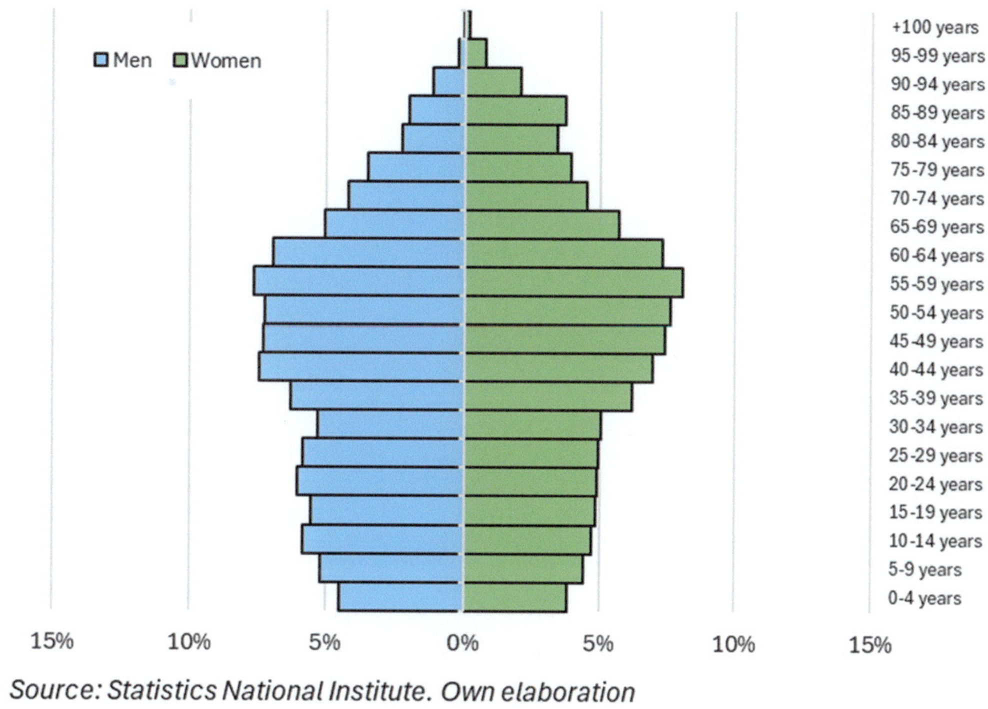


**Fig. 4.11** Population pyramid of Soria capital as of 1/01/2023 (% of total)

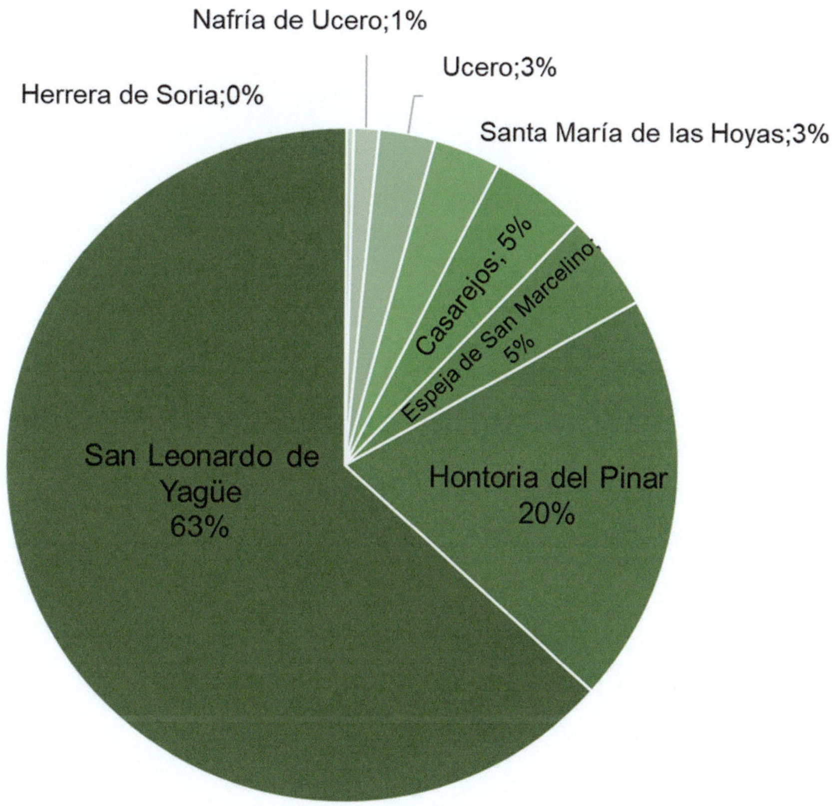


**Fig. 4.12** Distribution of the population in the CRLNP municipalities as of 01/01/2023 (% of total)

**Table 4.3** Health center, doctors' offices, and pharmacies in CRLNP

| | Health center | Doctors' offices | Pharmacies |
|---|---|---|---|
| Hontoria del Pinar | | 3 | 1 |
| Casarejos | | 1 | |
| Espeja de San Marcelino | | 5 | |
| Herrera de Soria | | 1 | |
| Nafría de Ucero | | 3 | |
| San Leonardo de Yagüe | 1 | | 1 |
| Santa María de las Hoyas | | 1 | |
| Ucero | | 1 | |
| **Total** | **1** | **15** | **2** |

Source: Junta de Castilla y León. Date: 2022

**Table 4.4** Evolution of education centers, students and professors in the studied area (1990–2022)

| | 1990 | 2010 | 2022 |
|---|---|---|---|
| Educational establishments | 6 | 4 | 4 |
| Non-university students | 560 | 518 | 309 |
| Teachers | | 72 | 57 |

Source: Junta de Castilla y León

The predominant economic sector is the service sector, and specifically since its declaration, the CRLNP has a consolidated tourism, despite not being a geopark or a tourist destination. It is an NPA and, therefore, its objective is conservation, education, enjoyment for citizens, and improving the quality of life of its inhabitants. Since 2006, it has been recognized by the Quality Tourism System of the Spanish Institute for Tourism Quality, based on its services and public use facilities and a management policy appropriate to its conservation objectives.

Massive visits began in the 1980s, coinciding with its declaration, and since then the interests of visitors have been very varied. Tourists are not a homogeneous group, but a very diverse public, which can be classified into three categories: mass tourism, cultural tourism, and active tourism (Serrano et al., 2020). Mass tourism is the most relevant in the Park, concentrated in recreational areas and near Ucero, and its objective is enjoyment, recreation, and socialization, with short visits that overcrowd specific sites of the Park, such as the hermitage of San Bartolomé. Cultural tourism is characterized by tourists interested in cultural and/or natural

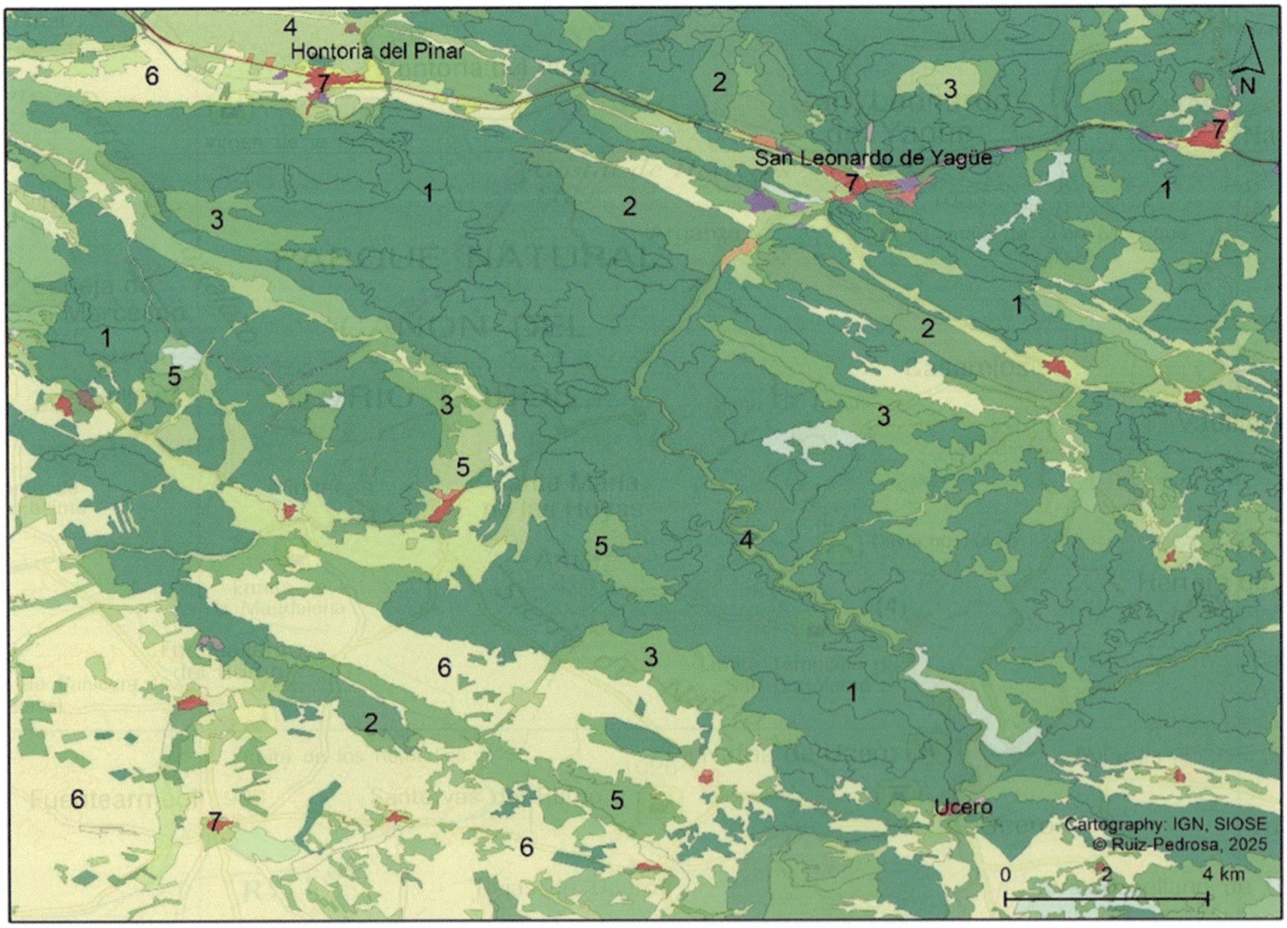


**Fig. 4.13** Land uses in the CRLNP. Legend: 1. coniferous forest; 2. mixed forest; 3. transitional shrubland; 4. grasslands; 5. sclerophyllous shrubland; 6. dryland farming; 7. urban core

**Table 4.5** Heads of livestock in CRLNP

| | Cattle | Goat | Rabbits | Horse | Sheep | Pig | Total |
|---|---|---|---|---|---|---|---|
| Hontoria del Pinar | 987 | 6 | | 40 | 1557 | 1400 | 3990 |
| Casarejos | 116 | | | 7 | 907 | | 1030 |
| Espeja de San Marcelino | 83 | 15 | 3 | 3 | 2468 | | 2572 |
| Herrera de Soria | | | | | 284 | | 284 |
| Nafría de Ucero | | | | | 1821 | | 1821 |
| San Leonardo de Yagüe | 161 | 57 | | 8 | 1734 | | 1960 |
| Santa María de las Hoyas | | 7 | | | 2312 | | 2319 |
| Ucero | | | | | 1150 | 840 | 1990 |
| Total | 1347 | 85 | 3 | 58 | 12,233 | 2240 | 15,966 |

Source: Agricultural Census (2009), Junta de Castilla y León

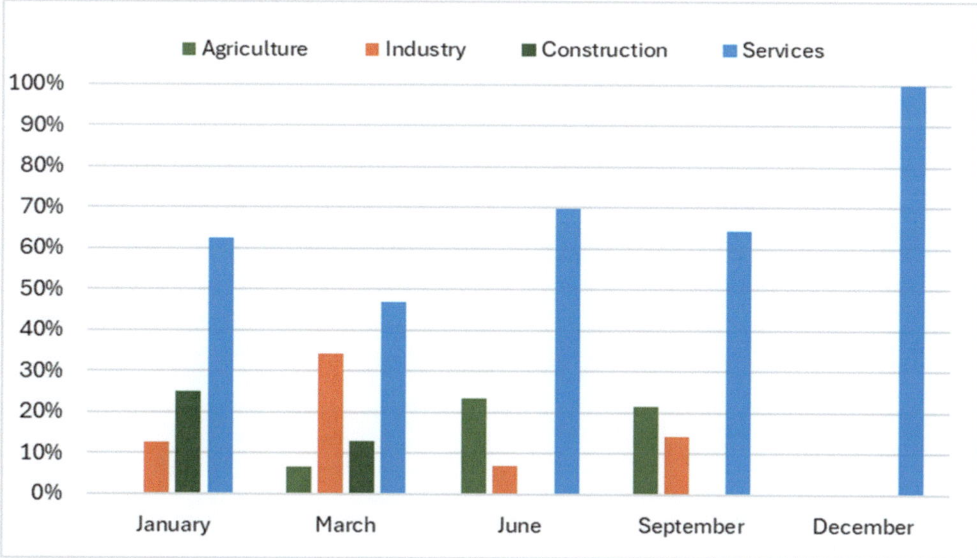


**Fig. 4.14** Contracts by sector in the study area (year 2023). (Source: SEPE)

heritage, much more aware of conservation and sustainable use, visiting multiple sites according to their objective, and interested in learning about the values of the Park, consuming interpretive and informational products. The active tourist, finally, seeks physical training and sports satisfaction, and is usually not interested in interpretative and knowledge proposals (hikers, runners, cyclists, climbers, etc.).

The CRLNP Lobos keeps two different records of the number of visitors: on the one hand, visitors who go to the Park Visitors Center, and on the other hand, through four pedestrian counters and one vehicle counter distributed throughout the park, one of the pedestrian counters located at the hermitage of San Bartolomé. The park house stood out recently, in 2021, for being the most visited in the entire region, with 38,801 visitors, despite the restrictions and limitations caused by COVID, which forced its closure in the high spring season and prevented the visit of tourists from other autonomous communities until late May. In 2022, visitors to the visitor center dropped to 32,748, still being the second most visited in Castilla y León. Therefore, the visitors center has been able to recover the visitor flow prior to COVID, which in 2019 was 38,007 (Fig. 4.15).

However, these good data are overshadowed by the registered visitors in the traffic counters located at the entrances to the Park, which show a total of 151,015 visitors in 2022, a very significant decrease compared to 2019, when it received 313,966 visitors. In 2019, only 12% of tourists visited the interpretation center, while in 2022 this percentage has increased to 21%. In short, the Cañón del Río Lobos has lost half of its pre-pandemic visitors, but these visitors are now more receptive to interpretation and come to the Visitor Center seeking information and advice. The tourism model is changing, from independent tourists to tourists who seek knowledge, in other words, from mass tourism to cultural tourism.

As for the typology of tourists, it is mainly family and social tourism, encompassed in contemplative nature tourism and recreation, whose focus is increasingly, as we have seen, on interpretation and learning. There is a second type of visitors, school groups and cultural associations, which in 2021 barely represented 2% of the total visitors to the Park, while before the pandemic they represented around 10%. The pandemic has disrupted this dynamic of organized visits and collaboration between the

**Fig. 4.15** Visitors to the Cañón del Río Lobos visitor center (2005–2022)

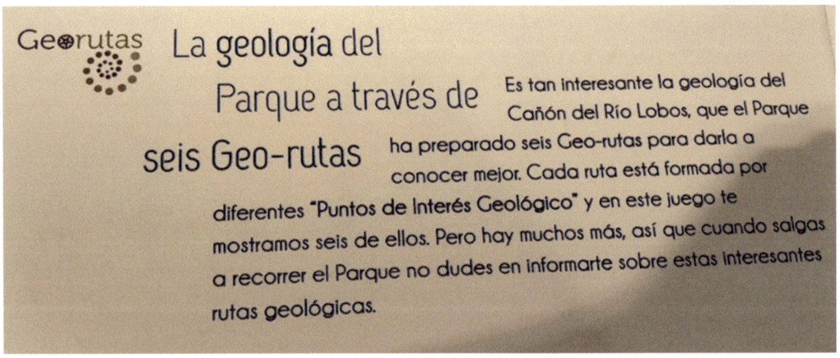


**Fig. 4.16** Promoted geo-routes in the Cañón del Río Lobos Visitor Center. The text says: *The geology of the park through six geo-routes. The geology of the Lobos River Canyon is so interesting that the park has prepared six geo-routes to help visitors learn more about it. Each route consists of different "Geological Points of Interest," and in this game we show you six of them. But there are many more, so when you go out to explore the park, be sure to find out about these interesting geological routes*

Park authorities and groups of people who used to visit for educational purposes. And finally, there is a last type of visitors attracted by the possibility of engaging in active or adventure tourism activities, such as horseback riding, cycling, and especially those related to caving, focused on a few caves in the Natural Park. In 2021, 25 authorizations were granted to carry out caving activities. The tourist cave of La Galiana is the main attraction for organized active tourism, through a concession to a private company (Serrano et al., 2020).

The flow of visitors is distributed throughout all months of the year, but with strong concentrations in August (18% of visitors in 2022), Easter (15%), some long weekends, and weekends in general. The months with the highest influx coincide with long holiday periods and weekends, with weekends accounting for 52.3% of annual visits, explaining the crowds of visitors on certain dates that allow for the characterization of mass tourism.

Regarding the origin of visits to the Visitor Center, with the 2022 data, almost all are of national origin (over 98%), with Madrid being the region that contributes the most visitors (36%) by a large margin, followed by Castilla y León (21%), Valencia (9%), and Catalonia (7%). Of the 2% of foreign visitors, 33% came from France, followed by the United Kingdom, Germany, and the Netherlands, each with 11%.

The Visitor Center of the Cañón del Río Lobos, located at the entrance to the Ucero Park, has two distinct areas: the ground floor for temporary exhibitions, and the second floor with the permanent exhibition on the Cañón del Río Lobos. The permanent exhibition focuses on the biodiversity of the Park, flora and fauna, with landforms and landscape taking a back seat. Geology is barely mentioned in a small section on karst hydrology, but there is no description of what karst is from a landform perspective.

Some geo-routes with different points of geological interest are promoted (Fig. 4.16), although these routes have not been designed and no information is provided about them at the Visitor Center. In general, both the interpretive panels at the Visitor Center and those in the Park, as well as the informational brochures, do not contain information related to geomorphology or an explanation of the karstic modeling that justifies the Park's designation.

This tourist development is reflected in the economic sector, not only in labor contracts by sectors as already mentioned but also in tourist infrastructure (Table 4.6). Thus, in the CRLNP there are a total of 37 rural accommodations, 13 tourist use houses, 4 hotels, 4 tourist apartments, 2 hostels, and a camping site. As for the hospitality sector, there are 19 restaurants and 4 cafes. Linked to active tourism, there is a company in San Leonardo de Yagüe. It is in this municipality where most tourist services are concentrated, followed by Hontoria, Casarejos, and Ucero.

**Table 4.6** Tourist services in the CRLNP

| | H | TR | H | TA | CAF | C | AT | R | TH | Total |
|---|---|---|---|---|---|---|---|---|---|---|
| Hontoria del Pinar | 1 | 8 | 1 | 1 | 1 | | | 4 | 2 | 18 |
| Casarejos | | 9 | | | | | | 2 | 1 | 12 |
| Espeja de san Marcelino | | 3 | | 1 | | | | | 1 | 5 |
| Nafría de Ucero | | 1 | | | | | | | 1 | 2 |
| San Leonardo de Yagüe | | 8 | 2 | | 3 | | 1 | 9 | 8 | 31 |
| Santa María de las Hoyas | 1 | 2 | | | | | | 1 | | 4 |
| Ucero | | 6 | 1 | 2 | | 1 | | 3 | | 13 |
| Total | 2 | 37 | 4 | 4 | 4 | 1 | 1 | 19 | 13 | |

Source: SIE, JCyL. The letters are: H, hostels; TR, rural tourism accommodations; H, hotels; TA, tourist apartments; CAF, cafeterias; C, campings; AT, active tourism companies; R, restaurants; TH, tourist use homes

Since 2023, the Cañón del Río Lobos has seen its tourist infrastructure strengthened due to regional funding amounting to almost half a million euros, from the Ministry of Environment of the Junta de Castilla y León and the funds for the improvement of Public Forests of Regional Interest. Several actions and interventions have been carried out in the Park, aimed at meeting the needs arising from an increasing influx of visitors to the Park. In the summer of 2023, a footbridge over the Río Lobos (Fig. 4.17) was inaugurated, which helps to relieve congestion on the main path, leading to the hermitage by creating a new accessible path throughout the year, regardless of river floods. The surface between the walkway and the hermitage was also improved, accessibility was enhanced by adding railings, and the pavement of the main entrance and a new access control area at Fuente Engomez were reinforced, improving parking.

In 2024, the new viewpoint of La Galiana (Fig. 4.18) has been inaugurated, after an expansion of the observation area by adding walkways forming a lookout point (Fig. 4.19), and the expansion of the parking lot. This viewpoint project, which is part of the Tourist Infrastructure Program in Natural Areas of Castilla y León, has sparked debate and critics of the project, as it involves an aggressive intervention with a high landscape impact on a cliff that naturally served as a viewpoint.

### 4.2.4 Cultural Heritage

The Cañón del Río Lobos Natural Park has a rich cultural heritage, both tangible and intangible, which has historically been linked to the spectacular landscape. In the socioeco-

**Fig. 4.17** New footbridge over the Rio Lobos on the path to San Bartolomé

**Fig. 4.18** New viewpoint at La Galiana, with interpretive panel and birdwatchers

nomic area of the Park, there are nine Cultural Interest Assets recognized by the Junta de Castilla y León, among which we can highlight four caves with rock art for their state of conservation and historical relevance, the monument of the Hermitage of San Bartolomé, the San Leonardo castle, and the Ucero castle.

Undoubtedly, the monument that symbolizes the Park is the Hermitage of San Bartolomé, with a prominent monumental character due to its artistic and historical value and its location at the entrance of the canyon. It has been recognized as a Historic Artistic Monument since 1983 and as a Cultural Interest Asset with Monument category in 2015. This Romanesque construction from the thirteenth century is characterized by its Templar symbolism, esoteric figures, and its location at the entrance of the Canyon, which adds a special landscape value (Fig. 4.20).

Equally monumental are the San Leonardo castle and the Ucero castle. The Castillo de San Leonardo (Fig. 4.21),

**Fig. 4.19** Footbridges at the New Galiana Lookout Point

**Fig. 4.20** Hermitage of San Bartolomé at the entrance to the Cañón del Río Lobos

in the southwest of the urban area and from the sixteenth century, is considered the first manorial castle in Spain in the bastioned style, although only its base remains. The Ucero castle (Fig. 4.22), from the thirteenth century, is a compound with triple walls and a well-preserved tower, located on a hill near Ucero, and is part of a monumental complex that also includes Celtiberian, Visigothic, and Roman remains, as well as the ruins of a Romanesque hermitage. Both castles are easily accessible, have parking, and interpretive panels.

Inside the karst caves of the Park and its surroundings, up to four cave paintings can be found. These are small-sized paintings (less than 11 cm) in red color depicting various figures: anthropomorphic, geometric representations, and elements of nature.

The Cañón del Río Lobos also has special relevance as one of the itineraries used by pilgrims to connect with the French Way of the Camino de Santiago, in addition to being the setting for movies and series such as "Fuenteovejuna" (1972), "The Three Musketeers: The Queen's Diamonds" (1973), and the series Curro Jiménez in 1977.

**Fig. 4.21** San Leonardo castle. Above: main entrance; below: inside the ruined castle

**Fig. 4.22** Ucero castle. Above: main façade; below: view of the keep from the inside

## 4.3 Cañón del Río Lobos Natural Park Geomorphosites Inventory

A total of 14 geomorphosites have been inventoried in the Cañon del Río Lobos Natural Park (Fig. 4.23). Each geomorphosite has also been assigned a geomorphological classification, as well as values of accessibility and interest (Table 4.7). The predominant geomorphological attribution is structural (five cases) and karstic (five cases), representing 71% of the total, due to the presence of folded and limestone materials. Fluvial geomorphosites and an example of interesting slope processes are less frequent.

These sites have been identified as potential geomorphosites based on their geomorphological and landscape characteristics. A descriptive sheet has been created for each site, detailing its intrinsic values (scientific content), natural dynamics, uses, impacts on the geomorphosites, and additional values (cultural, educational, touristic, use and management) (Tables 4.8, 4.9, 4.10, 4.11, 4.12, 4.13, 4.14, 4.15, 4.16, 4.17, 4.18, 4.19, 4.20, and 4.21 and Figs. 4.24, 4.25, 4.26, 4.27, 4.28, 4.29, 4.30, 4.31, 4.32, 4.33, 4.34, 4.35, 4.36 and 4.37). The descriptive sheet also includes information on the location and images of the geomorphosites.

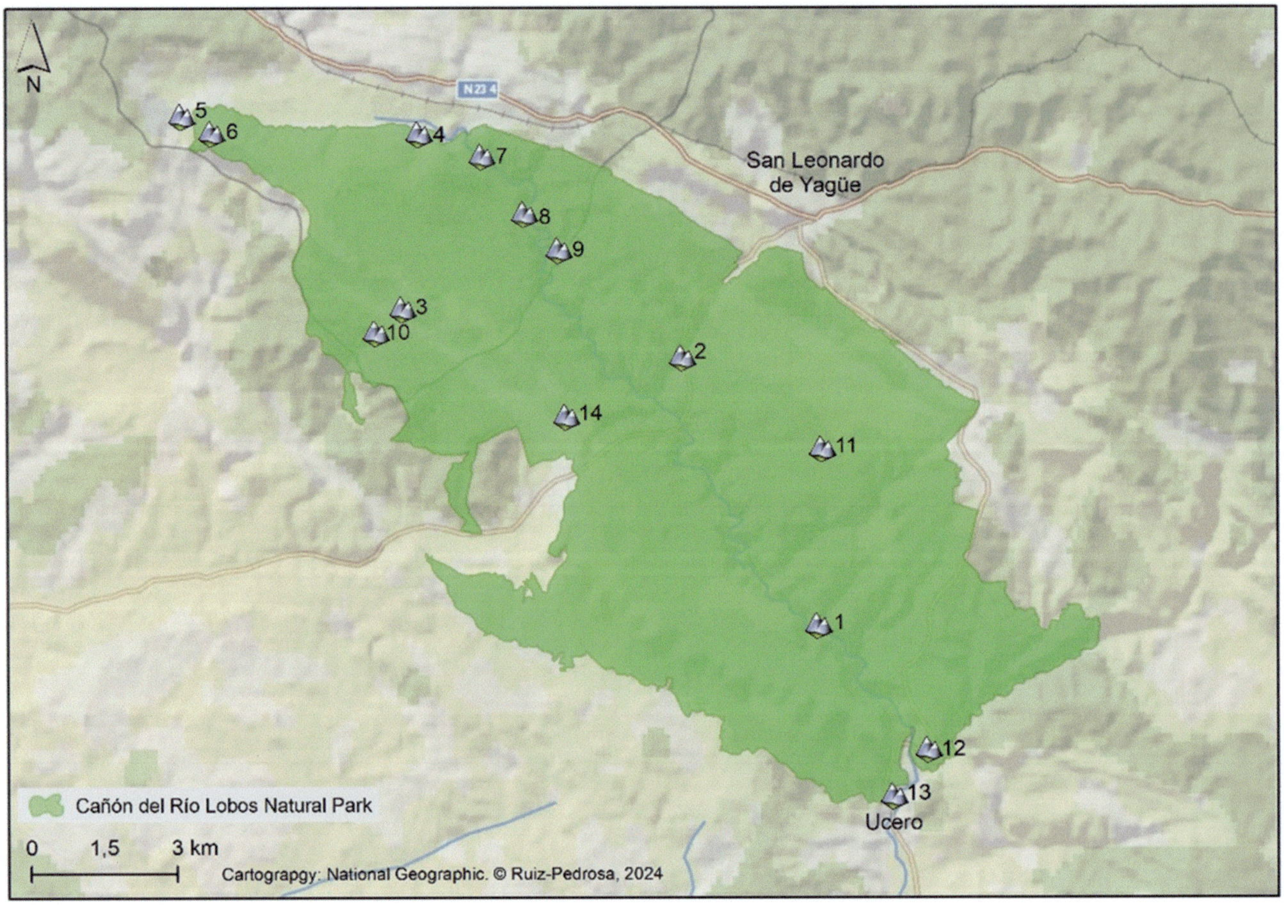


**Fig. 4.23** CRLNP inventoried geomorphosites

**Table 4.7** CRLNP inventoried geomorphosites

| N° | Geomorphological attribution | Type | | Accessibility | Interest |
|---|---|---|---|---|---|
| 1 | Fluviokarstic | Site | Representative | High | High |
| 2 | Fluvio-structural | Site | Exceptional | High | Medium |
| 3 | Structural | Site | Representative | Medium | High |
| 4 | Structural | Site | Representative | High | High |
| 5 | Slope | Element | Exceptional | Medium | High |
| 6 | Structural | Site | Representative | Medium | High |
| 7 | Karstic | Element | Representative | High | High |
| 8 | Structural | Site | Exceptional | Medium | Medium |
| 9 | Fluvial | Site | Representative | Medium | High |
| 10 | Structural | Site | Exceptional | High | High |
| 11 | Karstic | Site | Representative | Medium | Medium |
| 12 | Karstic | Element | Representative | High | High |
| 13 | Karstic | Element | Representative | High | Medium |
| 14 | Karstic | Element | Representative | Medium | High |

Geomorphosites: 1. Rio Lobos–San Bartolomé Canyon, 2. Arganza fault-line valley, 3. La Sierra syncline flank crest, 4. Virgen de la Cueva syncline flank, 5. Pico Navas slopeslide, 6. Pico Navas perched syncline, 7. Las Raideras river sink, 8. Hoyo de los Lobos syncline flank, 9. La Isla entrenched meander, 10. Costalago orthocline valley, 11. Las Tainas y el Torcajón pits and karstic area, 12. La Galiana karstic system, 13. Ucero spring, 14. Chorrón fluvial sink

**Table 4.8** Río Lobos–San Bartolomé Canyon geomorphosite

| Geomorphosite identification sheet | | | |
|---|---|---|---|
| Identification | **Name: Río Lobos–San Bartolomé Canyon** | **Site: Río Lobos canyon in San Bartolomé hermitage** | **N° 1**<br>Altitude: 970 m |
| Situation | Municipality: Ucero | Coordinates:<br>X: 494319.10; Y: 4622211.40 | |
| Geomorphological values | Type | Site<br>Representative | |
| | Genesis | Fluviokarstic. Karstic dissolution and river incision | |
| | Morphology: description, morphostructures, erosion | Meander, asymmetrical valley between escarpments carved in the Coniacian limestones. System formed by the holokarstic canyon with vertical walls (with traces of karstification levels), hanging horizontal caves (like the Cueva Grande), detrital and limestone fillings inside and in the Lobos river | |
| | Dynamics | Karstic and fluvial | |
| | Chronology | Miocene-Holocene | |
| | Main interest | Fluviokarstic (karst on walls and cave, fluvial in the valley) | |
| | Secondary interest | Cultural, slopes, and hydrological | |
| | Geomorphosite attribution | Fluviokarstic | |
| Additional values | Landscape | High, panoramic view of the canyon | |
| | Cultural elements | Hermitage, cave paintings, viewpoints | |
| | Economic | Active tourism | |
| | Social | Recreation | |
| | State of conservation | Medium: bad in the cave due to presence of garbage, graffiti, overcrowding... and good outside | |
| | Actual uses | Tourism, hiking, cycling, religious | |
| | Infrastructures | Track, bridge, path, footbridge over the river | |
| | Impacts | Trail, influx, overcrowding | |
| | Legal situation | Natural Park. Reserve Area | |
| Accessibility | Walking distance | 2.5 km from parking | |
| | Slope | No slope | |
| | Security | Good, marked trail | |

The five geomorphosites with structural attribution are representative of folded reliefs, including ridges of synclinal flanks and hanging synclines. N° 3, La Sierra syncline flank crest, and N° 4, Virgen de la Cueva syncline flank, represent the ridges of the SW and N flanks, respectively, of the Lobos river syncline. The Pico Navas perched syncline, geomorphosites N° 6, elevated to the NW, adds to the structural features the presence of karstic forms and processes, dominated by karren, slope dynamics, and the added cultural interest derived from the remains of the Celtiberian fortification wall. The Costalago orthocline valley (N° 10) has been classified as unique for its great structural, landscape, lacustrine, and fluvial interest, with active dynamics of incision, solifluction, and landslides.

The karstic or fluviokarstic geomorphosites are the most significant landscapes in the park, presenting karstic canyons, sinkholes, chasms, caves, and springs. The best-known and representative geomorphosite is the Río Lobos–San Bartolomé Canyon (N° 1), a system formed by a holokarstic canyon with vertical walls and traces of karstification levels, hanging horizontal caves like the Cueva Grande, and detrital and limestone fillings inside. It is also of great cultural interest due to the presence of the hermitage of San Bartolomé and cave paintings.

In addition to the fluviokarstic geomorphosites, there are also sites that exemplify the sinuous and meandering hydrographic network typical of the park. These are the geomorphosites of fluvial attribution. In particular, site N° 9, La Isla entrenched meander, is an example of a very sinuous entrenched meander, with steep walls, a flat bottom occupied by fluvial terraces, and a dry channel for most of the year. There are two geomorphosites attributed to both fluvial and structural aspects, where tectonics have influenced the river network. These are sites N° 2 and N° 8 (Arganza fault-line valley and Hoyo de los Lobos syncline flank, respectively), which are two straight valleys on a fault, interrupting the sinuosity of the park's river network.

The Pico Navas slopeslide (N° 5), located in the northwest corner of the park, is the only geomorphosite attributed to slope dynamics, standing out for its current dynamics, large size, and landscape content. It is an active rotational landslide formed by a body of blocks and a head with cracks, scarps, and uneven blocks easily visible.

**Table 4.9** Arganza fault-line valley geomorphosite

| Geomorphosite identification sheet | | | |
|---|---|---|---|
| Identification | **Name: Arganza fault-line valley** | **Site: Arganza valley** | **N° 2**<br>Altitude: 1010 m |
| Situation | Municipality: | Coordinates: | |
| | San Leonardo de Yagüe | X: 492595.64; Y: 4629526.95 | |
| Geomorphological values | Type | Site | |
| | | Exceptional | |
| | Genesis | Fluvial and structural. Fault-line valley. Fluvial incision on the erosion surface over the Coniacian limestones and Maastrichtian marls, following the NE-SW fault line | |
| | Morphology: description, morphostructures, erosion | Straight flat-bottomed valley. NE-SW fracture line. Two levels of fluvial terraces, one of them being a floodplain. Two alluvial cones. Allochthonous drainage from the Navaleno River | |
| | Dynamics | Fluvial | |
| | Chronology | Terraces, Holocene | |
| | Main interest | Fluvial and morphostructural landforms | |
| | Secondary interest | – | |
| | Representativeness | High | |
| | Geomorphosite attribution | Fluvial and structural | |
| Additional values | Landscape | Medium | |
| | Cultural elements | Mill, 7 ojos Bridge | |
| | Economic | Agricultural use | |
| | Social | Recreation | |
| | State of conservation | Poor, high degree of abandonment | |
| | Actual uses | Picnic area, mill, livestock use | |
| | Infrastructures | Two bridges, recreational area, canal | |
| | Impacts | Livestock, abandonment of arable fields | |
| | Legal situation | Natural Park. Compatible Use Zone | |
| Accessibility | Walking distance | 3 km from Arganza to recreational area | |
| | Slope | – | |
| | Security | High, marked path | |

**Table 4.10** La Sierra syncline flank crest geomorphosite

| Geomorphosite identification sheet | | | |
|---|---|---|---|
| Identification | **Name: La Sierra syncline flank crest** | **Site: La Sierra** | **N° 3** |
| | | | Altitude: 1070 m |
| Situation | Municipality: | Coordinates: | |
| | Hontoria del Pinar | X: 485386.74; Y: 4628791.57 | |
| Geomorphological values | Type | Site | |
| | | Representative | |
| | Genesis | Structural | |
| | Morphology: description, morphostructures, erosion | Karren. Landslides. Crest of the SW flank of the Rio Lobos syncline, with a NW–SE direction, a gently sloping ridge to the NE, and a steep and stepped front to the SW. The ridge is dominated by structural limestone pavements and karst depressions (sinkholes, uvala) with chasms and sinkholes. The front is stepped due to the stratigraphy. The landslides are active and generate scree slopes | |
| | Dynamics | Slope (landslides) and karst | |
| | Chronology | Tertiary, Quaternary | |
| | Main interest | Structural | |
| | Secondary interest | Karstic and slope | |
| | Geomorphosite attribution | Structural | |
| Additional values | Landscape | High, panoramic view | |
| | Cultural elements | Quarries, hut, watchtower | |
| | Economic | Livestock and forestry | |
| | Social | Recreation | |
| | State of conservation | Good | |
| | Actual uses | Livestock and forestry | |
| | Infrastructure | Hut, tower, and track | |
| | Impacts | House, abandoned quarry. Moderate | |
| | Legal situation | Natural Park. Limited Use Area of Special Interest | |
| Accessibility | Walking distance | 1.7 km from the Costalago refuge/5 km from Grillares | |
| | Slope | 120 m/150 metros | |
| | Security | Good, track | |

**Table 4.11** Virgen de la Cueva syncline flank geomorphosite

| Geomorphosite identification sheet | | | |
|---|---|---|---|
| Identification | **Name: Virgen de la Cueva syncline flank** | **Site: Virgen de la Cueva** | **N° 4** |
| | | | Altitude: 1050 m |
| Situation | Municipality: | Coordinates: | |
| | Hontoria del Pinar | X: 486118.71; Y: 4632367.83 | |
| Geomorphological values | Type | Element | |
| | | Representative | |
| | Genesis | Structural and fluvial | |
| | Morphology: description, morphostructures, erosion | North flank crest of the River Lobos syncline incised by the river, forming a meander cut into the Coniacian limestones that has eroded the flank, creating a steep north-facing slope and a south-facing ridge | |
| | Dynamics | Fluvial and slope | |
| | Chronology | Tertiary, Quaternary | |
| | Main interest | Structural | |
| | Secondary interest | Fluvial | |
| | Geomorphosite attribution | Structural | |
| Additional values | Landscape | High | |
| | Cultural elements | Hermitage, houses, viewpoint, Roman bridge | |
| | Economic | – | |
| | Social | Religious, recreation | |
| | State of conservation | Good with numerous current and past uses on the slope and the ridge | |
| | Actual uses | Recreation, religious, tourism | |
| | Infrastructures | Viewpoint, path | |
| | Impacts | Path, houses | |
| | Legal situation | Natural Park. Area of Special Interest Reserve | |
| Accessibility | Walking distance | Accessible by car | |
| | Slope | – | |
| | Security | High | |

**Table 4.12** Pico Navas slopeslide geomorphosite

| Geomorphosite identification sheet | | | |
|---|---|---|---|
| Identification | **Name: Pico Navas slopeslide** | **Site: Pico Navas** | **N° 5**<br>Altitude: 1276 m |
| Situation | Municipality: | Coordinates: | |
| | Hontoria del Pinar | X: 481762.51; Y: 4632531.46 | |
| Geomorphological values | Type | Site | |
| | | Singular | |
| | Genesis | Slope. Rotational landslide | |
| | Morphology: description, morphostructures, erosion | Rotational landslide generated in the Coniancese limestones in contact with the marls and marly limestones of the Turonian, resulting in a large displacement of the NW portion of the Rio Lobos syncline. The landslide has created a chaotic surface tongue, a block body, and a head with cracks and inclined blocks. The landslide is active | |
| | Dynamics | Slope. Active | |
| | Chronology | Holocene—active | |
| | Main interest | Slope. Active processes | |
| | Secondary interest | – | |
| | Geomorphosite attribution | Slope | |
| Additional values | Landscape | High, panoramic view | |
| | Cultural elements | Cross and nativity scene from 1929 | |
| | Economic | – | |
| | Social | Hiking | |
| | State of conservation | Good | |
| | Actual uses | Hiking. Abandoned quarries and mines | |
| | Infrastructures | – | |
| | Impacts | Population Center at the foot (possible risk) | |
| | Legal situation | Natural Park. Reserve Zone | |
| Accessibility | Walking distance | 3.5 km from Navas del Pinar | |
| | Slope | 300 m | |
| | Security | Medium, sections without a path | |

**Table 4.13** Pico Navas perched syncline geomorphosite

| Geomorphosite identification sheet | | | |
|---|---|---|---|
| Identification | **Name: Pico Navas perched syncline** | **Site: Pico Navas** | **N° 6**<br>Altitude: 1352 m |
| Situation | Municipality: | Coordinates: | |
| | Hontoria del Pinar | X: 482017.49; Y: 4632390.80 | |
| Geomorphological values | Type | Site | |
| | | Exceptional | |
| | Genesis | Structural | |
| | Morphology: description, morphostructures, erosion | Perched syncline with NW–SE direction and steep margins, culminating in a slope, elevated toward the NW. The culmination is shaped by karst processes, dominated by karren, along with dolines and uvalas | |
| | Dynamics | Tectonic and karstic | |
| | Chronology | Tertiary | |
| | Main interest | Structural | |
| | Secondary interest | Karstic | |
| | Geomorphosite attribution | Structural | |
| Additional values | Landscape | High, panoramic view | |
| | Cultural elements | Roman fort, ruined stone wall. Geodetic vertex | |
| | Economic | Livestock farming (sheep) | |
| | Social | Hiking | |
| | State of conservation | Good | |
| | Actual uses | Hiking, livestock farming | |
| | Infrastructures | Forest track | |
| | Impacts | Moderate: track and livestock | |
| | Legal situation | Natural Park. Reserve Zone | |
| Accessibility | Walking distance | 4 km from Navas del Pinar | |
| | Slope | 300 m | |
| | Security | Medium, sections without a path | |

**Table 4.14** Las Raideras river sink geomorphosite

| Geomorphosite identification sheet | | | |
|---|---|---|---|
| Identification | **Name: Las Raideras river sink** | **Site: Lobos river** | **N° 7** Altitude: 1030 m |
| Situation | Municipality: | Coordinates: | |
| | Hontoria del Pinar | X: 488623.40; Y: 4631736.65 | |
| Geomorphological values | Type | Site | |
| | | Representative | |
| | Genesis | Karstic | |
| | Morphology: description, morphostructures, erosion | Karstic river sink at the contact between Coniacian limestones and micrites, with a progressive filtration in the river deposits. Introduction into the limestones through sinkholes, with flow losses of up to 400 L/s (Segovia, 2008), which drains subaerially to the Ucero river | |
| | Dynamics | Karstic and fluvial | |
| | Chronology | Active | |
| | Main interest | Karstic | |
| | Secondary interest | Fluvial, hydrological | |
| | Geomorphosite attribution | Karstic | |
| Additional values | Landscape | Medium, forest | |
| | Cultural elements | – | |
| | Economic | – | |
| | Social | Recreation | |
| | State of conservation | Good | |
| | Actual uses | Hiking, recreation | |
| | Infrastructures | Forest track, path and boardwalk, parking | |
| | Impacts | Moderate: boardwalk and picnic area | |
| | Legal situation | Natural Park. Limited Use Area of Special Interest | |
| Accessibility | Walking distance | 1 km from Hontoria del Pinar | |
| | Slope | – | |
| | Security | High: forest track and path | |

**Table 4.15** Hoyo de los Lobos syncline flank geomorphosite

| Geomorphosite identification sheet | | | |
|---|---|---|---|
| Identification | **Name: Hoyo de los Lobos syncline flank** | **Site: Valle de río Lobos. Hoyo de los Lobos** | **N° 8** Altitude: 1020 m |
| Situation | Municipality: | Coordinates: | |
| | Hontoria del Pinar | X: 488325.58; Y: 4630684.73 | |
| Geomorphological values | Type | Site | |
| | | Exceptional | |
| | Genesis | Structural and fluvial | |
| | Morphology: description, morphostructures, erosion | Linear portion of the Lobos River valley, 560 m long, incised in the Coniacian limestones and oriented NNW–SSE. The valley has an alluvial terrace occupying the bottom and a fluvial accumulation level 5–10 m wide with substrate outcrops. Its genesis is attributed to fluvial incision along a NNW–SSE fracture line | |
| | Dynamics | Torrential-fluvial. Adaptation of fluvial erosion to the geological structure | |
| | Chronology | Pleistocene, Holocene—active | |
| | Main interest | Tectonic and fluvial | |
| | Secondary interest | – | |
| | Geomorphosite attribution | Structural | |
| Additional values | Landscape | Medium, forest | |
| | Cultural elements | Stone hut and stone steps in the riverbed | |
| | Economic | Forestry | |
| | Social | – | |
| | State of conservation | Good | |
| | Actual uses | Hiking, forestry | |
| | Infrastructures | Trail | |
| | Impacts | – | |
| | Legal situation | Natural Park. Limited Use Area of Special Interest | |
| Accessibility | Walking distance | 2.5 km from Hontoria del Pinar | |
| | Slope | – | |
| | Security | Good, trail | |

**Table 4.16** La Isla entrenched meander geomorphosite

| Geomorphosite identification sheet | | | |
|---|---|---|---|
| Identification | **Name: La Isla entrenched meander** | **Site: Lobos river valley** | **N° 9** |
| | | | Altitude: 1015 m |
| Situation | Municipality: | Coordinates: | |
| | Hontoria del Pinar/San Leonardo de Yagüe | X: 488953.64; Y: 4630015.53 | |
| Geomorphological valley | Type | Site | |
| | | Representative | |
| | Genesis | Fluvial. Incision | |
| | Morphology: description, morphostructures, erosion | Entrenched meander of high sinuosity, incised in the Coniacian limestones. Morphology with steep walls and flat bottom occupied by fluvial terraces (one level) and the channel | |
| | Dynamics | Fluvial | |
| | Chronology | Pleistocene, Holocene | |
| | Main interest | Fluvial | |
| | Secondary interest | – | |
| | Geomorphosite attribution | Fluvial | |
| Additional values | Landscape | Medium, forest | |
| | Cultural elements | – | |
| | Economic | Forestry | |
| | Social | – | |
| | State of conservation | Good | |
| | Actual uses | Hiking, forest | |
| | Infrastructures | Trail, blocks to cross the river | |
| | Impacts | – | |
| | Legal situation | Natural Park. Reserve Area | |
| Accessibility | Walking distance | 5 km from parking | |
| | Slope | – | |
| | Security | Good, trail | |

**Table 4.17** Costalago orthocline valley geomorphosite

| Geomorphosite identification sheet | | | |
|---|---|---|---|
| Identification | **Name: Costalago orthocline valley** | **Site: Costalago valley** | **N° 10**<br>Altitude: 1070 m |
| Situation | Municipality: | Coordinates: | |
| | Hontoria del Pinar | X: 485477.45; Y: 4628504.92 | |
| Geomorphological values | Type | Site | |
| | | Singular | |
| | Genesis | Structural and differential erosion | |
| | Morphology: description, morphostructures, erosion | Orthoclinal valley developed on the southwest flank of the Lobos river syncline, shaped by differential erosion. The valley opens in the sand stratum (Albian) between the conglomerates (Aptian) to the south and the marls and limestones (Turonian-Coniacian) to the north. The valley has peat fill in the bottom, with slopes affected by solifluction. On the north slope, there are landslides and active falls | |
| | Dynamics | Fluvial and slopes. Processes of incision, solifluction, and active falls | |
| | Chronology | Tertiary, Pleistocene, and Holocene | |
| | Main interest | Structural/landscape | |
| | Secondary interest | Lacustrine/fluvial/slopes | |
| | Geomorphosite attribution | Structural | |
| Additional values | Landscape | High, wide valley | |
| | Cultural elements | – | |
| | Economic | Livestock | |
| | Social | Recreation | |
| | State of conservation | Good, historical modifications (drainage, tracks) | |
| | Actual uses | Livestock, recreation | |
| | Infrastructures | Tracks, shelter, viewpoint, spring, and watering hole | |
| | Impacts | Low (tracks and livestock) | |
| | Legal situation | Natural Park. Limited Use Zone and Limited Use Zone of Special Interest | |
| Accessibility | Walking distance | Arrival by car | |
| | Slope | – | |
| | Security | High, road and track | |

**Table 4.18** Las Tainas y el Torcajón pits and karstic area geomorphosite

| Geomorphosite identification sheet | | | |
|---|---|---|---|
| Identification | **Name: Las Tainas and el Torcajón pits and karstic area** | **Site: Las Tainas and el Torcajón** | **N° 11**<br>Altitude: 1088 m |
| Situation | Municipality: | Coordinates: | |
| | Casarejos | X: 494355.51; Y: 4625901.67 | |
| Geomorphological values | Type | Site | |
| | | Representative | |
| | Genesis | Karstic | |
| | Morphology: description, morphostructures, erosion | Erosion surface on Coniacian limestones and micrites where a karst complex develops, consisting of exokarstic elements (structural karren, dolines, and captured karst depressions) and endokarstic elements (vertical development pits, such as Tainas and Torcajón, with a depth of 100 m) adapted to the structural conditions | |
| | Dynamics | Karstic | |
| | Chronology | Pleistocene and Holocene | |
| | Main interest | Karstic | |
| | Secondary interest | Landscape | |
| | Geomorphosite attribution | Karstic | |
| Additional values | Landscape | Low | |
| | Cultural elements | Tenadas | |
| | Economic | Livestock farming | |
| | Social | – | |
| | State of conservation | Good | |
| | Actual uses | Speleology, livestock farming, hiking | |
| | Infrastructure | Signage and fencing of the pits | |
| | Impacts | Moderate: tourism and livestock farming | |
| | Legal situation | Natural Park. Compatible Use Zone and Limited Use Zone | |
| Accessibility | Walking distance | 3.5 km from El Congosto | |
| | Slope | – | |
| | Security | Good, track | |

**Table 4.19** La Galiana karstic system geomorphosite

| Geomorphosite identification sheet | | | |
|---|---|---|---|
| Identification | **Name: La Galiana karstic system** | **Site: La Galiana caves** | **N° 12**<br>Altitude: 1040 m |
| Situation | Municipality: | Coordinates: | |
| | Ucero | X: 496386.52; Y: 4620070.87 | |
| Geomorphological values | Type | Site | |
| | | Representative | |
| | Genesis | Karstic | |
| | Morphology: description, morphostructures, erosion | Karst system characterized by the existence of predominantly horizontally developed cavities staggered on the slopes at 6, 15, 150, and 160 m above the current channel. Its interior morphology is conditioned by the structure (stratification and jointing), visible in the galleries, and presents a wide representation of speleothems (columns, flows, stalagmites, stalactites, curtains), as well as detrital fillings and crusts of high educational and scientific value | |
| | Dynamics | Karstic | |
| | Chronology | Tertiary, Pleistocene | |
| | Main interest | Karstic | |
| | Secondary interest | – | |
| | Geomorphosite attribution | Karstic | |
| Additional values | Landscape | High, viewpoint | |
| | Cultural elements | Rock art | |
| | Economic | Active tourism (speleology) | |
| | Social | Recreation | |
| | State of conservation | Medium: presence of graffiti, high influx | |
| | Actual uses | Speleology, hiking, tourism | |
| | Infrastructure | Road, path, parking, and viewpoint, adaptation for touristic use. Galiana Baja has regulated touristic use | |
| | Impacts | Human influx: Graffiti, wear, trampling, fire, overuse in Galiana High. Galiana Baja has access and adaptation for tourism | |
| | Legal situation | Natural Park. Limited Use Zone of Special Interest | |
| Accessibility | Walking distance | 100 m from parking | |
| | Slope | – | |
| | Security | High | |

**Table 4.20** Ucero springs geomorphosite

| Geomorphosite identification sheet | | | |
|---|---|---|---|
| Identification | **Name: Ucero springs** | **Site: Ucero** | **N° 13**<br>Altitude: 960 m |
| Situation | Municipality: | Coordinates: | |
| | Ucero | X: 496197.58; Y: 4620105.19 | |
| Geomorphological values | Type | Element | |
| | | Representative | |
| | Genesis | Karstic | |
| | Morphology: description, morphostructures, erosion | The Ucero river originates in La Galiana spring. Example of aquifer and groundwater circulation (infiltration through limestone) | |
| | Dynamics | Karstic | |
| | Chronology | Active. Pleistocene, Holocene | |
| | Main interest | Karstic | |
| | Secondary interest | Fluvial, hydrological | |
| | Geomorphosite attribution | Karstic | |
| Additional values | Landscape | Medium | |
| | Cultural elements | – | |
| | Economic | – | |
| | Social | Recreation | |
| | State of conservation | Good | |
| | Actual uses | Hiking, tourism | |
| | Infrastructure | Road, path, bridge, parking, recreational area | |
| | Impacts | Moderate: road construction, path improvement, visits | |
| | Legal situation | Natural Park. Compatible Use Zone | |
| Accessibility | Walking distance | Parking at the spring itself | |
| | Slope | – | |
| | Security | High | |

**Table 4.21** Chorrón fluvial sink geomorphosite

| Geomorphosite identification sheet | | | |
|---|---|---|---|
| Identification | **Name: Chorrón fluvial sink** | **Site: Barranco del Chorrón** | **N° 14**<br>Altitude: 1036 m |
| Situation | Municipality: | Coordinates: | |
| | Santa María de las Hoyas | X: 490041.15; Y: 4626340.59 | |
| Geomorphological values | Type | Element | |
| | | Representative | |
| | Genesis | Karstic | |
| | Morphology: description, morphostructures, erosion | Fluvial sink at the contact between the Albense sand strata (impermeable) and the limestones, with moderate dip toward the karstic system. It forms a horizontal and elongated mouth (following the stratigraphy) through which the water current enters | |
| | Dynamics | Karstic | |
| | Chronology | Pleistocene, Holocene—active | |
| | Main interest | Karstic | |
| | Secondary interest | Fluvial, hydrological | |
| | Geomorphosite attribution | Karstic-fluvial | |
| Additional values | Landscape | Medium (forest) | |
| | Cultural elements | Archeological site | |
| | Economic | – | |
| | Social | – | |
| | State of conservation | Good | |
| | Actual uses | Hiking | |
| | Infrastructure | Trail and path | |
| | Impacts | Medium visitor frequency | |
| | Legal situation | Natural Park. Limited Use Zone. Reserve Zone | |
| Accessibility | Walking distance | 1 km from Siete Ojos bridge | |
| | Slope | – | |
| | Security | High, path | |

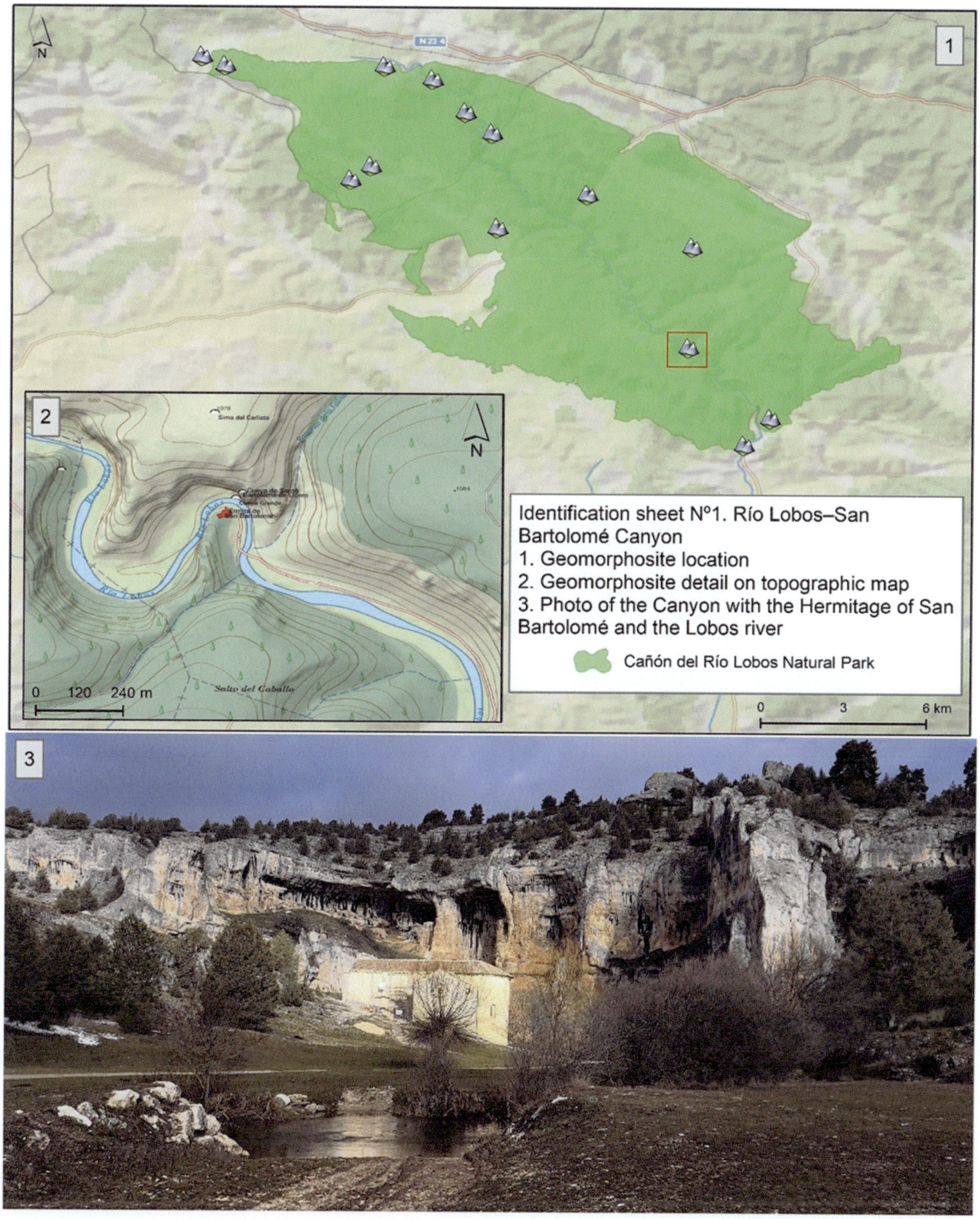


**Fig. 4.24** Río Lobos–San Bartolomé canyon identification sheet

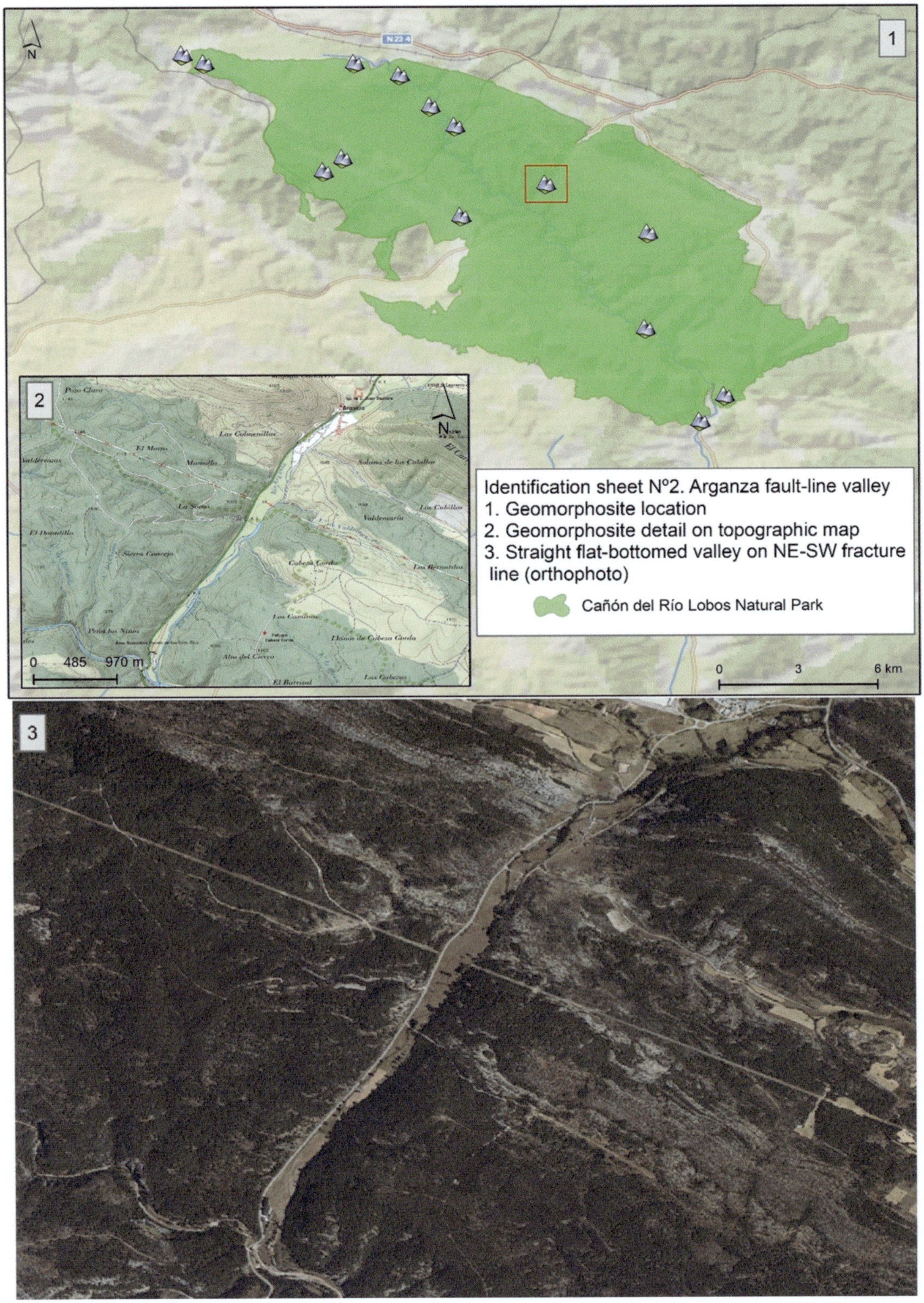


**Fig. 4.25** Arganza fault-line valley identification sheet

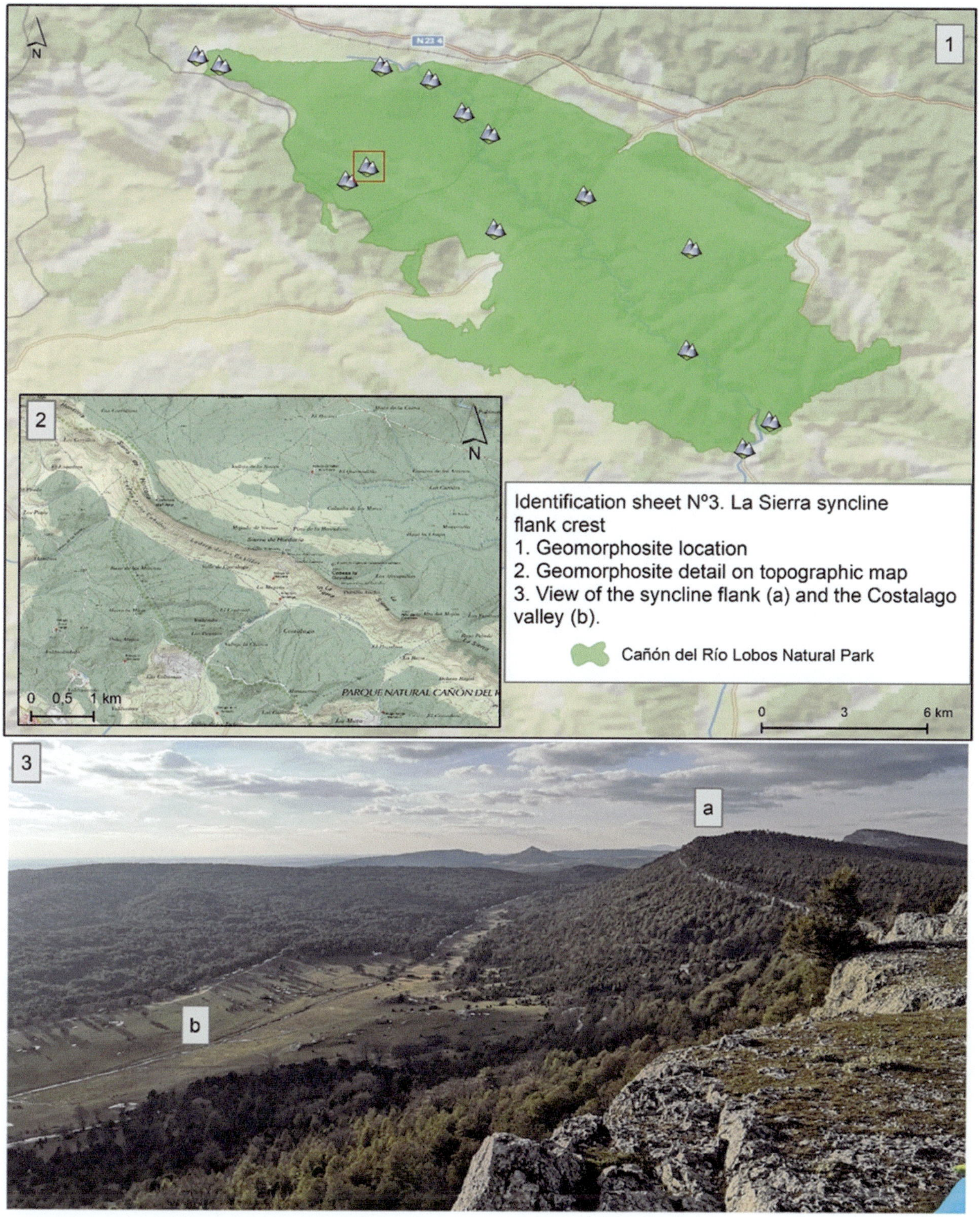


**Fig. 4.26** La Sierra syncline flank crest identification sheet

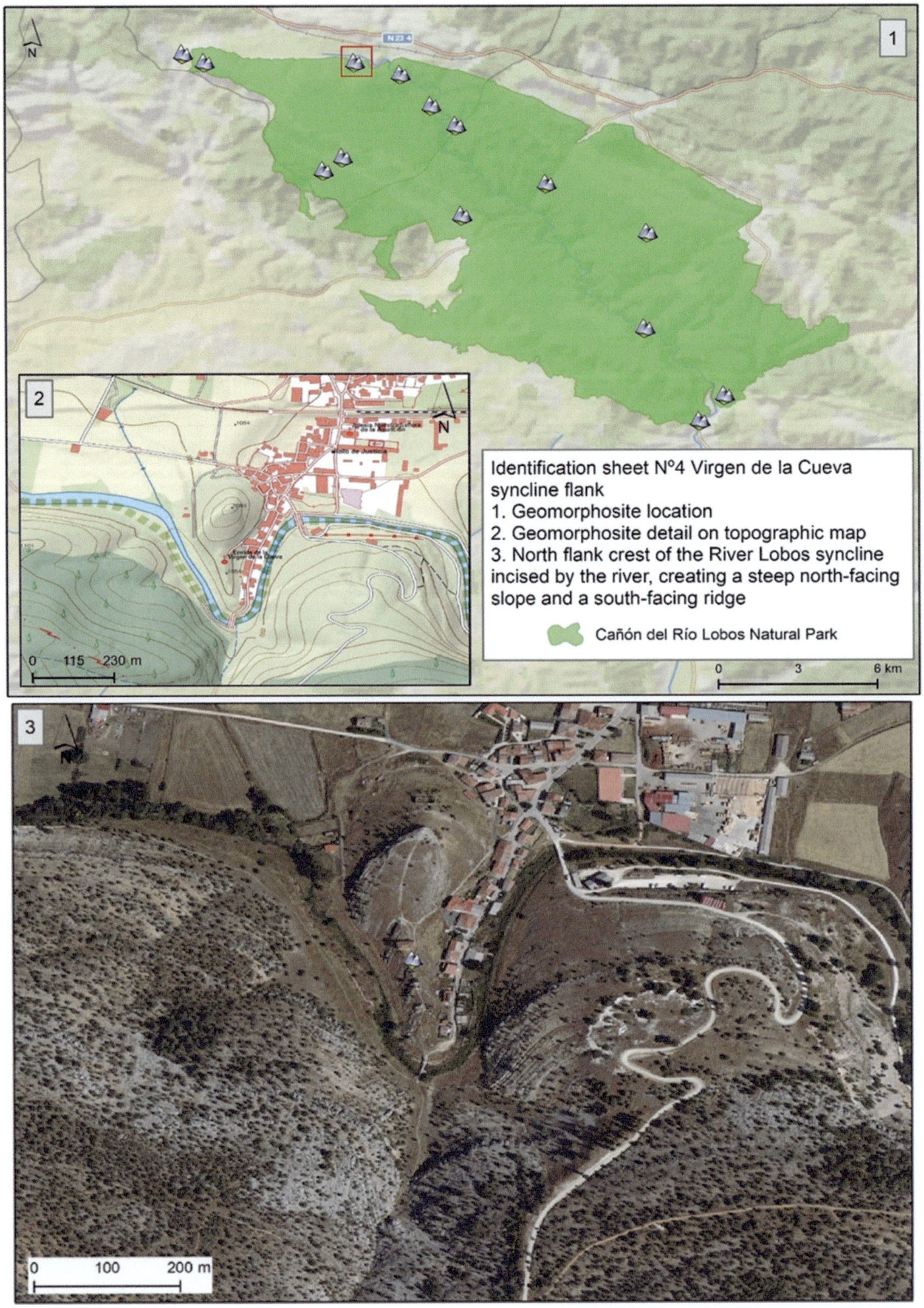


**Fig. 4.27** Virgen de la Cueva syncline flank identification sheet

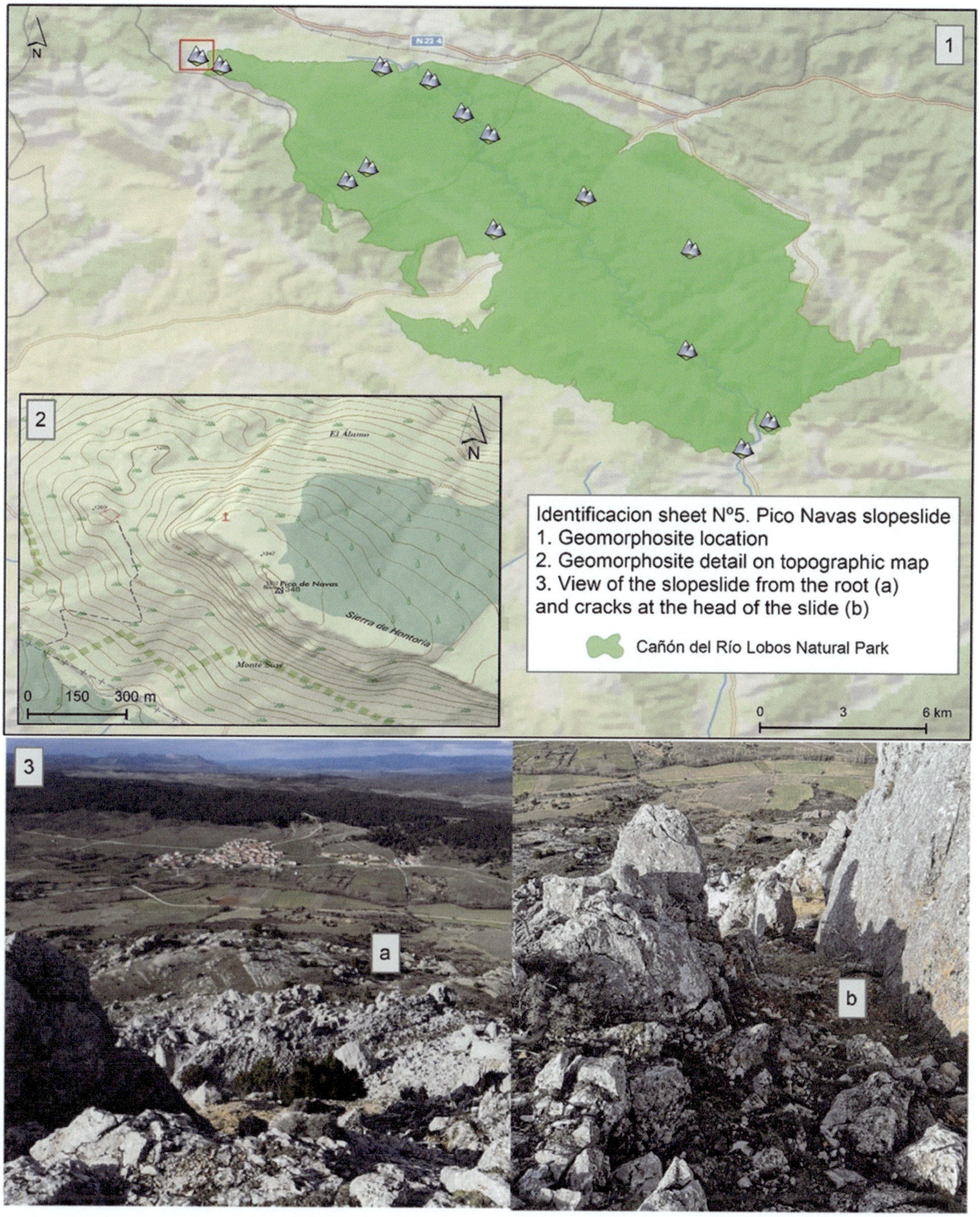


**Fig. 4.28** Pico Navas slopeslide identification sheet

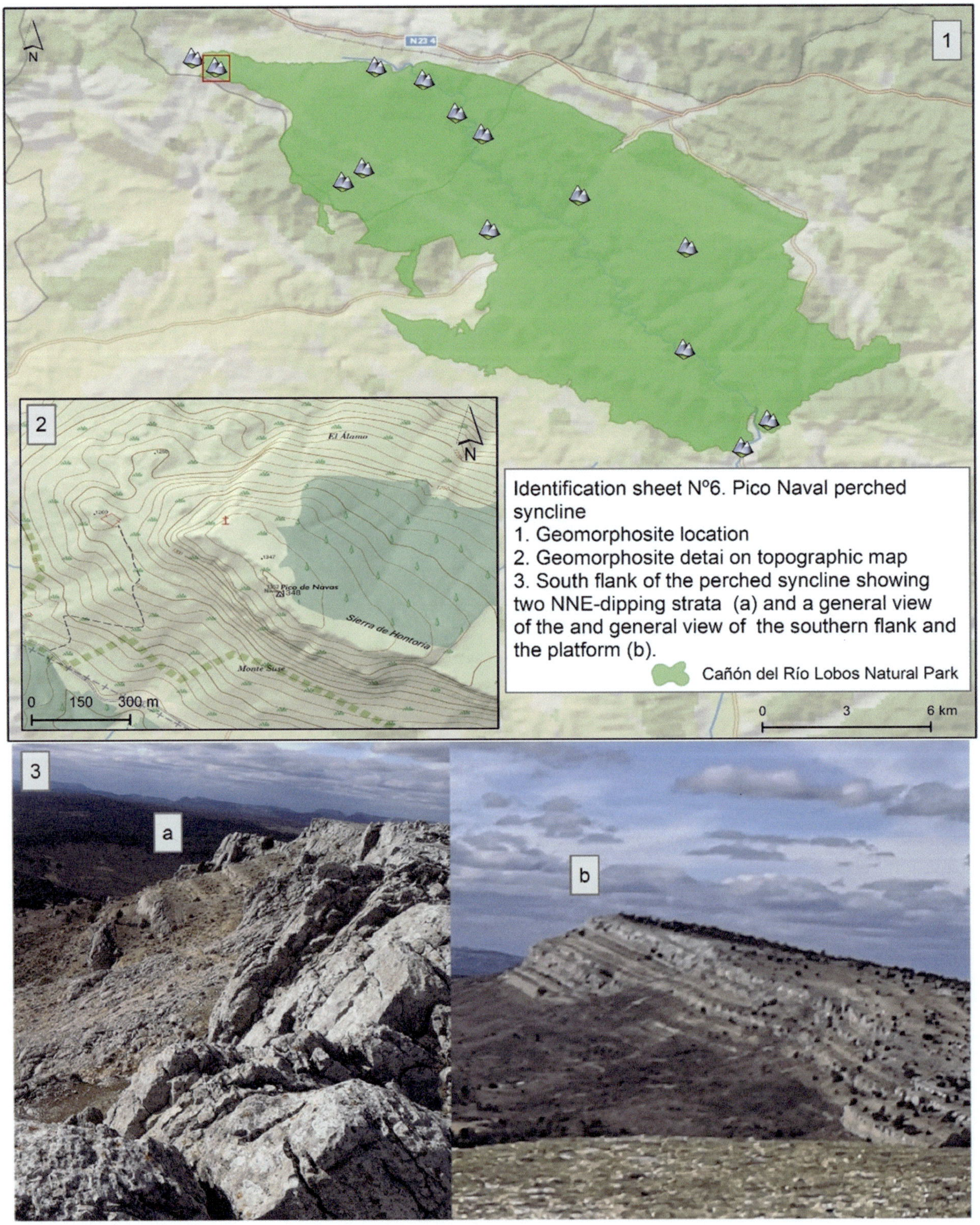


**Fig. 4.29** Pico Navas perched syncline identification sheet

**Fig. 4.30** Las Raideras river sink identification sheet

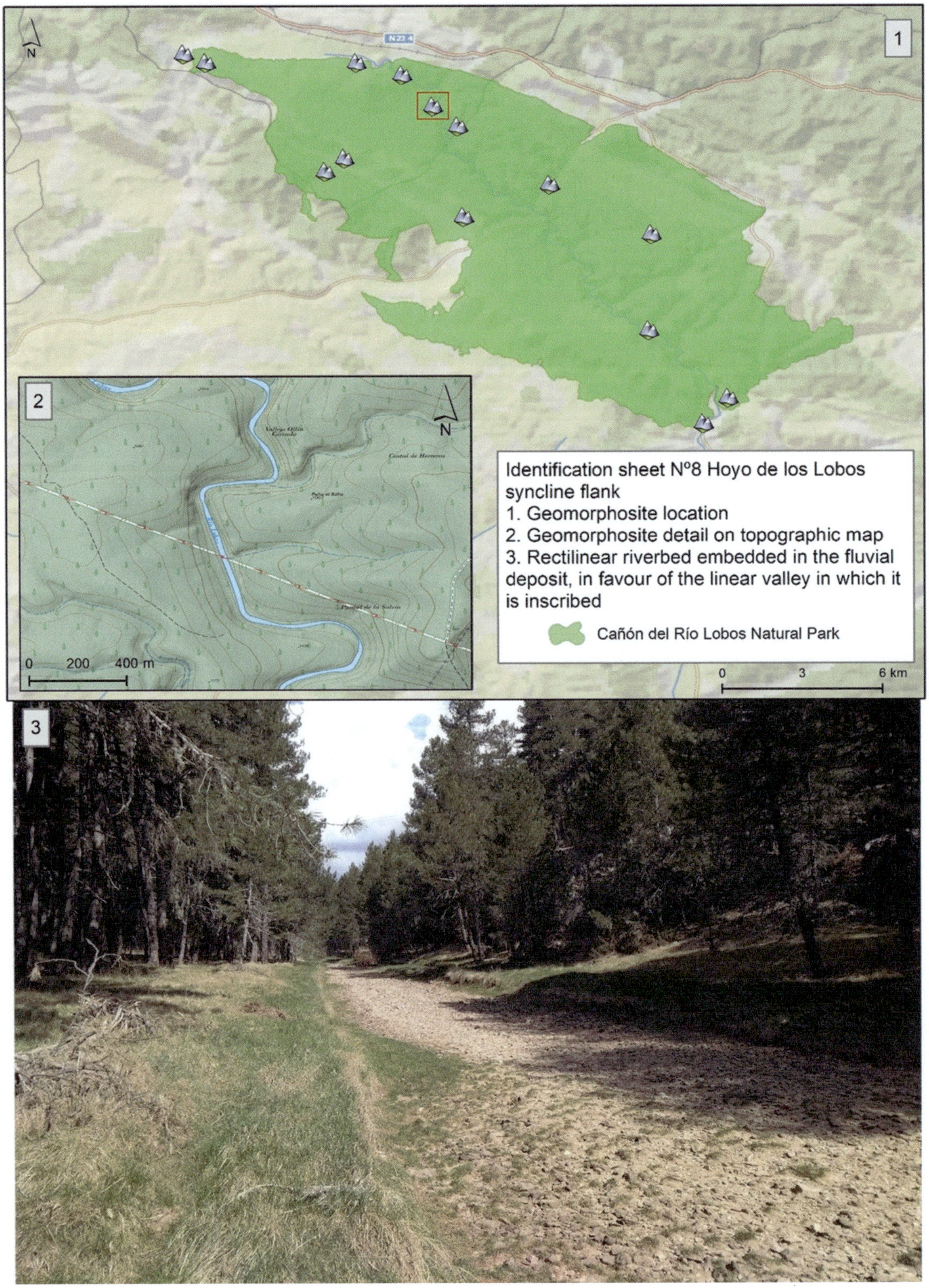


**Fig. 4.31** Hoyo de los Lobos syncline flank identification sheet

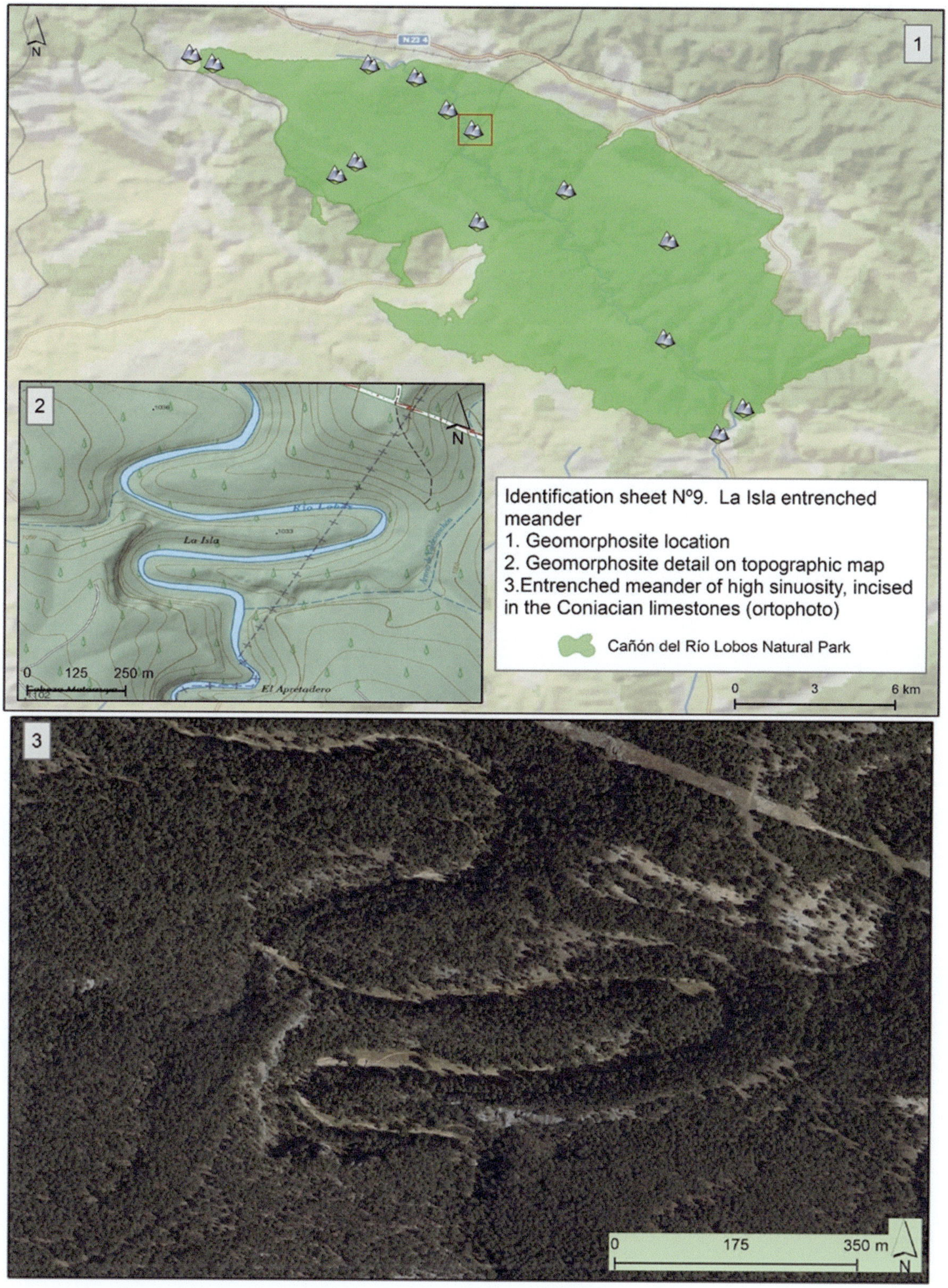


**Fig. 4.32** La Isla entrenched meander identification sheet

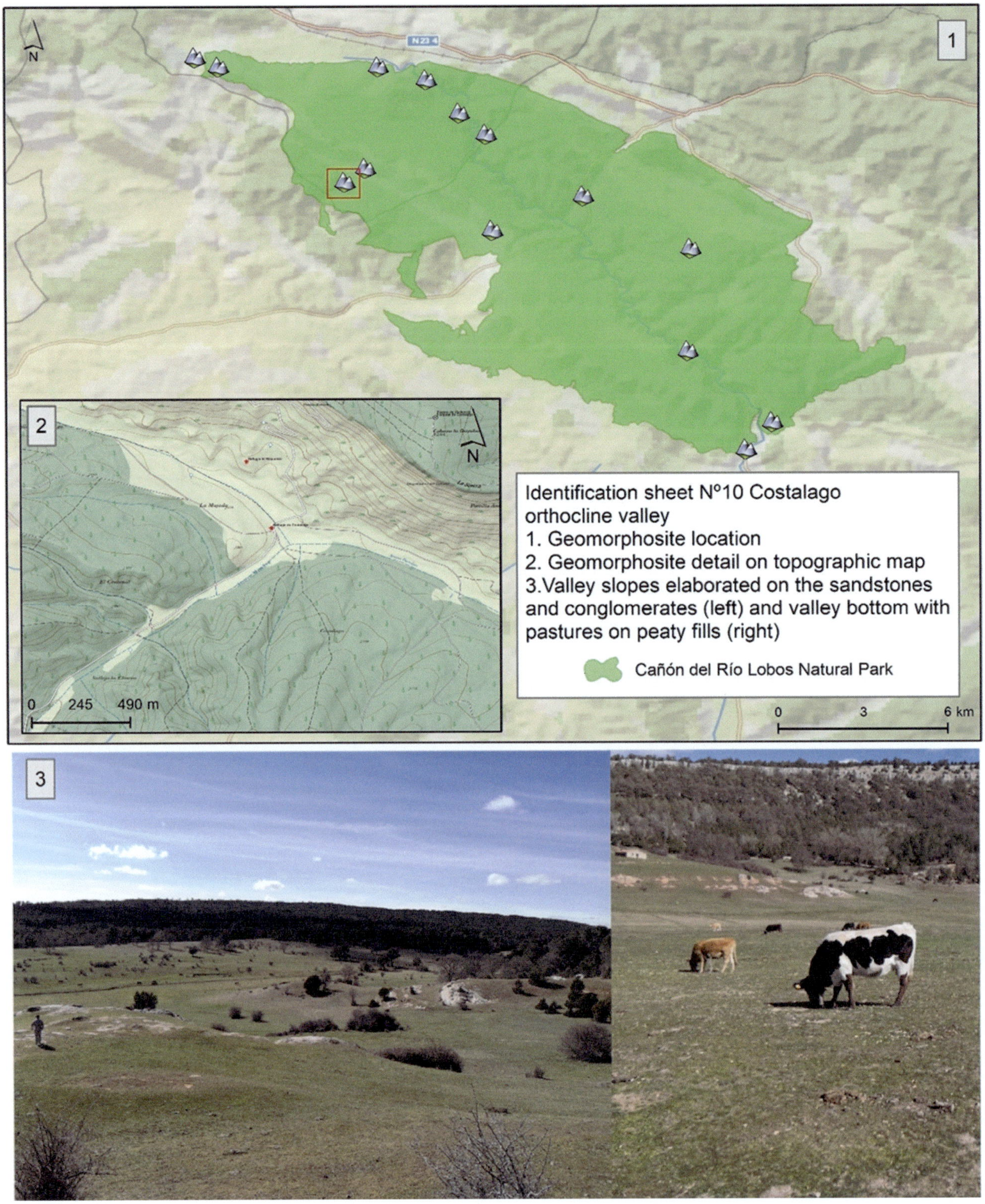


**Fig. 4.33** Costalago orthocline valley identification sheet

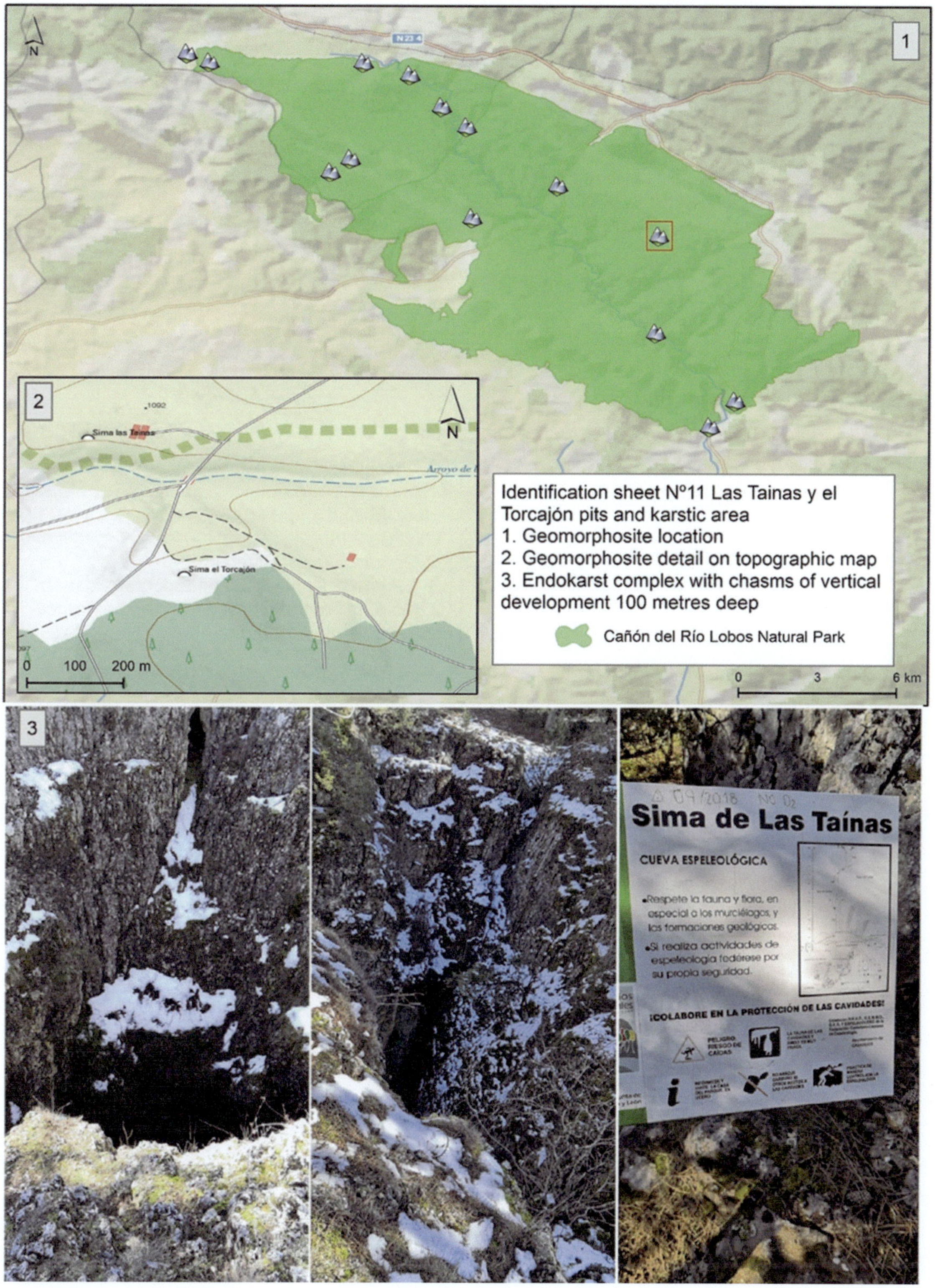


**Fig. 4.34** Las Tainas and el Torcajón pits and karstic area identification sheet

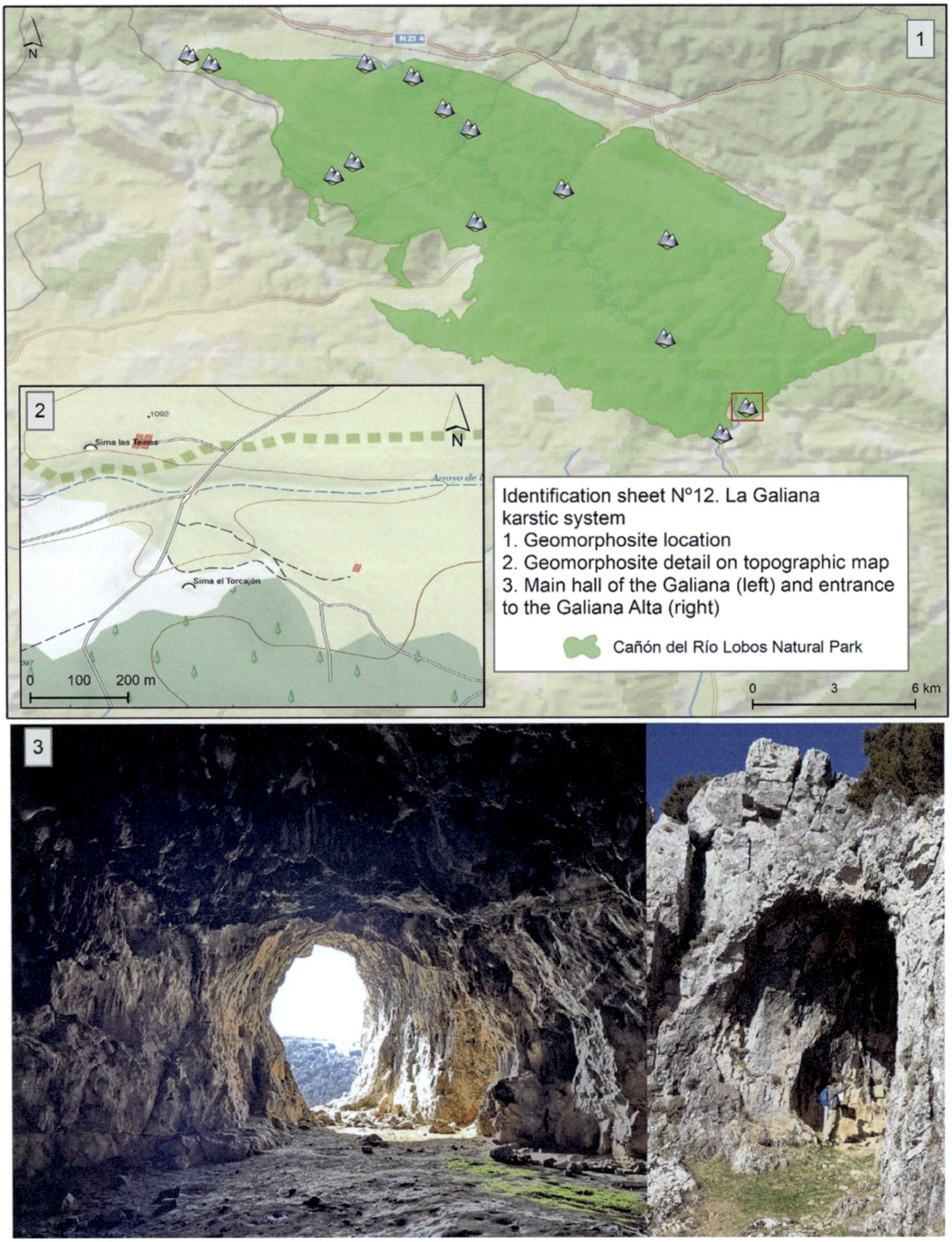


Fig. 4.35 La Galiana karstic system identification sheet

Fig. 4.36 Ucero springs identification sheet

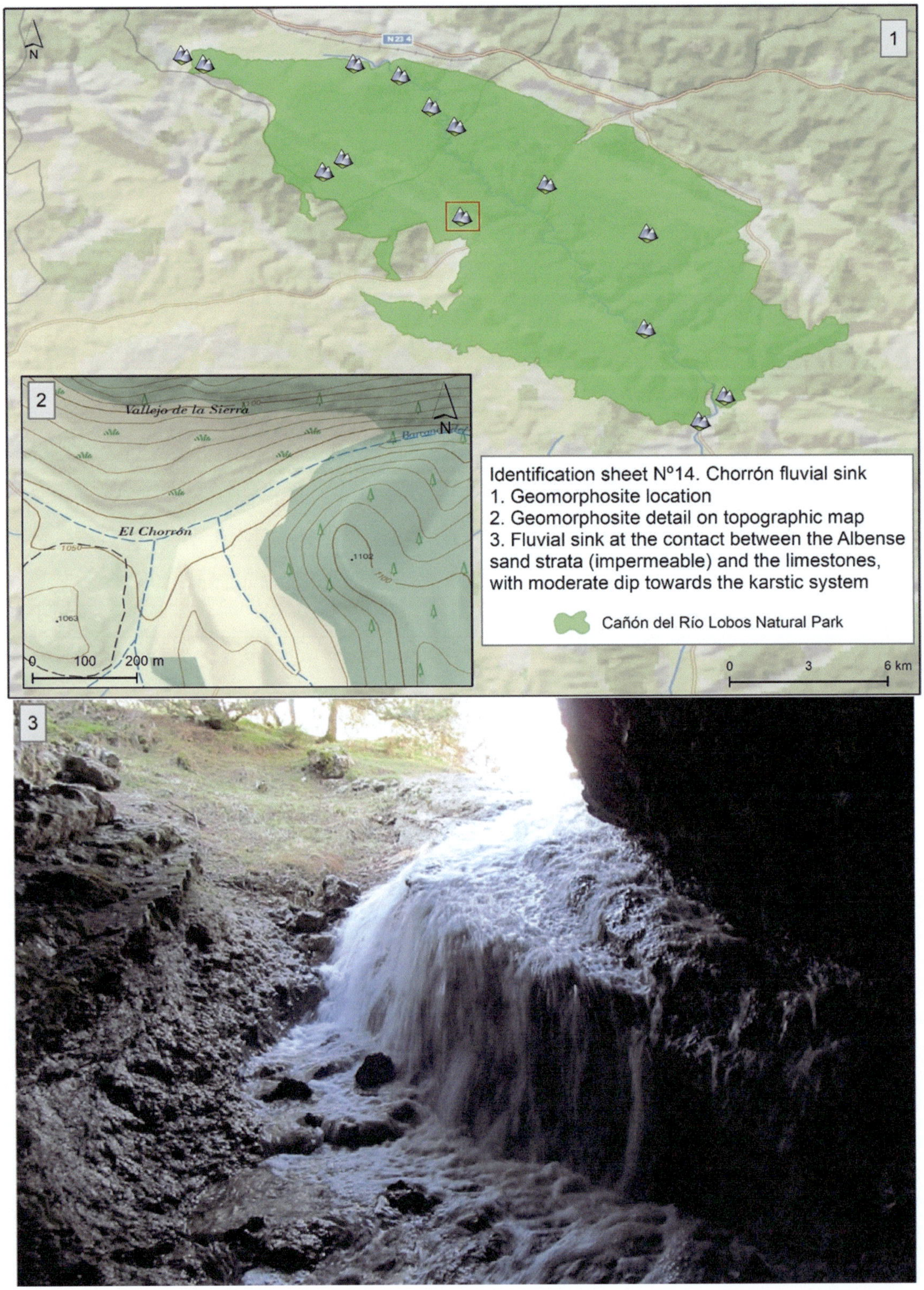


**Fig. 4.37** Chorrón fluvial sink identification sheet

# 5 Results

Once the geomorphosites of the two Natural Protected Areas (NPAs) under study have been inventoried, the methods of geomorphosite potential use assessment and touristic and didactic potential have been applied, allowing for a comparison between the geomorphosites of the two study areas and the application of the most appropriate management strategies in each of them based on their potential and needs.

## 5.1 Geomorphosites Potential Use Assessment

To determine the scientific and additional value of the inventoried geomorphosites, the first step is to apply the assessment method that will analyze their conditioning factors of use, geology, and geomorphology, and the additional values (see Chap. 2). Thus, results have been obtained for each geomorphosite and each of the three groups in both study areas, the summary and comparison of which can be seen in Table 5.1.

All Geomorphosites in the Sierra de la Paramera have a medium global value, with the lowest values in the Additional values and the highest in the **Conditioning factors for its use** (Table 5.2(1)). The highest values correspond to the geomorphosites located within the scope of the Red Natura area, as they have medium environmental protection, and most of them have a very high visibility value, since, as described, the granite, fluvial, and glacial landforms are perfectly recognizable in the field. Accessibility is a negative factor, as in most geomorphosites the path is rocky and unmarked, and the intensity of physical activity is medium in some of them, with marked slopes and long distances. The geomorphosites with the highest value in this section are N° 8 Adaja river spring and N° 11 Puerto de Mengamuñoz fault, characterized by high accessibility as both have current uses, the first for hiking and livestock and the Mengamuñoz fault as a mountain pass. They are followed by geomorphosites N° 2, 5, 7, and 10 with high value (Manqueospese Granite castle, Responsos Granite isolated rock, Serrota glacial cirques and Gavilanes and Chinitas granite system), all of them with low fragility and medium accessibility. All geomorphosites except the Serrota glacial cirques, where snow hinders visits in winter, are visitable all year round. Regarding the values of **Physical elements** (Table 5.2(2)), all geomorphosites have a medium value, except for N° 9 and 11, of structural attribution Garganta Honda fault-line valley and Puerto de Mengamuñoz fault—for not having examples of erosion, transportation, and sedimentation processes or weathering; nor active processes, biodiversity, or hydrological examples. On the contrary, the highest score is for geomorphosites N° 12, Picuezo river giant's kettle, as in addition to presenting several landforms, it is an excellent example of weathering, erosion, transportation, sedimentation, as well as presenting an active process and an excellent hydrological example with the Picuezo river. They are followed, with few points of difference, by four geomorphosites: N° 1, 2, 4, and 10, all of granite attribution and, despite the spectacularity of their forms, do not have hydrological examples or active examples. Common to the 12 geomorphosites is the scarcity of biodiversity, as it is an area affected by a forest fire and where it is difficult to visualize any faunal species.

Finally, in the Paramera and Serrota, the **Additional values** are high in two geomorphosites (N° 1 and 2), low in three geomorphosites (N° 4, 9, and 12), and the rest are



R. M. Ruiz Pedrosa, *Geomorphological Heritage as Didactic and Geotouristic Resources*, Geoheritage, Geoparks and Geotourism, https://doi.org/10.1007/978-3-032-06205-5_5

**Table 5.1** Summary of geomorphosites assessment in both NPA

| Sierras de la Paramera and la Serrota | | | | | Cañón del Río Lobos | | | | |
|---|---|---|---|---|---|---|---|---|---|
| N° | A | B | C | Global | *N°* | A | B | C | Global |
| 1 | M | M | H | **Medium** | *1* | H | H | H | **High** |
| 2 | H | M | H | **Medium** | *2* | H | M | M | **Medium** |
| 3 | M | M | M | **Medium** | *3* | H | H | H | **High** |
| 4 | M | M | L | **Medium** | *4* | H | M | H | **High** |
| 5 | H | M | M | **Medium** | *5* | H | M | M | **High** |
| 6 | M | M | M | **Medium** | *6* | H | H | H | **High** |
| 7 | H | M | M | **Medium** | *7* | H | H | L | **High** |
| 8 | H | M | M | **Medium** | *8* | H | H | M | **High** |
| 9 | M | L | L | **Medium** | *9* | H | M | M | **Medium** |
| 10 | H | M | M | **Medium** | *10* | H | H | M | **High** |
| 11 | H | L | M | **Medium** | *11* | H | M | M | **Medium** |
| 12 | M | M | L | **Medium** | *12* | H | M | H | **High** |
| | | | | | *13* | H | M | M | **Medium** |
| | | | | | *14* | H | M | M | **Medium** |

A: Conditioning factors for its use, B: Physical elements, C: Additional values. The numbers of the geomorphosites are corresponding with those previously described. Letters mean: L, low; M, medium; H, high

mediums (Table 5.2(3)). Except for those geomorphosites with livestock use, no geomorphosites have economic value. The same happens with cultural elements, with high value only in the Ulaca Semihorst and Manquospese Granite castle, which are the two geomorphosites with high value. Landscape values are generally mediums or highs, and anthropogenic influence is high in those geomorphosites with cultural elements and economic activity.

The assessment in the Cañón del Río Lobos is quite different, with nine geomorphosites having high global value and five with medium value. All 14 geomorphosites have high value, with very similar scores (between 22 and 26 out of 28) in terms of the **Conditioning factors for its use** (Table 5.3(1)), as a result of high accessibility, with good trails and short distances and low slopes; a high level of protection, as the geomorphosites are located in a Natural Park; all geomorphosites have sporadic or frequent use; visibility is high in most of them; and they are visitable all year round. In terms of **Physical elements** (Table 5.3(2)), six geomorphosites have high value and the rest have medium value. The geomorphosites with high value are N° 1, 3, 6, 7, 8, and 10—Río Lobos–San Bartolomé Canyon, La Sierra syncline flank crest, Pico Navas perched syncline, Las Raideras river sink, Hoyo de los Lobos syncline flank and Costalago orthocline valley—attributed to karstic, fluviokarstic, or structural processes. The Río Lobos–San Bartolomé Canyon (N° 1) stands out above the others, as the best example in the Park of a variety of fluvial and karstic landforms, erosion processes, transportation and sedimentation with the Lobos river, weathering of limestone rocks, as well as a high value of biodiversity and hydrology. The lowest values are for geomorphosites N° 4 and 11—Virgen de la Cueva syncline flank and Las Tainas and el Torcajón pits and karstic area—due to lower geological variety, scarce or no biodiversity, and lack of active processes. The rest of the geomorphosites, with medium value, are characterized by structural attribution, with fewer landforms, few or no examples of weathering, and medium biodiversity value.

The **Additional values** of the geomorphosites in the Cañón del Río Lobos are high in five geomorphosites, medium in eight, and low in one (Table 5.3(3)). The low value is for N° 7, Las Raideras river sink, with a medium landscape value and little anthropic influence, and without any cultural elements or economic value. The geomorphosites with high additional value are characterized by the presence of cultural elements, once again highlighting the Cañón in San Bartolomé, with the maximum landscape, cultural, anthropic, and economic value, followed by La Sierra syncline, Virgen de la Cueva syncline, and the Pico Navas syncline, all with high landscape value and medium cultural, anthropic, or economic values.

We have seen that all the inventoried geomorphosites in both NPAs have a medium or high value in terms of conditioning factors for its use, geomorphological value, and additional values. This information is valuable for understanding the potential generic use of geomorphosites, and for identifying strengths and weaknesses and proposing management actions. However, it falls short for a didactic and/or geotouristic application, for which it is necessary to consider specific parameters, as will be done in the following sections.

**Table 5.2** Paramera–Serrota geomorphosites potential use assessment

| Parameters | Criteria | 1 | 2 | 3 | 4 | 5 | 6 | 7 | 8 | 9 | 10 | 11 | 12 |
|---|---|---|---|---|---|---|---|---|---|---|---|---|---|
| 1. Conditioning factors for its use | Accessibility | 2 | 2 | 2 | 0 | 4 | 2 | 4 | 4 | 0 | 2 | 4 | 2 |
| | Fragility (risk of degradation) | 2 | 4 | 2 | 4 | 4 | 2 | 4 | 4 | 4 | 4 | 4 | 0 |
| | Seasonality | 4 | 4 | 4 | 2 | 4 | 2 | 2 | 4 | 4 | 4 | 4 | 4 |
| | Intensity of physical activity (distance and slope) | 2 | 4 | 4 | 2 | 4 | 2 | 2 | 4 | 4 | 4 | 4 | 4 |
| | Visibility of the geomorphosite | 4 | 4 | 4 | 4 | 4 | 4 | 4 | 4 | 2 | 4 | 2 | 4 |
| | Current uses | 2 | 0 | 0 | 0 | 2 | 2 | 2 | 2 | 0 | 2 | 2 | 2 |
| | Legislation | 2 | 2 | 2 | 4 | 2 | 4 | 4 | 4 | 4 | 2 | 4 | 2 |
| | **Total (up to 28)** | **18** | **20** | **18** | **16** | **24** | **18** | **22** | **26** | **18** | **22** | **24** | **18** |
| 2. Physical elements | Geology | 0 | 0 | 0 | 0 | 0 | 0 | 2 | 0 | 0 | 0 | 2 | 0 |
| | Rocks (igneous, sedimentary, and metamorphic) | 2 | 2 | 2 | 2 | 2 | 2 | 2 | 2 | 2 | 2 | 2 | 2 |
| | Superficial landforms | 4 | 4 | 2 | 2 | 2 | 4 | 4 | 2 | 2 | 4 | 2 | 2 |
| | Sedimentological cascade | 4 | 4 | 4 | 4 | 4 | 2 | 4 | 0 | 2 | 4 | 0 | 4 |
| | Weathering (chemical, physical, or biological) | 4 | 4 | 0 | 4 | 4 | 4 | 0 | 0 | 0 | 4 | 0 | 4 |
| | Active processes | 2 | 2 | 4 | 4 | 0 | 2 | 0 | 2 | 0 | 2 | 0 | 2 |
| | Biodiversity | 0 | 0 | 0 | 0 | 0 | 0 | 0 | 0 | 0 | 0 | 0 | 0 |
| | Hydrology | 0 | 0 | 4 | 0 | 0 | 0 | 2 | 4 | 2 | 0 | 0 | 4 |
| | **Total (up to 32)** | **16** | **16** | **16** | **16** | **12** | **14** | **14** | **10** | **8** | **16** | **6** | **18** |
| 3. Additional values | Landscape and aesthetic | 4 | 4 | 2 | 2 | 2 | 4 | 4 | 2 | 2 | 4 | 0 | 4 |
| | Cultural elements | 4 | 4 | 2 | 0 | 2 | 2 | 2 | 2 | 0 | 2 | 2 | 0 |
| | Anthropic influence | 4 | 4 | 4 | 2 | 4 | 0 | 0 | 4 | 0 | 0 | 4 | 0 |
| | Economic value | 0 | 0 | 0 | 0 | 0 | 0 | 2 | 0 | 2 | 0 | 0 | 0 |
| | **Total (up to 16)** | **12** | **12** | **8** | **4** | **8** | **6** | **8** | **8** | **4** | **8** | **6** | **4** |
| | **Global value (up to 76)** | **46** | **48** | **42** | **36** | **44** | **38** | **44** | **44** | **26** | **46** | **34** | **40** |

Geomorphosites: 1. Ulaca Semihorst, 2. Manqueospese Granite castle, 3. Picuezo river aluvial fan, 4. Granite rock thermal weathering, 5. Responsos Granite isolated rock, 6. Picos Zapateros granite mountain ranges, 7. Serrota glacial cirques, 8. Adaja river spring, 9. Garganta Honda fault-line valley 10. Gavilanes and Chinitas granite system, 11. Puerto de Mengamuñoz fault, 12. Picuezo river giant's kettle

**Table 5.3** Cañón del Río Lobos geomorphosites potential use assessment

| Parameters | Criteria | 1 | 2 | 3 | 4 | 5 | 6 | 7 | 8 | 9 | 10 | 11 | 12 | 13 | 14 |
|---|---|---|---|---|---|---|---|---|---|---|---|---|---|---|---|
| 1. Conditioning factors for its use | Accessibility | 4 | 4 | 4 | 4 | 2 | 2 | 4 | 4 | 4 | 4 | 4 | 4 | 4 | 4 |
| | Fragility (risk of degradation) | 0 | 2 | 4 | 4 | 4 | 4 | 4 | 4 | 4 | 4 | 4 | 0 | 2 | 4 |
| | Seasonality | 4 | 4 | 4 | 4 | 4 | 4 | 4 | 4 | 4 | 4 | 4 | 4 | 4 | 4 |
| | Intensity of physical activity (distance and slope) | 4 | 4 | 4 | 4 | 4 | 4 | 4 | 4 | 2 | 4 | 4 | 4 | 4 | 4 |
| | Visibility of the geomorphosite | 4 | 2 | 4 | 2 | 4 | 4 | 4 | 2 | 2 | 4 | 0 | 4 | 2 | 4 |
| | Current uses | 4 | 2 | 2 | 4 | 2 | 2 | 2 | 2 | 2 | 2 | 2 | 4 | 2 | 2 |
| | Legislation | 4 | 4 | 4 | 4 | 4 | 4 | 4 | 4 | 4 | 4 | 4 | 4 | 4 | 4 |
| | **Total (up to 28)** | **24** | **22** | **26** | **26** | **24** | **24** | **26** | **24** | **22** | **26** | **22** | **24** | **22** | **26** |
| 2. Physical elements | Geology | 2 | 4 | 4 | 2 | 2 | 2 | 2 | 0 | 0 | 2 | 0 | 0 | 2 | 2 |
| | Rocks (igneous, sedimentary, and metamorphic) | 2 | 2 | 4 | 2 | 2 | 2 | 2 | 2 | 2 | 2 | 2 | 2 | 2 | 2 |
| | Superficial landforms | 4 | 2 | 4 | 2 | 2 | 4 | 2 | 4 | 2 | 4 | 4 | 4 | 2 | 2 |
| | Sedimentological cascade | 4 | 4 | 2 | 4 | 4 | 4 | 2 | 4 | 4 | 4 | 2 | 2 | 4 | 4 |
| | Weathering (chemical, physical, or biological) | 4 | 0 | 4 | 0 | 2 | 4 | 2 | 0 | 0 | 0 | 4 | 4 | 0 | 0 |
| | Active processes | 2 | 0 | 4 | 0 | 4 | 4 | 4 | 4 | 2 | 4 | 0 | 2 | 0 | 0 |
| | Biodiversity | 4 | 2 | 2 | 0 | 2 | 2 | 2 | 2 | 2 | 2 | 2 | 2 | 2 | 2 |
| | Hydrology | 4 | 4 | 0 | 4 | 0 | 0 | 4 | 4 | 4 | 2 | 0 | 2 | 4 | 4 |
| | **Total (up to 32)** | **26** | **18** | **24** | **14** | **18** | **22** | **20** | **20** | **16** | **20** | **14** | **18** | **16** | **16** |
| 3. Additional values | Landscape and aesthetic | 4 | 2 | 4 | 4 | 4 | 4 | 2 | 2 | 2 | 4 | 0 | 4 | 2 | 2 |
| | Cultural elements | 4 | 2 | 2 | 4 | 2 | 4 | 0 | 2 | 0 | 0 | 2 | 2 | 0 | 2 |
| | Anthropic influence | 4 | 2 | 4 | 4 | 2 | 2 | 2 | 2 | 2 | 2 | 2 | 4 | 4 | 2 |
| | Economic value | 2 | 2 | 2 | 0 | 0 | 2 | 0 | 2 | 2 | 2 | 2 | 2 | 0 | 0 |
| | **Total (up to16)** | **14** | **8** | **12** | **12** | **8** | **12** | **4** | **8** | **6** | **8** | **6** | **12** | **6** | **6** |
| | **Global value (up to 76)** | **64** | **48** | **62** | **52** | **50** | **58** | **50** | **52** | **44** | **54** | **42** | **54** | **44** | **48** |

Geomorphosites: 1. Rio Lobos–San Bartolomé Canyon, 2. Arganza fault-line valley, 3. La Sierra syncline flank crest, 4. Virgen de la Cueva syncline flank, 5. Pico Navas slopeslide, 6. Pico Navas perched syncline, 7. Las Raideras river sink, 8. Hoyo de los Lobos syncline flank, 9. La Isla entrenched meander, 10. Costalago orthocline valley, 11. Las Tainas and el Torcajón pits and karstic area, 12. La Galiana karstic system, 13. Ucero spring, 14. Chorrón fluvial sink

## 5.2 Geomorphosites Didactic and Geotouristic Assessment

For the assessment of the didactic potential of geomorphosites, understood as their suitability to be incorporated into formal education as a didactic complement in the teaching of Earth Sciences, it is necessary to consult the curriculum of the country where they are located, as has been done in this book for the case of Spain (see Sect. 5.3.1.1). After consulting the contents of the subjects related to Earth Sciences in the Castilla y León curriculum, the results have been obtained for each geomorphosite in both study areas (Table 5.4).

The geomorphosites in both study areas share common characteristics in terms of their didactic potential. On the one hand, the main difference between geomorphosites with high and medium value is their suitability for primary education, as geomorphosites with high geomorphological specialization are not suitable, and the presence of cultural and hydrological elements is more important, adding didactic potential to the geomorphosites. Most geomorphosites have high value in secondary education, where specific contents are greater. And in higher education, all geomorphosites have high potential. Similarly, the relationship with other disciplines always presents the maximum value in all geomorphosites, as they are sites that, considered from a heritage and landscape point of view, are multidisciplinary in themselves. The natural and cultural values of geomorphological heritage bring together the disciplines of Geography, History, Biology, and Geology, and even Physical Education if the physical, recreational, and outdoor activity factor is considered.

In the case of Paramera and Serrota (Table 5.5), five geomorphosites have high didactic potential—N° 1, 3, 7, 8, and 12—and the rest have medium value. The highest score is for the Ulaca Semihorst, the Picuezo river alluvial fan, Serrota glacial cirques, Adaja river spring, and the Picuezo river giant's kettle, which have the highest cultural and hydrological value. All of them have medium or high value in secondary education, and high value in higher education, as well as in relation to other disciplines. The worst values are for the availability of educational material, as there is no type of sheet for students or guide for teachers on the geomorphology of Paramera and Serrota. There are no school trips to this NPA, not even from nearby schools. There are only sporadic visits by school groups to the Ulaca Hillfort, always focused on the cultural content of the archaeological site. And in the Serrota glacial cirques, there is some informative work that can serve as educational material for teachers.

The didactic potential is higher in the Cañón del Río Lobos (Table 5.6), mainly due to the presence of didactic materials for some of them and a greater suitability for primary education as there is a greater number of cultural elements. The highest score is, once again, for the Río Lobos–San Bartolomé Canyon, suitable for all educational levels and for which didactic material is available. The Cañón del Río Lobos carries out the School Visits Program to NPAs, with didactic materials available and the possibility of arranging outings with the guides from the visitor center, so all the geomorphosites have a high or medium value in this section. In addition to the didactic material offered by the visitor center, there are numerous studies on the Park, one of the most recent on geomorphology, didactics, and tourism published by the Pangea Research Group (Serrano et al., 2020), which can serve as a guide for teachers.

Regarding the geotourism potential, there is a significant difference between the two study areas. While in the Paramera area only one geomorphosite has high potential and more than half have low potential, in the Cañón del Río Lobos area there are five geomorphosites with high potential and only four with low potential (Table 5.7).

**Table 5.4** Summary of geomorphosites didactic potential in both NPAs

| Paramera–Serrota | Value (up to 24) | | Cañón del Río Lobos | Value (up to 24) | |
|---|---|---|---|---|---|
| 1. Ulaca Semihorst | 22 | High | 1. Río Lobos–San Bartolomé Canyon | 24 | High |
| 2. Manqueospese Granite castle | 12 | Medium | 2. Arganza fault-line valley | 16 | Medium |
| 3. Picuezo river aluvial fan | 18 | High | 3. La Sierra syncline flank crest | 16 | Medium |
| 4. Granite rock thermal weathering | 12 | Medium | 4. Virgen de la Cueva syncline flank | 16 | Medium |
| 5. Responsos Granite isolated rock | 12 | Medium | 5. Pico Navas slopeslide | 14 | Medium |
| 6. Picos Zapateros granite mountain ranges | 14 | Medium | 6. Pico Navas perched syncline | 16 | Medium |
| 7. Serrota glacial cirques | 20 | High | 7. Las Raideras river sink | 20 | High |
| 8. Adaja river spring | 20 | High | 8. Hoyo de los Lobos syncline flank | 16 | Medium |
| 9. Garganta Honda fault-line valley | 14 | Medium | 9. La Isla entrenched meander | 20 | High |
| 10. Gavilanes and Chinitas granite system | 16 | Medium | 10. Costalago orthocline valley | 22 | High |
| 11. Puerto de Mengamuñoz fault | 14 | Medium | 11. Las Tainas y el Torcajón pits and karstic area | 14 | Medium |
| 12. Picuezo river giant's kettle | 18 | High | 12. La Galiana karstic system | 18 | High |
| | | | 13. Ucero springs | 22 | High |
| | | | 14. Chorrón fluvial sink | 20 | High |

**Table 5.5** Paramera–Serrota geomorphosites didactic assessment

| Criteria | 1 | 2 | 3 | 4 | 5 | 6 | 7 | 8 | 9 | 10 | 11 | 12 |
|---|---|---|---|---|---|---|---|---|---|---|---|---|
| Primary education suitability | 4 | 0 | 2 | 0 | 0 | 2 | 2 | 4 | 0 | 2 | 0 | 2 |
| Secondary education suitability | 4 | 2 | 4 | 2 | 2 | 2 | 4 | 4 | 4 | 4 | 4 | 4 |
| Higher education suitability | 4 | 4 | 4 | 4 | 4 | 4 | 4 | 4 | 4 | 4 | 4 | 4 |
| Relation with other disciplines | 4 | 4 | 4 | 4 | 4 | 4 | 4 | 4 | 4 | 4 | 4 | 4 |
| Suitable material for formal education | 2 | 0 | 0 | 0 | 0 | 0 | 2 | 0 | 0 | 0 | 0 | 0 |
| Fieldwork techniques | 4 | 2 | 4 | 2 | 2 | 2 | 4 | 4 | 2 | 2 | 2 | 4 |
| **Total (up to 24)** | **22** | **12** | **18** | **12** | **12** | **14** | **20** | **20** | **14** | **16** | **14** | **18** |

Geomorphosites: 1. Ulaca Semihorst, 2. Manqueospese Granite castle, 3. Picuezo river aluvial fan, 4.Granite rock thermal weathering, 5. Responsos Granite isolated rock, 6. Picos Zapateros granite mountain ranges, 7. Serrota glacial cirques, 8. Adaja river spring, 9. Garganta Honda fault-line valley, 10. Gavilanes and Chinitas granite system, 11. Puerto de Mengamuñoz fault, 12. Picuezo river giant's kettle

**Table 5.6** Cañón del Río Lobos geomorphosites didactic potential assessment

| Criteria | 1 | 2 | 3 | 4 | 5 | 6 | 7 | 8 | 9 | 10 | 11 | 12 | 13 | 14 |
|---|---|---|---|---|---|---|---|---|---|---|---|---|---|---|
| Primary education suitability | 4 | 0 | 0 | 0 | 0 | 0 | 2 | 0 | 2 | 4 | 0 | 2 | 4 | 2 |
| Secondary education suitability | 4 | 4 | 2 | 2 | 2 | 2 | 4 | 4 | 4 | 4 | 2 | 4 | 4 | 4 |
| Higher education suitability | 4 | 4 | 4 | 4 | 4 | 4 | 4 | 4 | 4 | 4 | 4 | 4 | 4 | 4 |
| Relation with other disciplines | 4 | 4 | 4 | 4 | 4 | 4 | 4 | 4 | 4 | 4 | 4 | 4 | 4 | 4 |
| Suitable material for formal education | 4 | 2 | 2 | 2 | 2 | 2 | 2 | 2 | 2 | 2 | 2 | 2 | 2 | 2 |
| Fieldwork techniques | 4 | 2 | 4 | 4 | 2 | 4 | 4 | 2 | 4 | 4 | 2 | 2 | 4 | 4 |
| **Total (up to 24)** | **24** | **14** | **16** | **16** | **14** | **16** | **20** | **16** | **20** | **22** | **14** | **18** | **22** | **20** |

Geomorphosites: 1. Rio Lobos–San Bartolomé Canyon, 2. Arganza fault-line valley, 3. La Sierra syncline flank crest, 4. Virgen de la Cueva syncline flank, 5. Pico Navas slopeslide, 6. Pico Navas perched syncline, 7. Las Raideras river sink, 8. Hoyo de los Lobos syncline flank, 9. La Isla entrenched meander, 10. Costalago orthocline valley, 11. Las Tainas and el Torcajón pits and karstic area, 12. La Galiana karstic system, 13. Ucero spring, 14. Chorrón fluvial sink

**Table 5.7** Summary of geomorphosites geotourism potential in both NPAs

| Paramera–Serrota | Value (up to 100) | | Cañón Río Lobos | Value (up to 100) | |
|---|---|---|---|---|---|
| 1. Ulaca Semihorst | 80 | High | 1. Río Lobos–San Bartolomé Canyon | 95 | High |
| 2. Manqueospese Granite castle | 55 | Low | 2. Arganza fault-line valley | 60 | Low |
| 3. Picuezo river alluvial fan | 55 | Low | 3. La Sierra syncline flank crest | 75 | Medium |
| 4. Granite rock thermal weathering | 45 | Low | 4. Virgen de la Cueva syncline flank | 95 | High |
| 5. Responsos Granite isolated rock | 45 | Low | 5. Pico Navas slopeslide | 50 | Low |
| 6. Picos Zapateros granite mountain ranges | 60 | Low | 6. Pico Navas perched syncline | 90 | High |
| 7. Serrota glacial cirques | 70 | Medium | 7. Las Raideras river sink | 75 | Medium |
| 8. Adaja river spring | 65 | Low | 8. Hoyo de los Lobos syncline flank | 55 | Low |
| 9. Garganta Honda fault-line valley | 45 | Low | 9. La Isla entrenched meander | 75 | Medium |
| 10. Gavilanes and Chinitas granite system | 70 | Medium | 10. Costalago orthocline valley | 95 | High |
| 11. Puerto de Mengamuñoz fault | 60 | Low | 11. Las Tainas and el Torcajón pits and karstic area | 60 | Low |
| 12. Picuezo river giant's kettle | 60 | Low | 12. La Galiana karstic system | 75/85 | Medium–high |
| | | | 13. Ucero springs | 70 | Medium |
| | | | 14. Chorrón fluvial sink | 70 | Medium |

In the Paramera–Serrota geomorphosites, the worst values are found in the **Additional values**. Although observation conditions are high in all geomorphosites, which as it has been already mentioned are easily recognizable in the field, the scores for accessibility and security are average, given the lack of marked and tourist-friendly trails and paths. Additionally, in the study area, there is only one recreational area: the Molino del Quemado in the Picuezo river aluvial fan geomorphosite, although it was destroyed by the river floods after the fire. Not even the most visited site, which is the Ulaca Hillfort, with a relative local tourist importance, has a Visitor Center or a recreational area.

**Table 5.8** Paramera and la Serrota geomorphosites geotouristic assessment

| Geomorphosites | Scenic (up to 15) | Scientific (up to 10) | Cultural (up to 15) | Educational (up to 5) | Conservation (up to 15) | Added (up to 40) | Total (up to 100) |
|---|---|---|---|---|---|---|---|
| 1 | 15 | 10 | 15 | 5 | 10 | 25 | **80** |
| 2 | 5 | 5 | 10 | 0 | 10 | 25 | **55** |
| 3 | 5 | 5 | 5 | 5 | 10 | 25 | **55** |
| 4 | 5 | 5 | 0 | 0 | 10 | 25 | **45** |
| 5 | 0 | 5 | 5 | 0 | 10 | 25 | **45** |
| 6 | 15 | 10 | 0 | 0 | 10 | 25 | **60** |
| 7 | 15 | 10 | 5 | 5 | 10 | 25 | **70** |
| 8 | 10 | 5 | 5 | 5 | 10 | 30 | **65** |
| 9 | 5 | 10 | 0 | 0 | 10 | 20 | **45** |
| 10 | 15 | 10 | 10 | 0 | 10 | 25 | **70** |
| 11 | 10 | 10 | 0 | 0 | 10 | 30 | **60** |
| 12 | 15 | 10 | 0 | 5 | 10 | 20 | **60** |

Geomorphosites: 1. Ulaca Semihorst, 2. Manqueospese Granite castle, 3. Picuezo river alluvial fan, 4. Granite rock thermal weathering, 5. Responsos Granite isolated rock, 6. Picos Zapateros granite mountain ranges, 7. Serrota glacial cirques, 8. Adaja river spring, 9. Garganta Honda fault-line valley, 10. Gavilanes and Chinitas granite system, 11. Puerto de Mengamuñoz fault, 12. Picuezo river giant's kettle

**Table 5.9** Cañón del Río Lobos geomorphosites geotouristic assessment

| Geomorphosites | Scenic (up to 15) | Scientific (up to 10) | Cultural (up to 15) | Educational (up to 5) | Conservation (up to 15) | Added (up to 40) | Total (up to 100) |
|---|---|---|---|---|---|---|---|
| 1 | 15 | 10 | 15 | 5 | 10 | 40 | **95** |
| 2 | 5 | 10 | 5 | 0 | 15 | 25 | **60** |
| 3 | 15 | 10 | 10 | 5 | 15 | 20 | **75** |
| 4 | 15 | 15 | 5 | 5 | 15 | 40 | **95** |
| 5 | 10 | 5 | 5 | 0 | 15 | 15 | **50** |
| 6 | 15 | 15 | 10 | 5 | 15 | 30 | **90** |
| 7 | 5 | 10 | 10 | 0 | 15 | 35 | **75** |
| 8 | 5 | 10 | 5 | 0 | 15 | 20 | **55** |
| 9 | 15 | 10 | 5 | 0 | 15 | 30 | **75** |
| 10 | 15 | 10 | 10 | 5 | 15 | 40 | **95** |
| 11 | 10 | 10 | 5 | 5 | 15 | 15 | **60** |
| 12 | 15 | 10 | 10 | 5 | 5–15 | 30–40 | **75/85** |
| 13 | 5 | 5 | 5 | 5 | 10 | 40 | **70** |
| 14 | 10 | 5 | 10 | 5 | 15 | 25 | **70** |

Geomorphosites: 1. Rio Lobos–San Bartolomé Canyon, 2. Arganza fault-line valley, 3. La Sierra syncline flank crest, 4. Virgen de la Cueva syncline flank, 5. Pico Navas slopeslide, 6. Pico Navas perched syncline, 7. Las Raideras river sink, 8. Hoyo de los Lobos syncline flank, 9. La Isla entrenched meander, 10. Costalago orthocline valley, 11. Las Tainas and el Torcajón pits and karstic area, 12. La Galiana karstic system, 13. Ucero spring, 14. Chorrón fluvial sink

Nevertheless, Ulaca achieves the highest score, being the only geomorphosite in the Paramera–Serrota with high geotouristic potential (see Table 5.8) as it has the highest scenic, scientific, and cultural score. The educational value is low in most geomorphosites, understood here as interpretative potential and the existence of excursions, interpretation elements, or guided tours, with no examples of this in the Paramera and Serrota. The lowest geotouristic potential values are for those geomorphosites without cultural elements or interpretative aspects in which the scenic and/or scientific value is medium, as in geomorphosites N° 4, 5, and 9. In geomorphosites with medium value, such as the Serrota glacial cirques, Gavilanes and Chinitas and the Adaja river spring, an improvement in their additional values would be sufficient to increase their geotouristic potential to high.

Efficient tourism management and planning in the Cañón del Río Lobos make the geomorphosites geotouristic potential high in five of them, medium in five, and low in four (Table 5.9). The presence of recreational areas, good accessibility, and the existence of interpretation products in at least half of the geomorphosites, combined with high or medium scenic, scientific, and cultural values, make the Cañón del Río Lobos an ideal NPA for geotourism (for more details see Sect. 5.3.2.2). The Río Lobos–San Bartolomé Canyon, the Virgen de la Cueva syncline flank, and the Costalago orthocline valley score almost all the points, followed by the Pico Navas perched syncline and La Galiana karstic system. The lowest values are found in those geomorphosites with lower scenic value and worse accessibility, such as the Pico Navas slopeslide.

## 5.3 Dissemination and Management Proposals

The assessment results of the geomorphosites didactic and geotouristic potential, in addition to the intrinsic evaluation of these sites, provide us with information about the most appropriate actions in each site. While in the Sierras de la Paramera and La Serrota, there is no type of development, neither didactic nor geotouristic, the Cañón del Río Lobos Natural Park is a consolidated touristic destination since its declaration.

The **Sierras de la Paramera and La Serrota** do not have the necessary resources and infrastructure to design geotourism routes and receive visitors, but they do have an intrinsic didactic and heritage potential related to their geomorphological heritage. Therefore, the actions proposed for this NPA are focused on: the didactic application of its geomorphosites; the use of the geomorphological map for the dissemination of the abiotic heritage in the most unique space of the NPA, the Castro de Ulaca; citizen science, involving citizens in dissemination and heritage processes; and finally, the heritage and geomorphological map of La Serrota glacial cirques.

The **Cañón del Río Lobos Natural Park** suffers from overcrowding in very specific sites, while the rest of the Park and the inventoried geomorphosites do not receive the attention they equally deserve. The actions proposed for this Natural Park include diversifying the tourist offer through the design of georoutes and the management and planning of geotourism, with geomorphosites as tools to support territorial planning.

### 5.3.1 Didactics and Dissemination in Sierras de la Paramera and la Serrota

#### 5.3.1.1 Geomorphosites Didactic Potential: Study Case in Spain, Italy, and Switzerland

Geomorphosites didactic potential has been defined in this work as their suitability for teaching Earth Sciences disciplines within formal education. The potential for integration into the academic curriculum depends directly on the educational system of each country, which is structured around a set of recommendations, laws, and annual programming that teachers must adapt their classes to. National or regional laws in each country regulate the curricular contents, competencies, and learning objectives of each subject by academic year in compulsory and post-compulsory education levels, from primary education to preuniversity education, which may have different denominations in each country.

To analyze the integration of geomorphosites in the academic curriculum development, it is necessary to consult the contents and learning objectives that students must acquire by law in subjects related to Earth Sciences, such as Geology, Geography, Geomorphology, Natural Hazards, Hydrology, etc. Case studies have been conducted in Spain, Italy, and Switzerland.

The **Spanish education system** consists of general education and special education. General education (Table 5.10, “A. Spain”) includes early childhood education (0–5 years, non-compulsory), primary education (6–11 years, 6 years), compulsory secondary education (12–15 years, 4 years), and baccalaureate (16–17 years, 2 years). Vocational training and university education are also part of the general education system. Special regime teachings include artistic, linguistic, and sports education.

The education system is governed by the Organic Law of Education 3/2020 (LOMLOE) at the national level, except for university education, which is governed by the Organic Law of University Regime 2/2023 (LOSU). The system follows a decentralized state model, with educational competencies divided between the Government and the Autonomous Communities. The state law establishes the minimum requirements for each educational stage, while the Autonomous Communities develop their own curricula based on these guidelines. Thus, Spain has 17 different curricular designs, 1 for each Autonomous Community.

In Castilla y León, three laws regulate the curricular contents: Decree 38/2022 for Primary Education, Decree 39/2022 for Secondary Education, and Decree 40/2022 for Baccalaureate. These decrees specify the curriculum of each subject, indicating the competencies, evaluation criteria, and contents of each course.

The Italian education system, unlike the Spanish one, has a curriculum set at the national level. The State establishes general rules and essential educational levels through national guidelines, although each school and teacher can adapt their content based on the socioeconomic and territorial context of their educational center.

The **Italian education system** is organized into four main levels (Table 5.10, “B. Italy”): (1) the integrated system from 0 to 6 years (preschool); (2) the first cycle of education, which includes primary school (6–11 years, 5 grades) and lower secondary school (11–14 years, 3 grades); (3) the second cycle of education (14–19 years, 5 grades) consisting of upper secondary school and vocational training; and (4) higher education (+19 years), which includes university education, higher artistic, musical, and dance education, and technical institutes.

Decree N° 254 of 2012 establishes the national guidelines for the curriculum of early childhood education and the first cycle of instruction up to age 14. The curriculum areas are defined in these national guidelines, with Geography and Science included in the first cycle (primary and lower secondary education).

Specific curriculum contents for the second cycle of education, which includes upper secondary school (high school) and technical and vocational institutes, are detailed in Decree

**Table 5.10** Subjects related to Geomorphology in the Spanish, Italian, and Swiss educational systems

| Organization and legislation | | Subjects | |
|---|---|---|---|
| A. Spain | **Primary (6 academic years)** | Natural Sciences | 1st–5th year |
| | *Decree 38/2022, 29 September (Castilla y Leon)* | Social Sciences | 3rd–6th year |
| | **Secondary (4 academic years)** | Geography and History | 1st year |
| | *Decree 38/2022, 29 September (Castilla y Leon)* | Biology and Geology | 1st , 3rd, and 4th year |
| | **Baccalaureate (2 academic years)** | Geography | 2nd year |
| | *Decree 40/2022, 29 September ( Castilla y León)* | Geology and Earth Sciences | 2nd year |
| B. Italy | **Primary (5 academic years) + Secondary first grade (3 academic years)** | Geography | 1st–5th year (Primary) and 1st–3rd year (Secondary first grade) |
| | *Decree N° 254, 2012, National guidelines for primary and secondary first grade curriculum* | Sciences | 1st–5th year (Primary) and 3rd year (Secondary first grade) |
| | **Secondary second grade (5 academic years)** | Geography and History | 1st and 2nd year |
| | *Decree N° 211, 2010, National guidelines on the specific learning objectives related to the activities and teachings included in the curriculum plans for Secondary courses* | Natural Sciences (Earth Science) | 1st and 2nd year, 5th year in scientific itinerary |
| C. Switzerland | **Primary (6 academic years) + Secondary I (3 academic years)** | Geography | 1st , 2nd, and 3rd cycle |
| | *Plan d'étude romand (PER), May 27, 2010* | Natural Sciences | 1st , 2nd, and 3rd cycle |
| | **Secondary II (5 years academic)** | Geography | 1st–3rd year |
| | | | Baccalaureate schools: optional in 4th and 5th course |

N° 211 of 2010. The subjects to be studied each year and the number of hours is detailed in the annexes of each specialty. In the institutes, students aged 14–19 can choose from different educational pathways: artistic, classical, linguistic, musical, scientific, and humanities. History and Geography are studied in the first 2 years in all pathways, while Natural Sciences (Biology, Chemistry, and Earth Sciences) are usually studied in the first 2 years in most pathways and throughout the 5 years in others (Scientific and Humanities).

The **Swiss education system** (Table 5.10, "C. Switzerland") is structured into compulsory education, which consists of preschool, primary, and lower secondary education, and post-compulsory education with upper secondary level II, which is divided into vocational training and general education schools (ECG). The tertiary level is further divided into higher vocational training and higher schools or universities. As Switzerland is a federal state, education is the responsibility of the cantons, and there is no national law regulating the curriculum content of school subjects. The cantons and their municipalities are responsible for regulating and implementing compulsory education (primary and lower secondary). In upper secondary education, the responsibility is shared between the Confederation and the cantons. The cantons are responsible for the school curriculum, which means that there are 26 different education systems in Switzerland. On May 21, 2006, an inter-cantonal agreement for the harmonization of compulsory education (Harmos) was approved, with the aim of unifying compulsory education to allow greater student mobility. However, the Harmos agreement still does not establish the curriculum content that students must study at each stage. An example of harmonization is the curriculum of the French-speaking part of the country (Suisse romande), adopted on May 27, 2010 by the cantonal directors of Suisse romande (PER, Plan d'étude romand), which sets the pedagogical objectives for the seven French-speaking or bilingual cantons of western Switzerland and determines most of the didactic content of the 11 years of compulsory schooling.

Once compulsory education is completed, students can choose between Vocational Training Schools or General Education Schools. Within the latter, students can opt for General Culture Schools, which provide access to Higher Schools of professional practice, or for Baccalaureate Schools, whose objective is to prepare students for university studies. In both cases, geography appears as a discipline of study. Baccalaureate Schools are regulated by the Regulation of Gymnastic Studies of the Canton of Valais of June 10, 2009. However, this regulation does not establish the content that should be covered in the geography subject. General Culture Schools are regulated by the Regulation of General Culture Schools of August 25, 2021 of the Canton of Valais and, like the regulation of baccalaureate schools, does not establish the curriculum content of geography.

Analyzing **the learning contents in Castile and León**, Geography is introduced in secondary education, connected with History. In primary school, students learn about Geomorphology and Geology in the subjects of Natural Sciences and Social Sciences. The subject of Natural Sciences continues in secondary school with Biology, Geology, Physics, and Chemistry, but not with Geography. Natural Sciences focus on understanding the environment and promoting sustainable practices. Geography is part of

Social Sciences, which explore geographical, historical, economic, and sociological aspects. The goal of both subjects is for students to develop a sustainable relationship with their environment. In secondary education, Geography and History are combined into a single subject, alternating Geography and History classes each year.

The primary Natural Sciences curriculum includes the study of natural heritage, natural landscapes of Castilla and León, and geographical features. It also involves contact with the natural environment and outdoor activities, including excursions. In Social Sciences, students also study landscapes, physical elements, and geographical diversity of Europe. These topics continue in Secondary Education in Geography, specifically in the first year. In the Biology and Geology subject, students study a series of topics related to geomorphology and geology, including landscapes, landforms, rocks, minerals, and geological processes in the first, third, and fourth years. In the Baccalaureate, depending on the chosen educational itinerary, students take Geography and Geology and Earth Sciences subjects in the second year, which are based on the study of landforms and geomorphological analysis in Geography, and natural heritage and geodiversity in Geology and Earth Sciences (Table 5.11).

Regarding the Italian curriculum, in primary and first year of secondary school, Geography is presented as the cross-curricular subject par excellence, with the aim of teaching students about the relationships between human societies and the planet they inhabit, as well as the processes of transformation of the natural environment, whether natural or caused by human activity. Geography is necessary at this stage to develop the competence of active citizenship, which involves students approaching their closest territory. Through this approach and the study of the landscape, the goal is to raise awareness at this stage about the natural and cultural heritage inherited. The curriculum establishes a first approach to the territory through direct exploration. Field trips are therefore presented as a didactic necessity and obligation for the subject of Geography. In the Science subject, emphasis is placed on the processes of observation, research, and experimentation of the scientific method. This subject encourages students to ask questions, design experiments, and build models, and for research to be conducted outside the classroom, even in natural spaces.

The learning objectives of the first cycle of primary and secondary school for the subjects of Geography and Science focus on Geomorphology and Geology: studying the physi-

**Table 5.11** Content related to Geomorphology in the Castilla and León education system by levels and subjects

| | | |
|---|---|---|
| Primary | Natural Sciences | First year: Contact with nature through nearby natural spaces. Interest and enjoyment in outdoor activities. Care, respect, and empathy toward living beings and the environment they live in. Local natural heritage: use, enjoyment, care, and conservation |
| | | Second year: Natural landscapes of Castilla y León: characteristics and peculiarities |
| | | Third year: Natural heritage in Castilla y León. Protected Natural Areas of Castilla y León. Their use, care, and conservation. Most relevant geographical features. Basic classification of rocks |
| | | Fifth year: Basic classification of rocks and minerals. Basic geological processes of formation and shaping of landforms. Natural heritage as a good and resource, care, and conservation. Protected Natural Areas in Spain |
| | Social Science | Third year: Climate and landscape. Relationship between climatic zones and landscape diversity |
| | | Fourth year: Earth and natural disasters. Elements, movements, dynamics that occur in the universe and their relationship with certain physical phenomena that affect the Earth and impact daily life and the environment |
| | | Fifth year: The landscape: elements that form it and types of landscapes. Characteristics of the main landscapes of Castilla y León. The natural environment. The geographical diversity of Spain |
| | | Sixth year: The natural environment. The geographical diversity of Europe |
| Secondary | Geography and History | First year: Basic components of landforms. Processes of modification of the Earth's surface. Richness and value of natural heritage |
| | Biology and Geology | First year: Rocks and minerals. Classification of rocks: sedimentary, metamorphic, and igneous. Geodynamic and geochemical models. Earth movements |
| | | Third year: Internal and external geological agents. Modeling landforms. Characteristic reliefs of Castilla y León |
| | | Fourth year: Global effects of geosphere dynamics through plate tectonics. External and internal geological processes. Landforms and landscape: importance as resources and factors involved in their formation and modeling. Geological stratigraphic columns and geological histories reflecting the application of principles in the study of Earth's history |
| Baccalaureate | Geography | Second year: Spanish landforms. Peninsular and insular landforms units. Characteristics and location. Geomorphological factors. Physical factors and diversity of landscapes and ecosystems. Analysis of geomorphological, bioclimatic, edaphic, hydrological factors, and those related to human activity and prevention of associated risks |
| | Geology and Earth Sciences | Second year: Geological and environmental heritage of Castilla y León. Assessment of its importance and conservation of geodiversity. External geological processes (weathering, erosion, transportation, and sedimentation). Landforms. Influence of geological agents, climate, and properties and relative disposition of predominant rocks. Castilla y León landforms |

cal elements of the territory, landscapes, geological processes, geomorphological risks, and the importance of protecting natural heritage (Table 5.12). This continues in the second cycle of education, where all pathways involve learning about the relationships between the environment and the territory, describing morphological and climatic processes and conditions in Geography; in Natural Sciences, students learn about tectonics and landforms. Students in the Scientific and Humanistic pathways receive more specific training on mineralogy, petrology, orogeny, and global tectonics. Additionally, Decree N° 211 of 2010 establishes that for all pathways in the second cycle of secondary education, the contents of Earth Sciences (Natural Sciences) must be developed in coordination with the Geography subject, and teachers must adapt the contents according to the territorial context of the educational center.

In the Swiss education system, specifically in the Canton of Valais, the contents of Geomorphology for compulsory education are studied in the subjects of Geography, which is included with History in Human and Social Sciences, and Natural Sciences, which is included with Mathematics. In these cycles, students learn about natural and man-made elements, description of sites (hydrography, relief, vegetation, etc.) and natural hazards in the Geography subject (Table 5.13). In the Natural Sciences subject, students study natural phenomena such as volcanic activity and natural hazards. From Secondary II onward, there are no mandatory contents, so teachers have academic freedom. By consulting the websites of some educational institutions, we can find out, for example, the textbooks used, which are characterized by presenting a comprehensive content of Geography, addressing both concepts of human and physical geography

**Table 5.12** Content related to Geomorphology in the Italian education system by levels and subjects

| | | | |
|---|---|---|---|
| Primary and first grade secondary | Geography | Recognize and name the main physical geographical "objects" (rivers, mountains, plains, coasts, hills, lakes, seas, oceans, etc.). Recognize the characteristics of different European and global landscapes, comparing them in particular with Italian landscapes, significant physical elements, and historical, artistic, and architectural characteristics as natural and cultural heritage that must be protected and valued | |
| | Natural Sciences | Respect and appreciate the value of the social and natural environment. Recognize, through field research and concrete experiences, the main types of rocks and the geological processes that originated them. Understand the structure of the Earth and its internal movements (plate tectonics); identify the seismic, volcanic, and hydrogeological risks in your region | |
| Second grade secondary (Liceo) | Two first years | Geography and History | Understand the complex relationships between environmental conditions and socioeconomic and cultural characteristics. Describe transformation processes and morphological and climatic conditions. In the construction of didactic itineraries, the main themes to consider are landscape, environment, and society |
| | | Natural Sciences | Deepen the explanatory framework of Earth's movements and the geomorphological study of the structures that make up the Earth's surface (rivers, lakes, glaciers, seas, etc.) |
| | Scientific Liceo | Natural Science | References to mineralogy, petrology (rocks), and phenomena such as volcanism, seismicity, and orogeny are introduced, examining the transformations related to them. Complex meteorological phenomena and models of global tectonics are studied |

**Table 5.13** Content related to Geomorphology in Valais canton, Switzerland, education system in compulsory education by subjects

| | | |
|---|---|---|
| Compulsory education | Geography | First cycle: Categorization of elements, natural and man-made, in the environment (watercourses, vegetation, communication routes, buildings) |
| | | Second cycle: Description of the Site and reasons that could explain its location: natural reasons. Identification of certain characteristics of the Site (hydrography, relief, landscape, insolation). Appropriation and use of vocabulary and specific notions related to space and Geography: Zone, network, boundaries, borders, relief, hydrography, vegetation, etc. |
| | | Third cycle: Natural risks, particularly those related to the terrestrial Medium (earthquakes, volcanism, etc.). Climate change, including natural hazards related to atmospheric phenomena (floods, landslides, avalanches, etc.). Study of a resource: water and its problems (distribution and access, marine resources, etc.) |
| | Natural Sciences | First cycle: Observation of natural phenomena related to water (clouds, rain, snow, hail, floods, wind, sun, etc.) and experimentation with the transformation of liquid–solid–liquid water (snow, ice). Exploration of natural phenomena related to the present (storm, flood, drought, avalanche, bird migration, volcanic eruptions, etc.) and formulation of hypotheses to explain them |
| | | First cycle: Observation of natural phenomena related to water (clouds, rain, snow, hail, floods, wind, sun, etc.) and experimentation with the transformation of liquid–solid–liquid water (snow, ice). Exploration of natural phenomena related to the present (storm, flood, drought, avalanche, bird migration, volcanic eruptions, etc.) and formulation of hypotheses to explain them |

and topics that explain the relationships and interactions between the Earth and its population, although most teachers use their own class notes.

#### 5.3.1.2 Cultural Geomorphology: Geomorphological Map Applied to the Relationship Between Natural and Cultural Heritage in the Ulaca Hillfort

An integrative vision indicates that geomorphology is one of the components of the cultural heritage of a territory, and at the same time allows considering the relationships between cultural components and their context, expressed as "cultural Geomorphology" (Panizza & Piacente, 2003). In this approach, it is particularly important to determine the physical framework in which the cultural heritage is inscribed to fully understand the analyzed site, and geomorphological mapping is a basic tool for the inventory and assessment of geomorphological elements or geomorphosites. In most methodologies for the analysis and assessment of Geomorphosites, and particularly in the one developed by our group (Serrano & González-Trueba, 2005; Serrano et al., 2020), it is the first step to study the geomorphological heritage of any territory.

The Cerro del Castillo, where the site of the Castro de Ulaca is located, is situated on a smaller and intermediate stepped block, located between the main tectonic block and the valley (Fig. 5.1). The upper part of the semicorpus where the Castro de la Ulaca is located has a level physiography. It is the remains of the fundamental eroded surface (Moreno, 2008) before the tectonic uplift. This erosion of the materials involved the partial export of the previously generated alterites to the Alpine tectonics, which have their correlative deposits in the sandstones of the southern portion of the Duero Basin and the Amblés Valley, and the flattening of the granite outcrops. During the tectonic uplift, weathering and mechanical erosion on the high surface of the stepped block generated a circular depression with drainage toward the NE, indicating a possible inclination of the block toward the N-NE. In the culminating central depression, there are paleodrainage channels toward the NE, adapted to fracturing, and it is occupied by small sandy-filled alveoli where springs emerge.

In the depths of the Amblés valley, tertiary materials—Oligocene and Miocene arkoses—emerge, but the stepped block of Cerro del Castillo is formed only by granites. The granite has been extensively weathered throughout the Central System through hydrolysis processes that break down the rock into clays, evacuated by erosion, leaving mantles of alterites formed by grains of quartz and feldspars (Molina et al., 1987). These alterites cover large areas called "grus" and saprolites (Migón & Thomas, 2002; Sequeira et al., 2002) among which the granitic substrate emerges.

The granite landforms of Ulaca can be classified according to Migoń (2021), as a dissected plateau with representative granitic relief forms of medium and smaller size. Ulaca constitutes a characteristic granite landscape with important landforms and processes that intervene in the landscape configuration. The most characteristic forms of Cerro del Castillo are: nubbins, followed by tors, domes, half-domes, and pedestal rocks, among the larger ones; and gnammas, tafonis, and polygonal cracking among the microforms, the latter not represented on the map (for more detail, see Ruiz-Pedrosa & Serrano, 2023). In Ulaca, the granite landforms are distributed according to their location, and this has conditioned the uses and modes of exploitation by the Vettones (Fig. 5.2).

In Cerro del Castillo, there is a clear morphological difference between the more flattened summit portions and the slopes, where the alternation of curved and vertical fractures and the greater energy of the relief condition different landforms (Table 5.14).

The geomorphological map permits to do a landform inventory to know the distribution of the main landforms of the Castro de Ulaca and to establish the spatial relationship

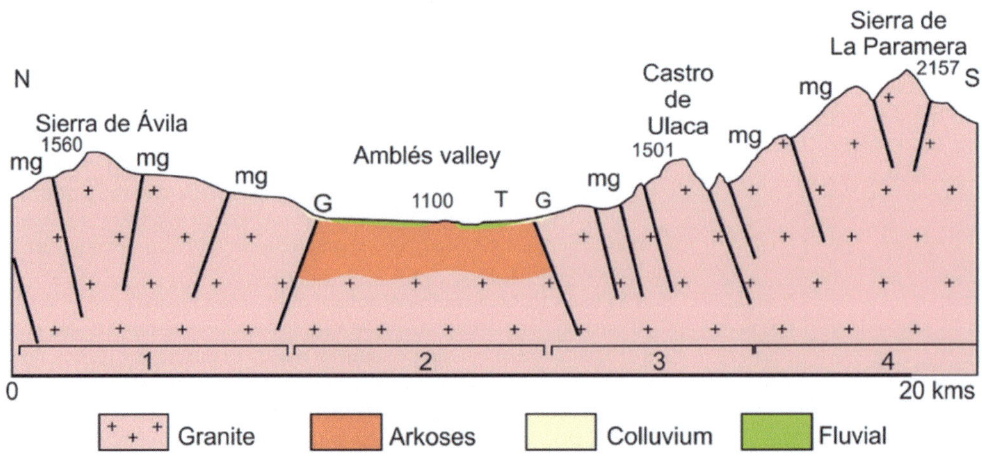


**Fig. 5.1** Morphostructural profile and location of the Ulaca Hillfort: (1) tectonic block of the Sierra de Ávila, (2) tectonic valley of Amblés, (3) Ulaca semihorst, (4) tectonic block of La Paramera, (mg) granite landforms, (g) slopes, (t) fluvial terraces

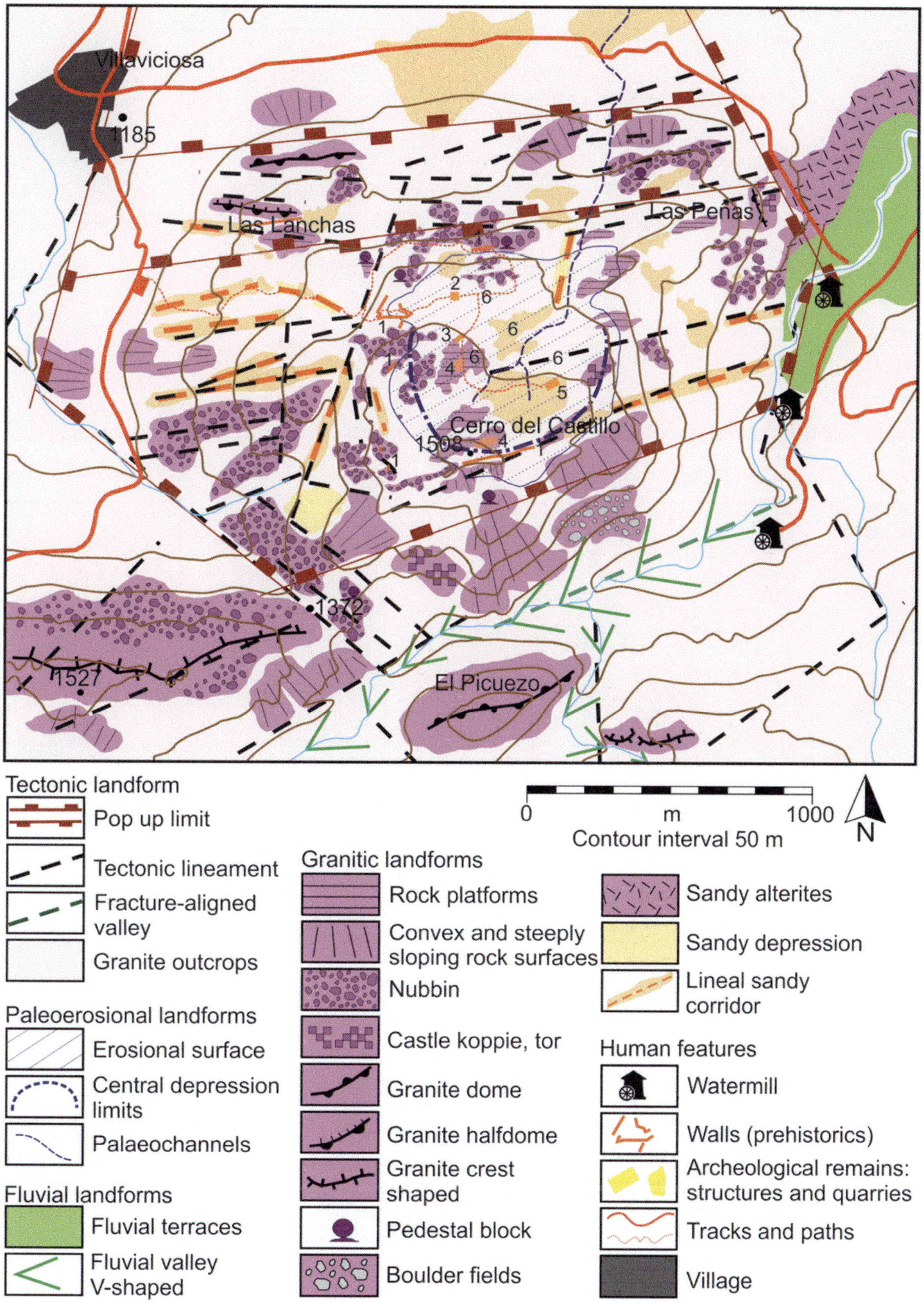


**Fig. 5.2** Cerro del Castillo geomorphological map

**Table 5.14** Granite landforms distribution in the Castro de Ulaca

| Unit | Physiography | Granite landforms | | Human uses main |
|---|---|---|---|---|
| | | Main | Secondaries | |
| Top | Flat and moderate slopes | Small weathering depressions. Sheet structure rock platforms | Pedestal rocks Tor | Settlements, quarries, ritual, and power building (altar and sauna) |
| Northern stepped blocks slopes | Steps and plains with hard slopes | Nubbins Boulder fields and Pedestal rocks | Tor Rock platform, half-domes, sandy depressions | Complex walls, settlements, and cattle meadows |
| Southern, eastern, and western fault-line valleys | Rectilinear slopes, hard slope ~24° | Subvertical rock platforms and convex and steeply slope rock surface | Half-domes, domes, boulder fields, and Nubbins | Defensive |

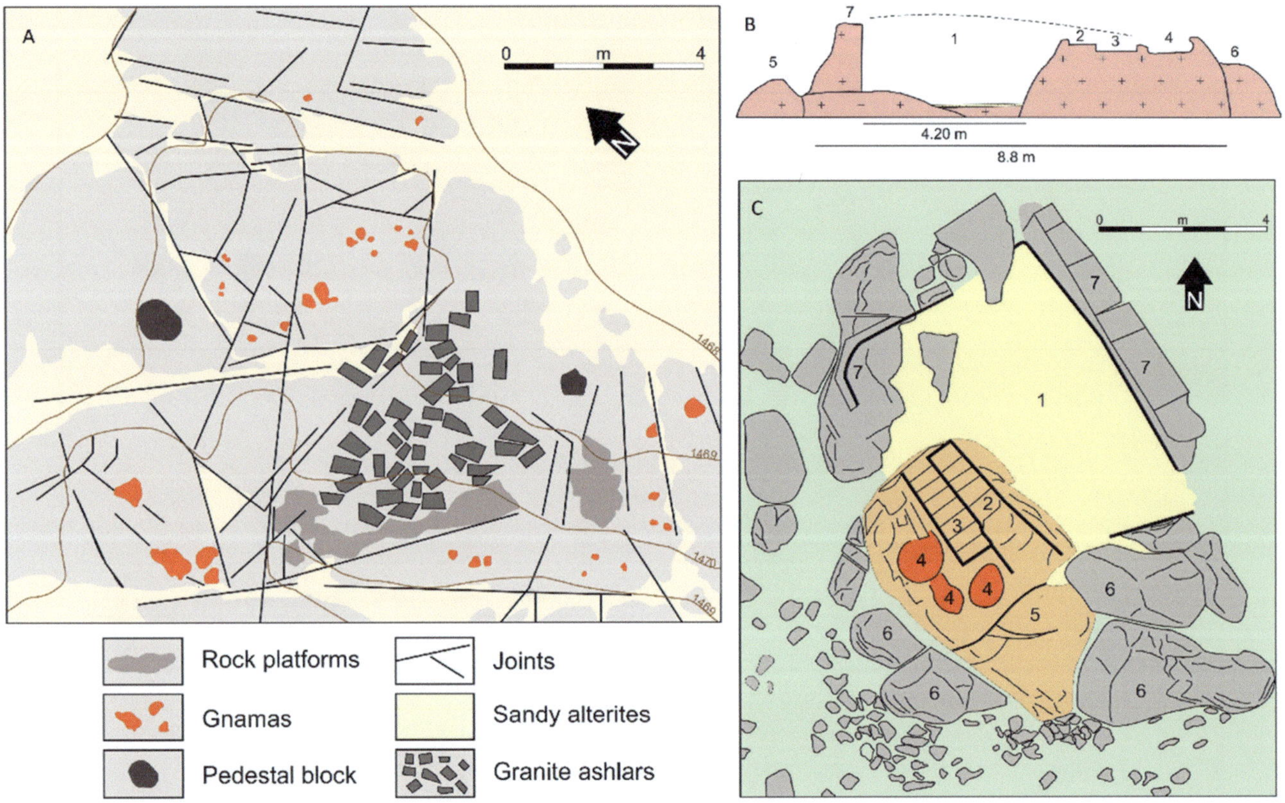


**Fig. 5.3** (**a**) Human and natural features relationship and the quarry components, (**b**, **c**) the sanctuary and the altar: (**b**) Profile of the tor worked by humans, (**c**) Map. Numbers mean: 1. Sanctuary main hall, 2. Ledges carved by human in the rock, 3. Steps carved by humans in the rock, 4. Gnammas used for sacred rites, 5. Natural boulder used as shrine, 6. Natural boulders, 7. Granite carved as sanctuary wall

between landforms, geomorphological units, and cultural remains.

The sandy depression infill by quartz sands and fines are places with wetter availability, and the springs are in the contact between granite and sandy accumulation. The small depressions are occupied by a wall of ancient houses, and on the central depression are located all the small settlements scattered by the site. In a depression near the spring is located the most important building, the power tower built with big granite ashlars obtained in the quarries.

The rock platforms and, in some cases, the convex and steeply slope rock surfaces are used as quarries to obtain building materials. The granite rock platform characterized by the sheet structure and subhorizontal curve joints, where the separation between joints is short (50–100 cm), was used to obtain suitable regular blocks, depending on the building supplies. Today, the features to work the stones are evident, and granite ashlar of different sizes are scattered around the exploitation front (Fig. 5.3).

On the top of the block, above the sandy depressions and where the granite outcrops are more common, symbolic and ritual elements are located directly carved on the rock. A small tor was worked to be a ritual sacrifice altar. The gnammas on the top and slopes of the small tor served to spill and channel the blood streams of the slaughtered specimens. The tor has been deeply reworked, emptied in the eastern side to build a chamber and stairs carved in the rock to reach the upper area and the gnammas. It is a good example where the natural landforms and human work are closely related. Another ritual structure built only 160 m away is explained as a sauna (Álvarez-Sanchís, 1999; Almagro Gorbea, M & Álvarez Sanchís, 1993). The isolated granite blocks have been emptied and worked to build different chamber and elements for sauna use.

The topography and granite landforms have been used as defensive walls. The big walls built around all sites are linked to nubbins, half-domes and tors where it was not necessary to build a wall. At the northeastern area, the separation between two tors allowed to open a useful door in the wall. Finally, the structural flats located to the north of the main hill and the sandy channel linked at the top with the fluvial valleys are favorable sites for the emergence of springs. The existence of channel, flats, and springs favored the location of the main access at NW and the livestock use.

The Ulaca map shows the dominance of granitic landforms, the processes inherited and their relationship with cultural elements, as well as the adaptation to the environment of this society, forced to occupy specific sites, the high half-horst, for socioeconomic and defensive reasons, but adapting to the geomorphological conditions (dwellings, defenses, gates) or developing skills on the ground (granite carving, stonemasonry). The 12 granitic landforms represented, compared to other structural, fluvial, or hillside ones with less representation, point to the potential of a granitic landscape characterized by the diversity of granitic geoforms. They are not elements of high scientific value, geomorphosites on an international, national, or regional scale, but they are of local value due to their relationship with archaeological remains. They constitute a cultural resource of tourist and educational orientation complementary to the archaeological elements.

When cultural spaces occupy large areas and prominent physiographies, landforms allow us to understand and interpret a part of the past, and it becomes necessary to incorporate this knowledge not only into spatial and archaeological analysis but also to apply it in two separate facets: to the management of the territory; and to interpretation and dissemination as a cultural service for visitors. One of the ways of bringing this information closer to the visitor is to incorporate the information into routes and walks by means of interpreters (the most effective system of cultural transfer), documentation, signage, or virtual guides, closely related to the archaeological remains. At Ulaca, the succession of sites allows for a high degree of effectiveness in incorporating these forms of information into the established routes.

The incorporation of geomorphological elements in the management of cultural spaces is necessary to guarantee their protection against potential visitor flows and infrastructure development, as well as to guarantee the survival and use of this cultural service for local societies and visitors.

#### 5.3.1.3 Dissemination and Citizen Science: Study Case in the Ulaca Hillfort

The Natural Protected Areas (NPAs) are home to a natural, cultural, educational, and tourist value that sometimes goes unnoticed by the inhabitants of their towns and villages. So much so that they are often unaware that they live in an NPA, or what are the natural and cultural features that have led to its declaration and protection.

This disconnection between protection figures and the inhabitants of NPAs leads to a serious problem of lack of interest, apathy, or mistrust toward public environmental bodies. Even more seriously, it implies a lack of knowledge of their values, rooted in the erroneous idea that the area does not deserve protection, which leads to a complicated process of acceptance and patrimonialization of geomorphosites. Geomorphosites are closely interrelated with human uses and cultural values that generate a reciprocal enrichment of both the cultural elements and Geomorphosites (Panizza, 2001, 2003; Serrano et al., 2020). These are the added values, which, in some cases, as in Ulaca, revalue the landforms due to the close relationship between the Geomorphosites, the landforms, the cultural remains, and the landscape.

The aim of this work is to investigate the knowledge and opinion of the inhabitants of the Sierra de la Paramera about the natural heritage of the area around the Ulaca site and the NPAs. In this context, the aim is to find out the attitudes of the local inhabitants toward the management of the NPAs and the archaeological site, as well as to give value to the geomorphological heritage in itself and in relation to the cultural elements through activities of interpretation and dissemination. In this territory there is cultural heritage planning focused on tourism and providing visitors with historical and archaeological information, but no attention has been paid to the natural heritage, despite its early assignment of the figure of NPA, more than 20 years ago, neither for the management nor for the visitors nor for the local population. For this reason, we propose a route through Ulaca Hillfort, where cultural and natural elements are combined for a global understanding of the landscape.

This interpretation activity took place in the Cerro del Castillo, where the Ulaca Hillfort is located. Prior to the activity, fieldwork was carried out for the geomorphological analysis of the Sierra de la Paramera and for the identification and evaluation of its geomorphosites. The qualitative method of Identification of Places of Geomorphological Interest of Valladolid was applied, which considers the intrinsic aspects, i.e., geomorphological, those of use and management of the places and the NPA, and the cultural or added aspects. In addition, a geomorphological map has been drawn up at a scale of 1:20,000 based on a bibliographic and cartographic review and fieldwork (Ruiz-Pedrosa & Serrano, 2023). Thirteen geomorphosites have been inventoried in the Sierra de La Paramera, one of which is the Castro de Ulaca, which stands out not only for its geomorphological heritage but also for the special relevance of its cultural heritage.

The activity consisted of a field trip with volunteer local inhabitants, therefore, the strategy used for the dissemination of natural heritage in this activity was that of heritage interpretation, which is characterized by the transmission of knowledge through first-hand experience and illustrative means (Tilden, 1957), in contact with a territory that is close to the participants.

To find out what knowledge the participants had about the Sierra de la Paramera and its granite modeling, a pre-activity questionnaire was carried out to find out whether they were aware of the environmental protection status of the Sierra, whether they had been to the Paramera and Ulaca before, and whether they were familiar with terms related to granite landscape and the morphostructure of the area in which we were. These same questions were asked at the end of the activity with the aim of evaluating the usefulness of the explanations given on the route, comparing the absorption of information, evaluating the interest of the activity, and the change in the attitude of the participants toward natural heritage.

The surveys were answered by a total of 32 participants, all of them inhabitants of the villages of La Paramera and the city of Ávila, of different ages (between 18 and 60 years old), with different levels of education (from compulsory studies to higher education) and whose professional profiles were far from any job related to geoheritage or landscape, except for three assistants, one of them a geographer and two employees of the Iruelas Valley Nature Reserve.

Along the route (Fig. 5.4), structural geomorphology, granite landscape, and the relationship between natural elements and cultural uses were explained in different stops (for more details, see Ruiz-Pedrosa & Serrano, 2024). To reinforce the explanations, various illustrations and diagrams were used as support material.

The explained contents and detailed results of the surveys can be read in Ruiz-Pedrosa and Serrano (2024). The results of the post-activity questionnaire show the didactic utility of the explanations given during the route. There was clear learning of all the concepts, most of which were unknown, and all participants knew what a dome, a fault, a pedestal rock, and a tor were by the end of the activity, as magnificent and numerous examples of all of them could be seen along the route. Therefore, these landforms are also the most useful for the dissemination of geomorphosites in a first approach. The concepts least assimilated were horst and graben, which is understandable since they represent morphostructures that are not as easily visualized as the granite landforms.

Furthermore, the responses to the question "Do you think Paramera and Serrota should be declared a Natural Park, why?" given at the end of the activity show a better collective awareness of the value and importance of natural and cultural heritage, their need for protection, and their potential as a tourist attraction. Thus, the objective of the experimental activity, which was precisely to convey these values of appreciation for natural heritage, was achieved by all participants, and all provided valuable and reasonable justifications for the need to protect this Natural Space.

The inhabitants of the NPAs must be aware of the natural heritage that surrounds them in order to be included in the management and research processes carried out in them. This participation is beneficial not only for the local population, favoring their link with the natural heritage and therefore their conception of heritage and desire for conservation, but also for the researchers themselves, as local people can actively participate by contributing their knowledge and experience. In order to achieve this, the local inhabitants, nowadays partially detached from the natural environment, traditional uses and land use due to the depopulation of the mountains, must have a knowledge of the natural values and outstanding elements that will enable them to value the territory as a collective and outstanding asset.

The high participation in this activity, despite its promotion only among the local population of the NPA and close to the capital of the province, and its focus on geomorphology, demonstrates the remarkable interest in Geography, natural heritage and landforms, contrary to what may appear to be the case. This was also expressed by the participants in the final comments, valuing the geographical and geomorphological approach given to a space in which historical and archaeological explanations predominate.

Interpretation and dissemination activities thus demonstrate the necessary collaboration between the local community and the scientific community. If the local inhabitants respond positively when invited to participate in these processes and dissemination activities, the scientific community often forgets how valuable the participation of the local population is. The residents of the Sierra de la Paramera who participated in the activity were able to understand the geomorphological processes that have shaped the landscape in which they live, not only giving meaning to those landforms they know—tors, nubbins, domes, pedestal rocks—but also contributing their own knowledge in local terminology and legends about granite landforms, or complementing the appreciation and understanding of the territory through a global understanding of the natural and cultural elements. Geomorphosites, then,

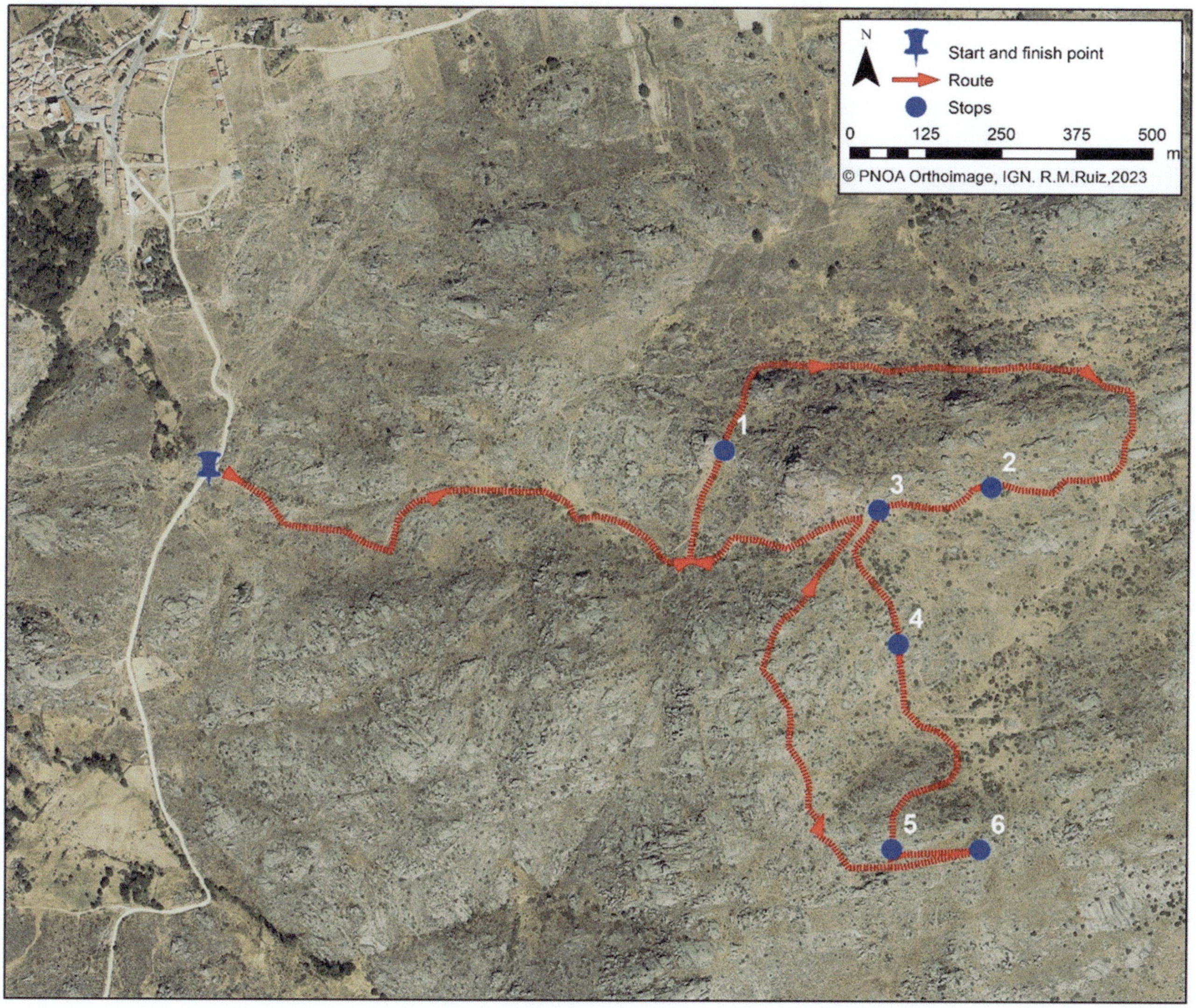


**Fig. 5.4** Route and stops made during the activity. At the top left, the village of Villaviciosa (Solosancho), which leads to the car park set up for the ascent to the Ulaca Hillfort, which is the starting and finishing point of the route. Stop 1: west entrance to the hillfort; Stop 2: north and east sector; Stop 3 and 4: northwestern sector, at the altar of sacrifices and ritual sauna; Stop 5 and 6: quarries in rock platforms, next to the Vetonean village and unfinished walls

prove to be a useful tool in the understanding of natural heritage and landscape, fundamental sites for the interpretation of geomorphological heritage.

### 5.3.2 Geotourism in Cañón del Río Lobos Natural Park

#### 5.3.2.1 Geomorphosites and Georoutes in Cañón del Río Lobos National Park

Cañón del Río Lobos gemorphosites are attractive territorial resources for visitors to the Park, not only for their intrinsic geomorphological value but also because they have a high component of landscape, aesthetic, cultural, and educational interest. It is evident that they have tourism potential, especially for those visitors who are not limited to recreational and aesthetic enjoyment of the natural heritage, but seek to acquire knowledge, understand the landscape, its dynamics, and the complex relationship of natural elements (Ruiz-Pedrosa et al., 2024).

The results of applying the tourism potential assessment method indicate that, out of the 14 geomorphosites inventoried in the park, 4 (28%) have high tourism potential, 6 (44%) have medium potential, and 4 (28%) have low potential.

Geomorphosites with high tourist potential are characterized by their scenic and cultural values, good conservation status and good accessibility. The most valuable geomorphosite is Río Lobos–San Bartolomé Canyon (N° 1), where the scenic and cultural content generates a high-value space, being the most attractive but also the most frequented due to its multiple attractions, accessibility, and proximity to parking. The Costalago orthocline valley (geomorphosite N° 10) is another highly attractive site, with a combination of elements, such as a viewpoint, valley, equipment, and a high landscape value, making it a great tourist attraction. Additionally, geomorphosites N° 4, the Virgen de la Cueva synclinal flank, and N° 6, the Pico Navas perched syncline, combine cultural and aesthetic elements in unique environments, giving them a high tourism potential. These sites are valuable territorial resources with high geomorphological and didactic value, as well as significant geotouristic potential.

The majority of the geomorphosites (44%) are considered to have medium value, defined by a combination of aesthetic beauty and scenic interest. Some of these sites have lower value due to their specialization (N° 7, N° 13, and N° 14), as well as their remoteness and poor accessibility. Additionally, four geomorphosites (28%) are classified as having low value, primarily due to poor accessibility, moderate scenic values or their specialization, which detract from their attractiveness to tourists. These sites also have low educational value, despite being in a good state of conservation.

The objective of geotouristic maps is to provide educational materials that explain the abiotic features of the area, with the aim of enhancing the level of recreation, culture, and education in outdoor activities within natural protected areas. Based on the Río Lobos Natural Park inventory and geomorphosite tourism assessment, a geotouristic map was created. This map includes information about visible geomorphological elements and provides details for hikers such as trails linking geomorphosites, shelters, springs, and elements of natural and cultural significance. The resulting map serves as a valuable tool for promoting geotourism and providing support in the field for local guides, monitors and tourist hikers who want to interpret nature and understand the landscape through direct knowledge of the terrain (Fig. 5.5).

The routes are designed for both tourists and hikers, as well as for educational trips at various levels. They are centered around the concept of walking, learning about natural heritage elements along the way, and enjoying direct contact

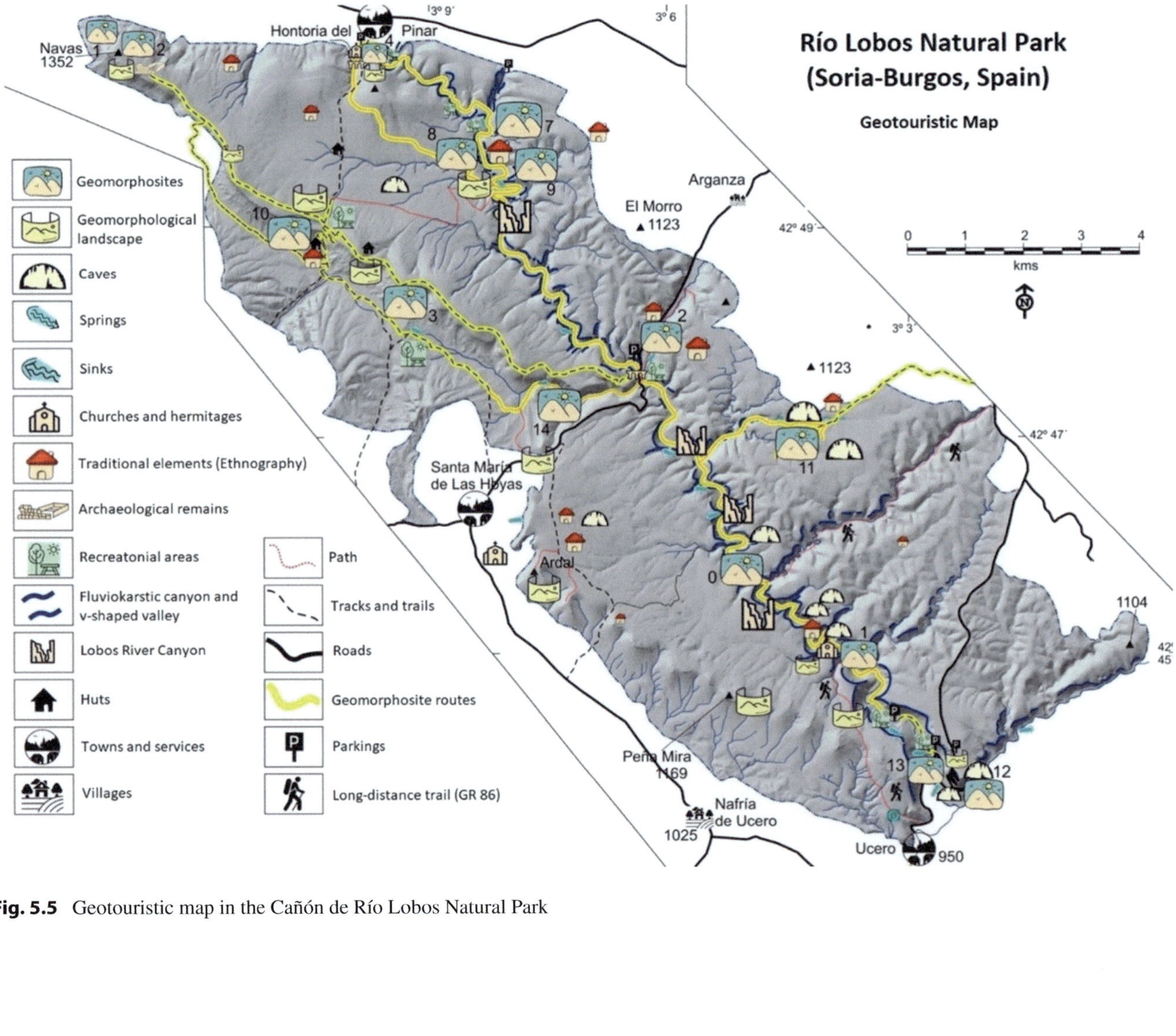


**Fig. 5.5** Geotouristic map in the Cañón de Río Lobos Natural Park

with nature and landscapes. The aim of these itineraries is, therefore, to combine leisure and education of tourists and students, integrating the interpretation of abiotic elements and landscapes with the experience of walking through the RLNP, from one geomorphosite to another. These geotourism routes serve as integrated tools for action, reflection, and feeling in an exceptional natural environment, such as the Río Lobos Natural Park. Five routes are indicated, which encompass several of the geomorphosites inventoried and allow a variety of geomorphological landscapes to be explored. The five routes run along marked trails and tracks in the park, ensuring accessibility. Each route is described in terms of type (traverse, circular route), distance and slope, difficulty, accessibility, the geomorphosites that are traversed, cultural elements present along the route, itinerary, and additional information. Detailed maps for each route have also been prepared (for more details, see Ruiz-Pedrosa et al., 2024).

In areas such as the Río Lobos Natural Park, the geotourism offerings are limited to well-known and frequently visited sites, leading to overcrowding at the most accessible and diverse geomorphosites, such as the canyon and the San Bartolomé hermitage. There is a lack of commitment to diversifying the routes offered and promoting lesser known but equally interesting geomorphosites from a natural, geomorphological, and scenic point of view.

Diversifying the routes could lead to a change in the tourism model in Río Lobos Canyon. Instead of predominantly short-term visitors, who only visit the canyon and the hermitage, without staying overnight, a multiday tourism model could be promoted, offering a wider range of places to visit and encouraging visitors to stay longer and explore the park from a broader perspective, encompassing natural—abiotic and biotic—and cultural heritages. This could have a positive impact on the economy of the municipalities in and around the park, fostering local development initiatives through the promotion of its natural heritage. The five georoutes cover all geomorphosites and the major cultural sites, expressed in a straightforward manner on the geotouristic map. Georoutes serve as a tourist resource that combines leisure and education, showcasing the park's natural and cultural heritage and diversifying the tourist experience while highlighting lesser known but scientifically, pedagogically, and geotouristically significant geomorphosites.

#### 5.3.2.2 Geotourism, Management and Public Use in Cañón del Río Lobos Natural Park

The declaration of NPAs aims to conserve areas of special value that are often at risk of degradation. NPAs serve also as tools for local development, providing both direct and indirect employment, as well as permanent and seasonal opportunities, leading to tensions between economic and conservation objectives. NPAs contribute to the local and regional economy and the tourism industry, but a better balance is needed between the objectives of preservation, public use and visitor enjoyment, always with a focus on preservation (Timmons, 2019; González-Domingo et al., 2021).

A significant issue in many NPAs is the concentration of visitors in specific areas, leading to environmental degradation and a diminished visitor experience (Manning et al., 2002; Carballo et al., 2019; González-Domingo et al., 2021; Mateos et al., 2020). Unequal overcrowding is a common management challenge, with some areas within NPAs experiencing high levels of visitors while others remain underutilized. Understanding these patterns can assist managers in territorial planning and directing visitors to less crowded and ecologically vulnerable areas for a more enjoyable experience.

When NPAs attract many visitors, overcrowding can occur in two ways affecting the entire NPA or being concentrated in specific sites within the NPA. In both scenarios, overcrowding at specific sites, especially natural landmarks, poses management challenges and conflicts of interest between biodiversity conservation, tourism development, and visitor appreciation and enjoyment. Numerous studies have identified overcrowding as the primary threat to ecosystems, habitats, and landscapes (e.g., Leung et al., 2019; Carballo et al., 2019), leading to degradation of the visitor experience, infrastructure overload, damage to nature (habitats, ecosystems, and species), and threats to cultural assets and heritage. In these cases, carrying capacity or limits of acceptable change are key tools in NPAs planning and management applied globally (Manning et al., 2002; Spenceley & Snyman, 2017; González-Domingo et al., 2021). While preventive, exceeding visitor limits and dispersing them to other areas within NPAs may only displace problems rather than solving them. However, NPAs managers and other stakeholders (such as park rangers, tourist agencies or guides, mountain guides) require tools to promote sustainable mobility to other areas within NPAs.

Geomorphosite mapping and assessment are effective tools for disseminating and explaining landscapes and abiotic elements of NPAs at different levels. They are essential for understanding the territory and promoting geotourism in socioeconomically disadvantaged areas. It is crucial to identify and document geomorphosites at a local level, even if they may lack national or international significance, as they can provide cultural surveys as part of ecosystem or nature services (Gray, 2018; Gray et al., 2024) to surrounding territories, often in peripheral areas. Geomorphosites serve as unique landscape elements in natural environments with high natural and scenic appeal, making them potential tourist attractions. This can help to reduce the pressure on iconic sites and raise public awareness of NPAs through responsible tourism and public use.

In the Natural Park of Río Lobos, there exists a strong attraction to the canyon, which receives most part of the visitors and which suffers a high massification, whereas the rest of the areas of the Park are poorly frequented and unknown for most of the visitors.

The objectives here are: (1) to understand the geomorphology of the Natural Park; (2) to identify and assess geomorphosites with high tourist appeal and use them as unique features to guide the spatial management of the Natural Park, especially to reduce overcrowding in popular areas; (3) to analyze geomorphosites as territorial resources for geotourism; and (4) to integrate geomorphosites with tourist potential into the Park's attractions, enabling their utilization by the public in a way that aligns with their characteristics and potential.

The CRLNP receives over 300,000 tourists annually, many of whom do not fully understand the complex evolution of the landscape but still appreciate and retain its beauty. The hermitage of San Bartolome and its surroundings (viewpoint, cave, valley, canyon) attract more than 300,000 visitors, along with over 20,000 vehicles accessing from Ucero. This area is visited by 81% of the natural park's visitors, while less than 19% visit the canyon between Valdecea and Siete Ojos. The rest of the RLNP receives very few visitors. Tourism is concentrated in specific areas and time periods, with 49.5% of annual visits occurring in August, Easter, and weekends. Most visitors are urban (34% from Madrid) and national (98%). All visitors look for natural and scenic values of the RLNP but the canyon is the main attractive and the potential values of other places and elements are not knowledge. Tourism is therefore highly concentrated in space and time, as 49.5% of annual visits are concentrated in August, Easter week, and weekends, and are mostly urban (34% of visitors come from Madrid) and domestic (98%).

The Geomorphosites inventoried in the Park, which are diverse in terms of attribution and complexity, can diversify and expand the offer of the Natural Park, relieving congestion in the most crowded sites and expanding their proposals. If geomorphosites are introduced as the basis of geotourism, they become an important resource for local and regional development (Reynard, 2003, 2008; González-Amuchástegui et al., 2013; Kubalíková & Kirchner, 2016; Kubalíková, 2019; Dóniz-Páez et al., 2023), which in the NPA must be responsible and sustainable, so the study of geoheritage is the first step for its incorporation into public use. In the Cañón de Río Lobos, this study has been carried out through the elaboration of the geomorphological map (Fig. 5.6) and the inventory of geomorphosites (see Sect. 5.3).

A second step that it is not common in the present conditions is its presentation both in the management plans of the NPAs as well as in the strategic development plans (Cocean & Cocean, 2017).

Natural parks have various sectors with high geotourism potential capable of attracting visitors, such as CRLNP. Planning based on integrated use of NPAs is essential. Geomorphosites will only be useful if they are known, rigorously presented, and integrated with natural, geographical, historical, and cultural aspects that define the NPAs (Panizza & Piacente, 1993; Serrano & González-Trueba, 2005; Reynard, 2009). To maximize their sustainable geotourism potential, geomorphosites need to be recognized for their intrinsic and added values (Kubalíková, 2013, 2019; González-Amuchástegui et al., 2013; Newsome & Dowling, 2018; Coratza & Hobléa, 2018). Planning should include zoning of geomorphosites based on exploitation and optimization, environmental impacts generated by tourism, and the natural risks that may affect the geomorphosite or visitors (Pralong, 2005; Kubalíková & Kirchner, 2016). This information should be incorporated into territorial plans of NPAs for effective management and public use. Geomorphosited must also be considered in natural space planning to ensure proper cataloging and incorporation into public use and flow management.

Geomorphosites can be utilized in various ways within the NPA. One approach is to diversify visits by directing visitors away from the most crowded areas toward other natural environments such as habitats, forests, geosites, or landscapes connected to the geomorphosites. Another method is to expand the range of recreational, educational, and geotourist activities in two perspectives: (1) to diversify the visited places and (2) to promote local development in the surrounding areas, stabilizing the population and creating new territorial dynamics, with the participation of local communities. To fully leverage the potential of geomorphology and geomorphosites as an attractive geotourism resource, it is essential to incorporate them into appropriate planning and management instruments, such as Special Plans (Use and Management Plans in Spain), which are currently not included in the CRLNP or in most NPAs.

Geomorphosites should be utilized as tools for territorial planning and management to aid in the conservation of the natural environment and the development of sustainable public use. This involves more than just creating inventories of geomorphosites and outstanding features; it requires actively participating in visitor flow management, setting usage guidelines, and protecting geoheritage and geodiversity as part of natural diversity. A comprehensive understanding of geomorphosites from various perspectives is crucial, as is integrating them into local culture and visitor experiences. Geomorphosites cannot be valued if they are not known, and effective conservation and management are not possible without proper valuation.

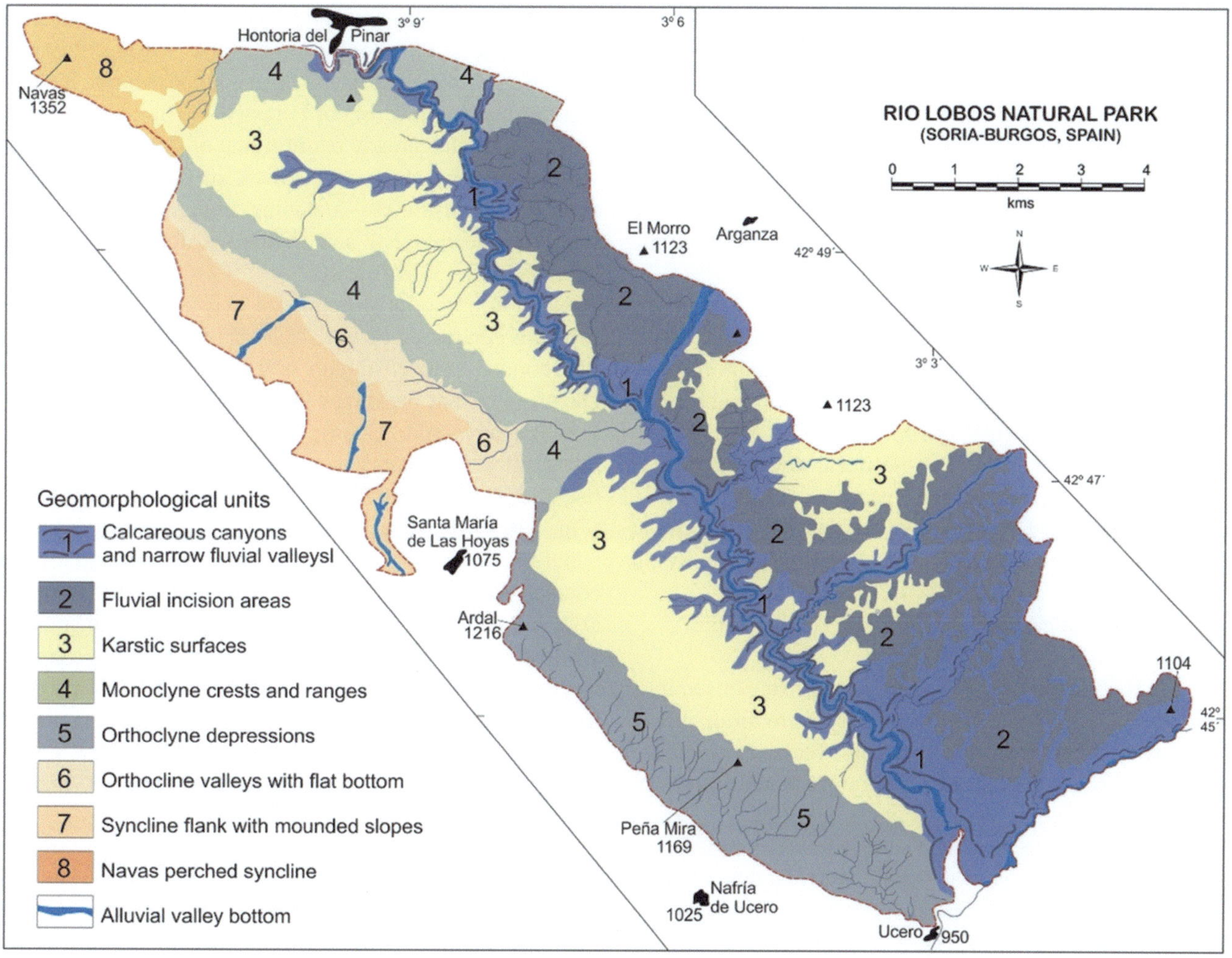


**Fig. 5.6** CRLNP geomorphological map

### 5.3.3 Geomorphology, Geoeducation and Field Trips: Study Case in Spanish High Schools

In July 2023, a survey on field trips aimed at teaching Geomorphology and Geology content was launched. The survey targeted educational centers throughout Spain, both public and private, offering compulsory secondary education and high school. The target audience of the survey were teachers from two departments: Geography and History, and Biology and Geology, as they are responsible for the subjects that cover physical geography and geology content. The form was open until January 31, 2024, and an intense dissemination effort was carried out:

- Schools of Geographers, Historians, Biologists, and Geologists from all Autonomous Communities
- Education unions at national and regional levels
- Regional and provincial Teacher Training Centers
- Dissemination through social media on educational dissemination profiles with a large following (>500 followers)
- Official emails published in the directory of all Autonomous Communities, sending the survey by mail to all educational centers, both public and private.

As of the survey closing date, 1163 responses had been obtained. The questions were as follows:

- In which Autonomous Community do you work?
- What is your education background?
- Public or private/concerted teaching center
- Years of experience
- Which department do you belong to? Geography and History or Biology and Geology?

- Does your school organize field trips to Natural Spaces focused on learning about landforms, Geomorphology, Geology, Hydrography, Climatology, Landscape, etc.?
- If the previous answer is yes,
  - Which department organized these field trips?
  - With which grade levels have you conducted field trips?
  - Do you think field trips are useful for reinforcing the knowledge taught in class?
- If the previous answer is no,
  - What is the reason? Lack of time during the course, I consider them not useful from a didactic point of view, lack of financial resources, lack of educational materials to prepare the trip, other.
- If you had all the necessary resources, would you like to organize field trips to Natural Spaces?
- If the Geography Department of a University proposed to take your students on a field trip to a Natural Space, would you accept?
- Comments, suggestions, or reflections on this topic

Of the 1663 responses, 15% were from private or semiprivate schools, while the majority, 85%, were from public schools (Fig. 5.7a). In terms of teachers' years of experience (Fig. 5.7b), 46% were teachers with more than 15 years of experience, followed by 23% with less than 5 years. Therefore, the most represented sample consists of teachers with a long academic career and employees of public schools. Regarding the Department (Fig. 5.7c), the representation is equitable, with 53% of Biology and Geology teachers and 47% of Geography, History, and Art History teachers.

Most of the teachers were biologists (34.51%) or historians (19.36%), with geographers representing only 11.88% of the responses and geologists 4.99% (Fig. 5.8). In some representation, graduates in Art History is around 8% and in Environmental Sciences, with 5.77%. In the "Other" section, we can include teachers with a joint degree in Geography and History (3.70%), Doctors (2.75%), and the rest with a variety of backgrounds related to humanities and health or Earth sciences (Philosophy and Letters, Nursing, Pharmacy, Chemistry, Biochemistry and Biomedicine, Marine Sciences, Agricultural Engineering or Forestry).

These results already confirm one of the hypotheses previously raised, which is the lack of specialists in Earth Sciences in institutes—including geographers and geologists—with Biologists and Historians dominating, each in their respective departments.

Regarding the most important question of whether field trips are conducted or not, 75% of the teachers responded yes, compared to 25% who said no (Fig. 5.9a). As for the Department that organizes them, 70% responded that it is the Biology and Geology department, 29% for the Geography and History department, and 1% of the teachers responded that both departments jointly organize them (Fig. 5.9b).

Of the teachers who responded that they do field trips, the course with the most field trips is 1st year of ESO (secondary), followed by 3rd and 4th year (Fig. 5.10). The numbers are lower in Bachillerato (preuniversity studies), especially in 2nd year of Bachillerato, a course focused on preparing for university entrance exams where time is very limited. The same occurs in 2nd year of ESO, a course where Biology and Geology are not studied, and in the subject of Geography and History only History is studied.

In response to the question, "Do you think field trips are useful for reinforcing the knowledge explained in class?" the vast majority, with 97.7%, answered yes, while the rest believe it depends on the student's behavior and implication (Fig. 5.11).

Regarding the bias by regions, the percentage of teachers who DO conduct field trips in their school has been represented cartographically (Fig. 5.12).

The region that conducts the most field trips is Aragon, with a 90% affirmative response, followed by Asturias and Cantabria with 84%, Canary Islands with 83%, and Pais Vasco with 80%. Between 70% and 79%, in descending

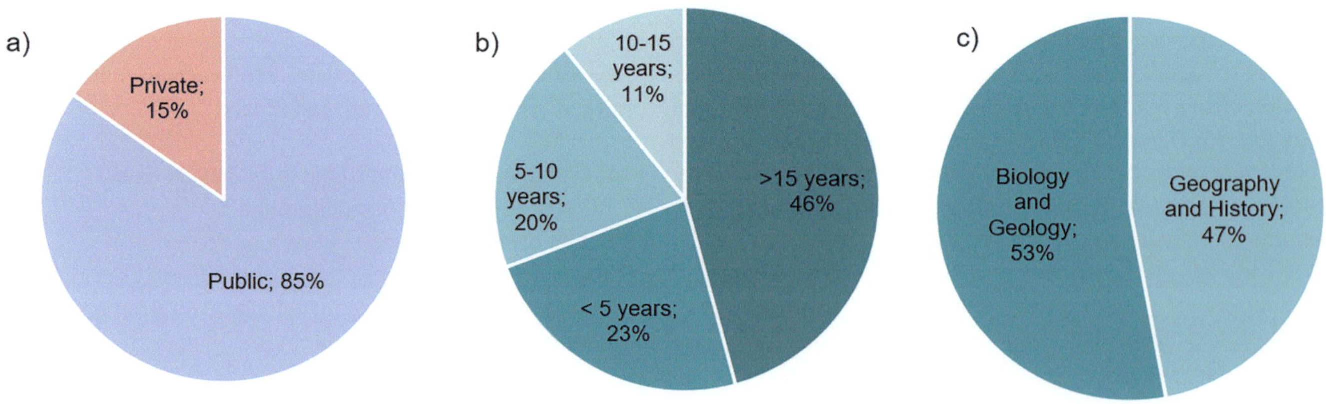


**Fig. 5.7** Public/private center, years of experience, and Department of the surveyed teachers

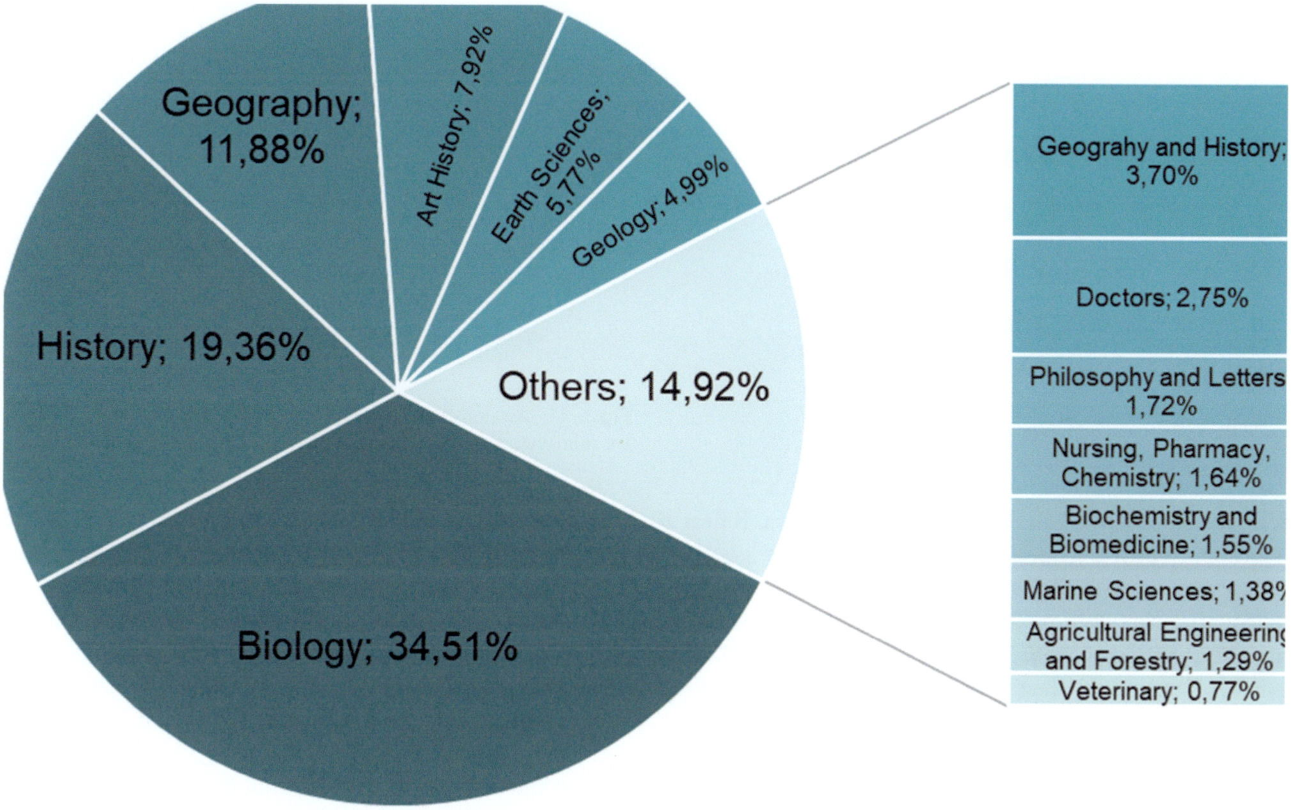


**Fig. 5.8** Education background of the surveyed teacher

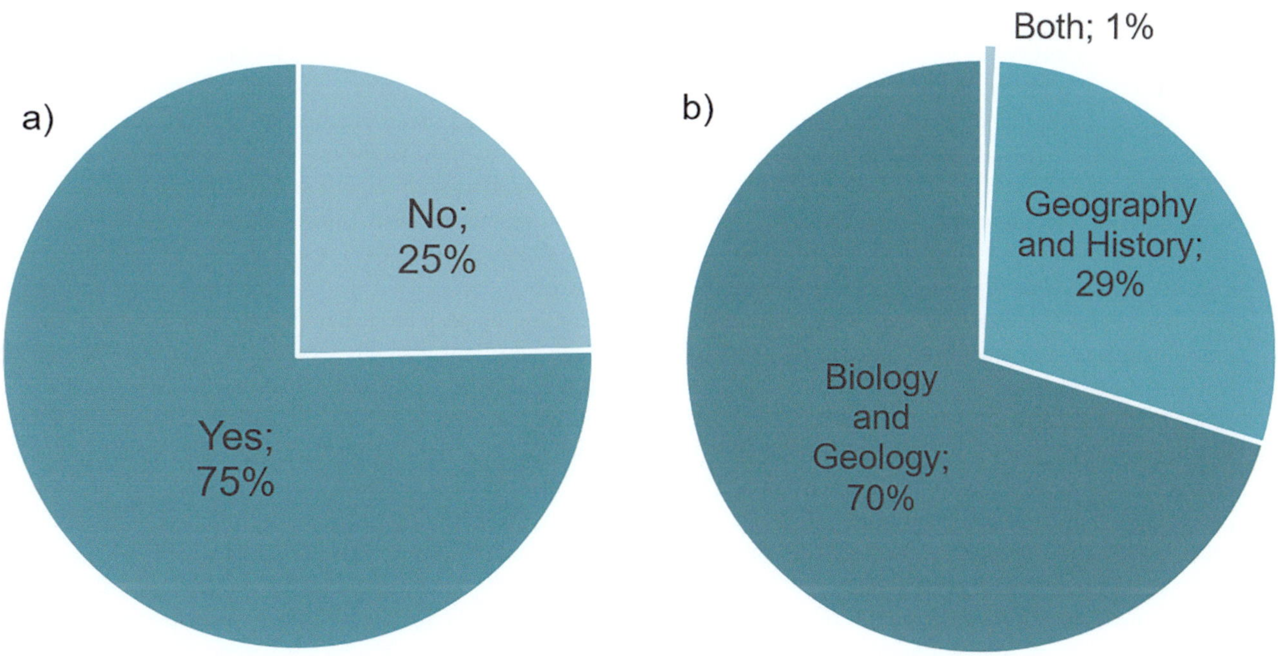


**Fig. 5.9** Answer to the question: Do you go on field trips? If so, from which Department they are organized?

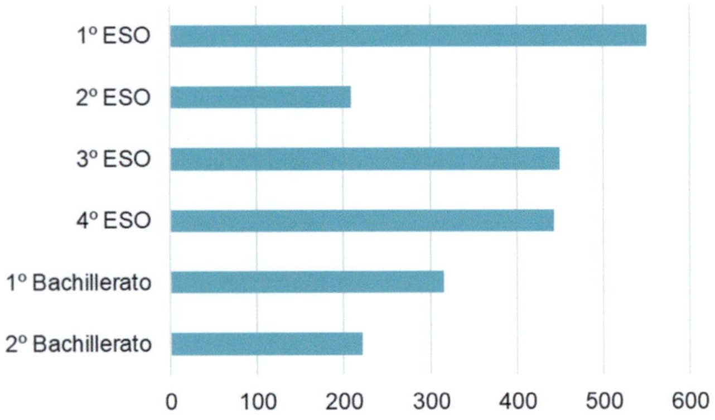


**Fig. 5.10** Courses in which field trips are conducted

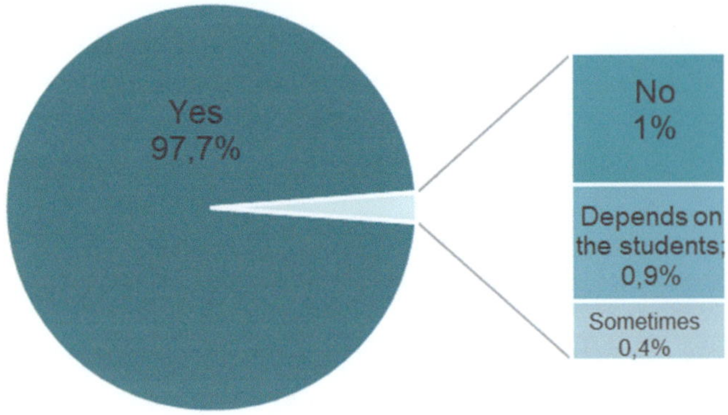


**Fig. 5.11** Response to the question "Do you think field trips are useful for reinforcing the knowledge explained in class?"

order, are Galicia and Murcia, La Rioja, Andalucia, Balearic Islands and Castilla y León, Madrid and Cataluña.

The regions with the worst data are Valencia, Navarra, and Melilla. And with the lowest data, Castilla-La Mancha, with 63%, and Ceuta, with 50%. These territorial imbalances in the tendency to conduct field trips to Natural Spaces may be due to several factors:

- More field trips are carried out in those regions with more NPAs
  - The region with the highest percentage of NPAs in its autonomy is the Canary Islands, with 40.6%, being one of the communities that conducts the most field trips. Additionally, it consists of small islands, which facilitates transportation.
  - In Castilla-La Mancha, with a vast size, the NPAs decreases to 7.4%, making transportation significantly more difficult. Ceuta and Melilla do not have any protected areas.
  - La Rioja is the second region with the most NPAs after the Canary Islands; however, it is not among the regions that conduct the most field trips. Cataluña, Andalucia, and Cantabria follow, with over 28% of their area protected, and all of them have a high percentage of field trips.
  - At the bottom are Aragon, Murcia, Castilla-La Mancha, and Extremadura, with less than 7% of their area protected, and all except Castilla-La Mancha have a high percentage. In fact, Aragon, with only 3.5% of its area protected, is the region that conducts the most field trips according to the survey results.
- Since no region directly finances field trips, families have to cover the costs of transportation and, if necessary, accommodation. Therefore, the per capita GDP influences the implementation of field trips.
  - Consulting the INE data for 2022, the autonomous communities with a per capita GDP above the national average are, from highest to lowest, Madrid, País Vasco, Navarra, Catalonia, Aragon, Balearic Islands, and La Rioja. All of them, except Navarra, have more than 75% of teachers who do conduct field trips.
  - However, Andalucia, Cantabria, Asturias, and the Canary Islands are below the national average, especially the Canary Islands and Andalucia at the bottom of the GDP ranking, and yet they lead in the implementation of field trips.
- The autonomous region size and, therefore, the proximity to NPA facilitate field trips
  - The regions with smaller surface area have the highest percentage of field trips: Balearic Islands, Cantabria, Pais Vasco, Canary Islands, Asturias, and Murcia.

Another bias to consider is the possible difference between public and private schools, assuming that, in general, students in private schools have a more favorable socioeconomic situation. However, the percentages of field trips are very similar, with slightly higher rates in public schools: 77% of responses were affirmative, compared to private school teachers, with 74% responding that they do conduct field trips.

Continuing with the teachers who responded that they do not conduct field trips, they were asked about the reasons, with lack of time in the course development being the overwhelming reason, followed by lack of economic resources and teaching materials (Fig. 5.13). In other options, teachers mentioned student misbehavior, not being useful based on the students attending the trip, and other reasons related to organization, distance, and bureaucracy.

In response to the question "*If you had all the necessary resources, would you like to organize field trips to Natural Spaces*?", 97.4% of teachers answered yes. Similarly, 90.6% were willing to collaborate with the university if the Geography Department proposed a field trip (Fig. 5.14). The remaining 10% responded that it depends on the time and alignment with the curriculum, cost, distance, physical activity, and the students attending the trip.

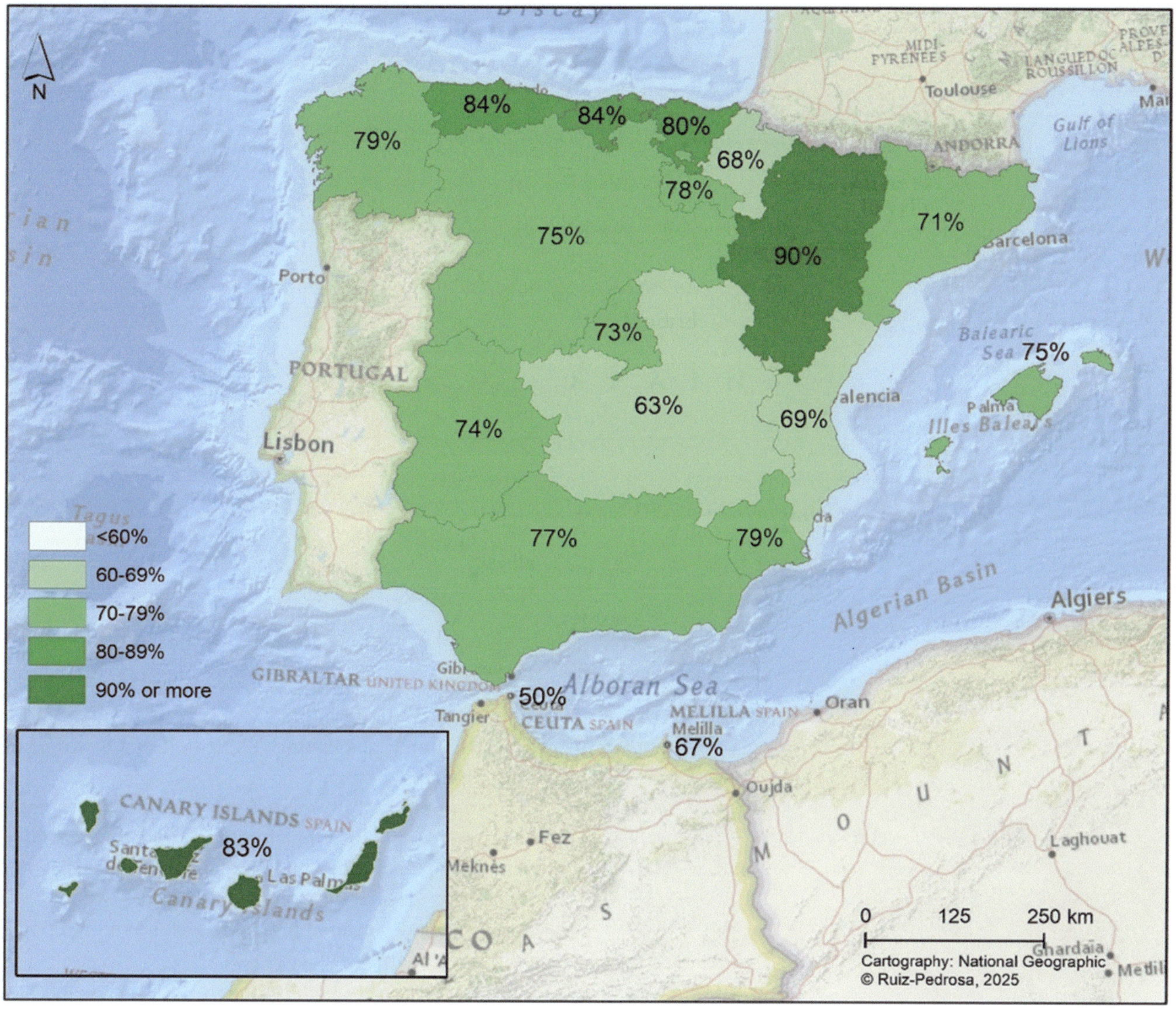


**Fig. 5.12** Percentage of teachers who do conduct field trips to NPAs

Finally, an open question was left for comments, suggestions, or reflections on field trips, in which teachers could express themselves freely. A total of 238 comments have been received, from which valuable information has been obtained about the situation of educational centers and the problems faced by teachers when trying to organize field trips. The conclusions that can be drawn are as follows:

- All teachers agree that field trips are a necessary and valuable methodological tool. Field trips reinforce the content learned in the classroom, strengthen the connection with natural heritage, and are aligned with the objectives of the LOMLOE and the Sustainable Development Goals. Many teachers cannot imagine their teaching without field trips.
- The main national impediment to conducting field trips is the lack of time due to a rigid curriculum, especially in Bachillerato (preuniversity), which hinders coordination with other subjects. Field trips require instructional time that is not available.
- Regional governments play an important role in facilitating or hindering field trips. In Cataluña, there are subsidized proposals for Learning Fields and Environments guided by professionals. In Cantabria, a new elective subject called Natural Heritage of Cantabria has been implemented, which is expected to support field trips to Natural Spaces. The Region of Murcia organizes training field trips for teachers as part of the Teacher Training Centers activities, usually two or three trips on Saturdays aimed at enabling teachers to conduct the same trips with their students later on. On the other hand, some communities are cutting funding and support for these trips. In Castilla y León, there used to be free environmental education programs in NPAs, with stays of 3–5 days, such as the Active Classrooms program; now stu-

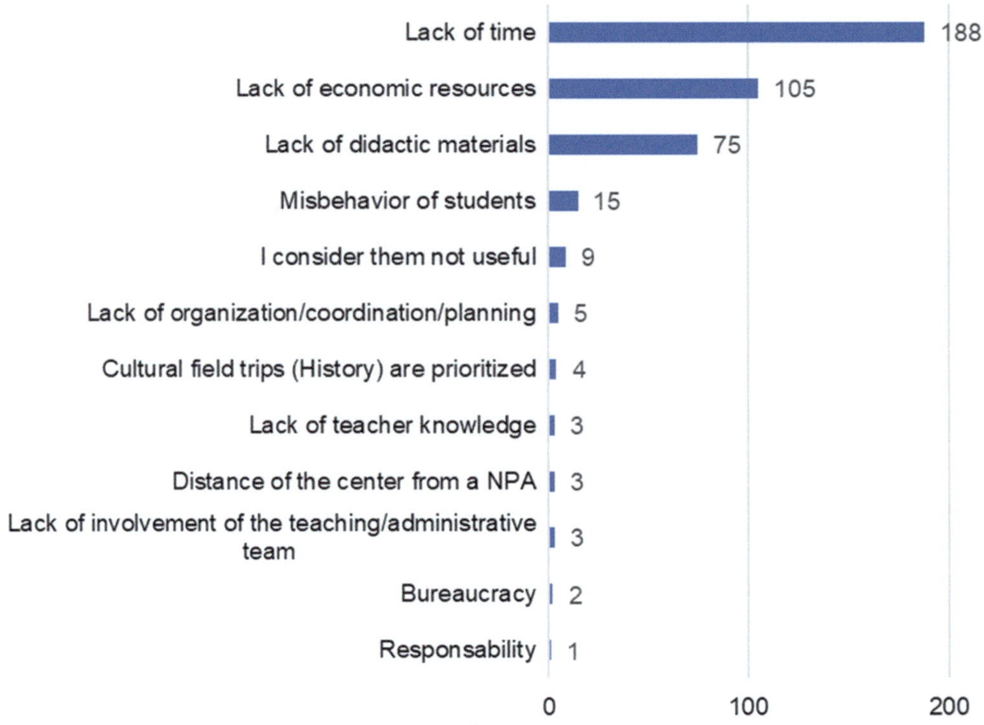


**Fig. 5.13** Reasons why field trips are not conducted

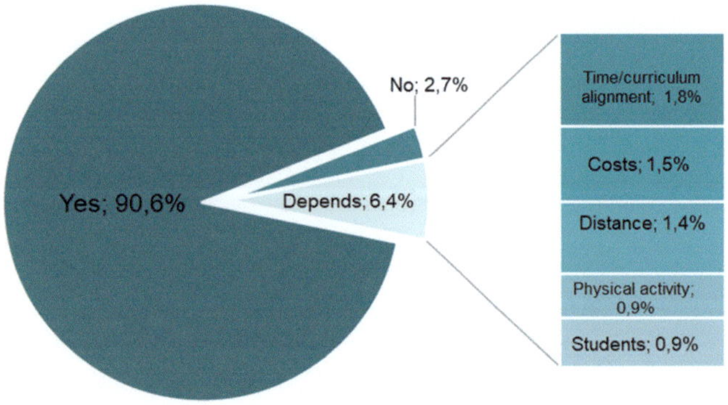


**Fig. 5.14** Answers to the question: If the Geography Department of a University proposed to you to take your students on a field trip to an NPA, would you accept?

dents have to pay, making it difficult for those from disadvantaged families to attend.

- The presence of disruptive students complicates field trips, along with the legal responsibilities that teachers must assume in case of any unforeseen events.
- History and Biology subjects and content have more weight in the curriculum than Geography and Geology. Additionally, there are more historians and biologists than geographers and geologists in high schools, as evidenced by participation results by field of study. This leads to a lack of training in Geomorphology and Geology among teachers.
- Cultural field trips, international programs like the Erasmus program, and recreational excursions are given priority over field trips to natural environments in schools. Some teachers express dissatisfaction with the promotion of international exchanges over field trips to nearby environments.
- The economic factor is also problematic. There are no subsidies to pay for transportation, and many students cannot afford it. Additionally, schools have been greatly affected by the increase in transportation costs in recent years. Teachers would appreciate subsidies and initiatives and proposals from universities.
- The size of the school and/or the group of students is crucial for conducting field trips. In small groups of students, generally less than 20, it is very difficult to afford a large bus. On the other hand, in schools with several large groups per grade, multiple buses and support teachers are needed, which complicates organization.
- The location of the school determines the possibility of conducting field trips. In schools in the Canary Islands and the Balearic Islands, due to the small size of the islands and the greater presence of Natural Protected Areas, teachers find it easy to organize trips. The same applies to schools located in rural areas near natural spaces. Conversely, in schools located more than 2 h away from a Natural Protected Area, field trips become almost impossible. Teachers express the need for supportive educational materials, such as field notebooks, to prepare for field trips. They state that they would appreciate the presence or proposal of educational itineraries by specialists, such as university professors or guides. Additionally, they believe that the presence of this external personnel makes students value field trips more.

# 6 Discussion

Geomorphosites and, in general, geoheritage and geomorphological heritage, are far from enjoying satisfactory integration in educational and touristic systems. Despite their high potential as an educational and touristic resource, as already justified, and the need to protect and preserve these sites, there are too many problems and obstacles to overcome, intrinsic to geographical education and geotourism.

On the one hand, we find a completely denigrated geographical education worldwide, with very diverse approaches. It has been proved in the educational systems we have analyzed, finding significant differences between Spain, Italy, and Switzerland. While in Spain our educational system is deeply regulated, with specific curricular contents established by law at a regional level in all educational levels, in Italy, legislation establishes general contents at a national level, with educational centers and teachers having to adapt their contents to the socioeconomic and territorial reality of their students. Something similar happens in Switzerland, where mandatory contents are only defined for primary education, with a decentralized system in which cantons and teachers have a great decision-making power. Common to the three countries is the idea of regional or proximity education, where the objective should be to strengthen the bond between students and the territory in which they live, while aiming to educate in a geography of globalization.

Despite these differences in the system and legislation, curricular contents are similar in Spain and Italy, which, unlike Switzerland, are part of the European Union and adhere to its education recommendations. The most relevant EU recommendation in educational legislation is the Council Recommendation of May 22, 2018 on key competences for lifelong learning, which includes eight competences: (1) Literacy; (2) Multilingualism; (3) Mathematics and science, technology, and engineering; (4) Digital competence; (5) Personal, social, and learning to learn competence; (6) Civic competence; (7) Entrepreneurship competence; (8) Cultural awareness and expression competence. The competence related to Geomorphosites is Science, defined as the ability to explain the natural world through observation and experimentation, as well as the student's ability to explain the influence of humans on the natural environment. However, these recommendations are limited in terms of teaching Earth Sciences, as they do not provide more details on the definition of the natural world or address awareness of natural heritage, while cultural awareness is mentioned. Nevertheless, both Spain and Italy have emphasized, at least in their curricula, the study of Earth Sciences.

In both countries there is a strong emphasis on the concept of Natural Heritage and the study of landscapes and landforms, not only at a regional and national level but also at a European scale. However, the distribution of Earth Sciences content among different subjects is interesting. In Italy, students learn Geography from the early levels of primary school, while in Spain, there is the subject of Natural Sciences in primary school, and Geography as a subject is not introduced until secondary school, always linked to History. The content of this subject in Spain mainly focuses on human geography, with basic concepts of landforms and natural heritage introduced only in the first year of secondary school. It is in the subject of Geology where most of the content related to Earth Sciences and Physical Geography is studied.

It is not until high school in Spain that a separate Geography subject is introduced, focusing on an in-depth study of Physical Geography, although it is optional. Therefore, in the Spanish educational system, Geography focuses on its human and socioeconomic aspects. In contrast, Italian students learn Physical Geography from the early years of primary education and, in secondary school, despite having various pathways, they study Physical Geography in the first 2 years. Additionally, in Italy, Geology is not a separate subject but is part of Natural Sciences. Based on this configuration, it seems that in Spain more importance is given to Geology, while in Italy, Geography is emphasized. However, it is known that in Italy, Geography is considered part of basic and general culture, as in higher



R. M. Ruiz Pedrosa, *Geomorphological Heritage as Didactic and Geotouristic Resources*, Geoheritage, Geoparks and Geotourism, https://doi.org/10.1007/978-3-032-06205-5_6

education, the university, there is a degree in Geology but not in Geography. There are only Geography master's programs, so those who consider themselves geographers have had a background in Sciences or Humanities and then specialized in Physical or Human Geography, recently through official master's programs.

However, the reality is that the curriculum is not always respected, and both subjects have minimal weight in the Italian and Spanish educational systems. It is common for teachers who teach Geography to be historians, and those who teach Geology to be biologists, so the content of Physical Geography and Geology is often overlooked due to lack of time in the course or lack of geological and geomorphological knowledge of the teachers, as some authors have pointed out (Orion, 2007; Rebelo et al., 2011) and as studied in Spanish high schools and in the United States by the National Geographic Society. The result is confusion among students, who do not know what Geography or Geology is, and even less achieve the learning objectives related to Geomorphology, geology, landforms, landscapes, orogeny, and so on. Ultimately, students' education in Earth Sciences is very limited, making it difficult to raise awareness of abiotic Natural Heritage. In this context, it seems more than necessary to strengthen the role of Earth Sciences in the educational system. The incorporation of geomorphosites in educational programs can undoubtedly be of great help.

In Switzerland, the situation is very different. We know that there is no national legislation, and each canton establishes its own content and learning objectives. After analyzing the curriculum in the French-speaking cantons (see Sect. 5.3.1.1), it is clear that it is much more flexible and general than in Spain or Italy. Teachers have academic freedom starting from non-compulsory education. Switzerland is the only country analyzed where Geography is present in both compulsory education (primary and secondary I) and post-compulsory education, which is reflected in universities. It is also the country where Geography is given more weight, with specific campuses and faculties in Geosciences, such as the Faculty of Geosciences and Environment at the University of Lausanne, where students can specialize in Geography, Environmental Sciences, or Geology, and pursue master's degrees in Geography, Environmental Sciences, and Earth Sciences. In Italy and Spain, Geography at the university level is linked to the Humanities, confined to faculties of Philosophy and Letters. In Italy, there are some degrees related to Geography, such as Human Sciences of the Environment, Territory, and Landscape (University of Milan) and Geographical Sciences for the Environment and Health (University of Rome), with a predominantly human focus on Economic, Political, and Urban Geography, and Physical Geography courses are optional. A similar situation exists in Spain, where Geography is often linked to history in many faculties (such as the University of Salamanca), with a greater emphasis on Human Geography and increasingly on Geographic Information Systems.

Therefore, it is worth reflecting: Is legislation synonymous with learning success? The data show the opposite. The latest Eurostat data on the school dropout rate in 2023 shows one of the highest rates for Spain at 13.7%, followed by Italy at 10.55%. Switzerland has a much lower rate at 5.5%. Therefore, the rigidity of the Spanish educational system does not seem to motivate students in their studies, whether in Geography or any other subject, while a more flexible system like Switzerland's offers better results. Furthermore, in countries where Geography and related Earth Sciences are emphasized throughout the education system, there is a greater supply and demand for related university studies.

Previously, we were talking about a school Geography of proximity, with field trips, experimentation, field research, didactic itineraries, and outdoor activities being one of the learning objectives of the analyzed educational systems. It is precisely in this proximity geography where geomorphosites and field trips have a place in the educational system, with geomorphosites acting as learning sites for the physical and cultural elements of students' immediate environment. However, field trips are becoming increasingly unaffordable for educational centers, teachers, students, and their families. The rising cost of transportation, the lack of time in the course development, rigid curriculum content impossible to fulfill, and the prevalence of field trips with cultural content are some of the reasons (Sect. 5.3.3). This issue is not limited to high schools but also affects universities, where budget cuts are making it difficult to conduct field trips in some Geography departments of Spanish universities.

Many questions arise. How do we intend to educate in Geography, Natural Heritage, conservation, and sustainability without students going to the field? Why, if the law recognizes the need and even the obligation of field trips, are the necessary resources not allocated? Why is there more priority given to visiting a museum, a city, or even an amusement park, rather than a Natural Space? At what point has a discipline as essential to life as Geography become a boring subject of memorization in schools?

The aim of this book is not to answer all these questions, but it aims to make a contribution to Geography didactics after reviewing the state of the art, research, and fieldwork conducted, and especially through the proposed method and the results obtained in the study areas. We aim to highlight the didactic potential of geomorphological heritage and advocate for the need to continue working in this direction for quality geographical education, forming citizens in contact with their Natural Heritage, who are aware of its value and the need for protection. The proposed method, initially designed for the Spanish educational system and its educational levels (primary, secondary, and high school), has proven to be useful for evaluating the didactic potential of Geomorphosites in educational systems of various countries, regardless of their configuration and legislation. The proposed method not only evaluates the scientific and additional values of Geomorphosites but also the conditions for their use—such as accessibility, fragility, seasonality, visibility, current uses, and legislation—as other authors have done. Focused on formal education and not previously evaluated in other works, an in-depth study of the educational legislation of the countries under study is added to the method, analyzing the curricula of subjects related to geography. Thus, by applying the method, Geomorphosites can be classified according to their suitability for the contents of different educational levels: primary, secondary, and higher education.

After a detailed analysis of the legislation and the contents and learning objectives established in the curricula, the method is easily adaptable to different levels, as has already been done in Italy and Switzerland. The results obtained in the three countries show that geomorphosites can be a valuable educational tool to support teachers in teaching Earth Sciences and help students not only in learning but also in raising awareness and appreciation of their Natural Heritage.

The relationship between geomorphosites and curriculum content by levels can be very useful in the academic field, as it allows for closer links between compulsory education and university. Although inventories of Geomorphosites and evaluation work are usually carried out by university working groups, researchers can help develop educational materials and design itineraries for teachers and students at primary, secondary, and higher education levels. Thus, the method is presented as a tool for selecting suitable sites for field trips in schools and institutes of education, already common at higher levels, instilling in younger students the importance of connecting with the territory, understanding the relief and landscape, and respecting the natural environment. This, in turn, will lead to a greater appreciation of Geography among students.

The unfortunate situation of geographical education in the education system is reflected, of course, in society as a whole. The limited teaching of Geography and Earth Sciences prevents the appreciation of geomorphological heritage by the population, and some authors have pointed out the need to involve young people in geoconservation (Worton & Gillard, 2013). According to Prosser (2019), most people do not understand Geosciences, geoheritage, or Geography, and therefore it is necessary to connect communities with Protected Natural Areas and geoheritage. This is especially important in areas with declining demographics, very low population densities, and abandonment of rural uses. Thus, support for environmental education and geotourism activities in Natural Protected Areas (NPAs) requires close communication with local communities and initiatives to develop effective management of geotouristic resources, incorporating Geomorphosites, as the involvement of the local population, in addition to promoting local development, improves the acceptance of conservation measures (Cocean & Cocean, 2017; Bouzekraoui et al., 2018a, 2018b; Pijet-Migoń & Migoń, 2019).

Within environmental education and education in nature, dissemination activities such as those carried out in the Sierra de la Paramera play a significant role in raising awareness about Geoheritage. This dissemination activity is based on heritage interpretation, which was defined by Tilden (1957) as "an educational activity that aims to reveal meanings and interrelationships through the use of original objects, first-hand experiences, and illustrative mediums, rather than simply transmitting factual information." This idea of heritage interpretation originated in the National Parks of the United States, in natural environments where it is necessary to guide visitors to discover the natural world, and the activity is always led by a professional interpreter who not only knows the medium in depth but also must be able to convey their feelings about the natural medium. However, this idea emerged outside formal education, and therefore it is important, as emphasized throughout this research, to differentiate between formal and nonformal geographic education. While there are many definitions and initiatives of nonformal geographic education (environmental education, dissemination, interpretation, mediation), formal school education is often overlooked. Therefore, geographic education should start in the formal sphere, in school education, through Earth Science didactics, and continue in the nonformal sphere of dissemination and interpretation.

Returning to the importance of interpretation and dissemination activities, it is necessary for residents of NPAs to know and understand the abiotic natural heritage that surrounds them and to be included in the management and research processes carried out in these areas. This participation is beneficial not only for the local population, fostering their connection with the natural heritage and therefore their conception of heritage and desire for conservation, but also for researchers themselves, as the local population can actively contribute their knowledge and experience. This is precisely the goal of Geography, the ability not only to see but to understand the territory, which in this case extends to abiotic heritage and Geomorphology. And this understanding should not be limited to the scientific community but should be shared with the local population, who generally know their territory, the landforms, but do not understand how they were formed, the reasons for those elements or resources, their location, or their traditional uses.

Interpretation is a very useful tool to increase the knowledge and awareness of the local population about the need to conserve and manage Geoheritage, not only by the authorities but the sensitization process is capable of awakening conservationist attitudes in the population (Tilden, 1957; Coratza et al., 2023; Rodrigues et al., 2023). To ensure the success of this process, it is necessary for the scientific community to approach the local population with an easy-to-understand language, with messages that promote dissemination and with a certain sentimental connotation that awakens in the public a feeling of appreciation for their natural heritage (Nesci & Valentini, 2020). Not only the local population can benefit from this type of activities, but interpreted natural trails, as a form of citizen science, allow the scientific community to undertake research of interest to the entire population from an innovative perspective (Kelly et al., 2020), enabling the transfer of knowledge between both groups. Other experiences to link researchers and local residents have been carried out to raise awareness and share cultural values and landscape among those involved in the NPAs (Tormey, 2019).

It is necessary for local residents, who are currently partially disconnected from the natural environment, traditional uses, and land management due to the depopulation of rural areas, to have knowledge of the natural values and outstanding elements that allow them to value the territory as a collective good. The high participation in the dissemination activity carried out in the Paramera, despite its promotion only among the local population of the NPA and near the provincial capital, and its geomorphological focus, demonstrates the significant interest in Geography, natural heritage, and geographical features, contrary to what might seem. This was also expressed by participants in the final comments, appreciating the geographical and geomorphological approach given to a space where historical and archaeological explanations predominate. Interpretation and dissemination activities thus demonstrate the necessary collaboration between the local community and the scientific community. If local residents respond positively when invited to participate in these processes and activities, the scientific community often overlooks how valuable the participation of the local population is. The residents of the Sierra de la Paramera who participated in the activity were able to understand the geomorphological processes that have shaped the landscape in which they live, not only giving meaning to the landforms they know—tors, nubbins, domes, pedestal rocks—but also contributing their own knowledge in local terminology and legends about the granite landforms, or complementing the appreciation and understanding of the territory through a global understanding of natural and cultural elements. Geomorphosites are once again revealed as a useful tool in understanding Natural Heritage, being fundamental sites for the interpretation of geomorphological Heritage.

And, just like with formal education, the necessary resources are also not allocated for nonformal environmental education. We have seen that environmental education in Spain is characterized by the precariousness of environmental education technicians, in a regional management model that delegates to private companies (Serantes et al., 2015). Specifically in Castilla y León, Environmental Education programs have been characterized by delays in approval processes, and by a supposed specialization or exclusive dedication to environmental education in the school environment, as stated in the III Environmental Education Strategy. While, as we have mentioned, geographical education in the school environment leaves much to be desired, the situation regarding dissemination among the population is even worse, and although Castilla y León has a decent network of NPAs with good facilities, it seems not to be enough. Without the necessary resources, it is therefore impossible to advance in environmental education, interpretation, and dissemination of Geoheritage.

Linked to dissemination and nonformal education is geotourism, which we have defined as tourism based not only on the abiotic heritage of a territory but also on its cultural elements, and always aiming to improve its identity, respect its heritage, and the well-being of its residents. The aesthetic, landscape, cultural, and didactic dimension of geomorphological heritage and geomorphosites make them territorial resources with geotouristic potential, but also tools for their promotion, management, and planning (Kubalíková, 2013; Serrano et al., 2020). Geomorphosites with aesthetic values and highly active landscapes or dynamics (volcanoes, glaciers, canyons) at any scale include the beauty of nature, embody the cultural conception of the romantic journey, in search of the beautiful and the sublime, and are therefore linked to geotouristic activity. Landforms are, therefore, a global touristic resource (Leung et al., 2019; Mishra, 2020).

The inventory and subsequent planning of geomorphosites geotouristic uses is based on three basic aspects: the exploitation and optimization of geomorphosites; the potential environmental impacts generated by tourism; and the natural risks that can affect the geomorphosite or visitors (Pralong, 2005; Kubalíková & Kirchner, 2016). The potential use of geomorphosites should be articulated on these three bases and included in the territorial plans of the NPAs. However, geomorphosites must also be active figures in the planning and management of natural spaces, oriented both toward their cataloging and their incorporation into public use and mass management.

Since one of the main impacts generated by tourism is overcrowding, which is accompanied by the deterioration of the NPA, but especially of geomorphosites as elements of special value, the first step for any territory conservation strategy is the decongestion of the most visited sites (McCool & Kohl, 2017; Leung et al., 2019; Carballo et al., 2019). In these cases, the key instrument in the planning and management of NPAs is carrying capacity or limits of acceptable change, applied in numerous parks around the world

(Manning et al., 2002; Leung et al., 2019), in addition to sustainable public use of the rest of the NPA areas.

The inventory of geomorphosites should be based on broad and global criteria that integrate natural, landscape, and cultural aspects, which is valid for exceptional environments as well as for NPAs of smaller dimensions and more modest values, such as the Cañón del Río Lobos Natural Park. Geomorphosites can be a powerful tool for local development based on geotouristic resources, although it has been identified that proper management and sustainable use are necessary to prevent deterioration of geoheritage resources (Newsome et al., 2012).

Cooperation with research institutions is important for providing activities to disseminate abiotic heritage (Pijet-Migoń & Migoń, 2019; Ruiz Pedrosa & Serrano, 2023), but also with local communities as part of tourism activities. In rural areas, often with few inhabitants and high aging rates (as is the case of Río Lobos), local communities have little capacity to identify new public uses, and do not see geotourism, currently very seasonal, as a resource for new jobs and activities (Bachiller, 2014; Raeisi et al., 2022; Zouros, 2007). In a rural area affected by depopulation, this can be addressed by promoting geotourism guides and nature interpreters with geographical training, who understand geomorphological, cultural, and territorial aspects. This training increases knowledge of the territory, promotes education, and contributes to the geoconservation of the NPA, overcoming the weaknesses detected in the Cañón del Río Lobos Natural Park (Bachiller, 2014).

Strategies for incorporating geomorphosites into the management of moderate-sized NPAs with varied uses, with overcrowded sites versus large unvisited but equally interesting spaces, involve a multitude of initiatives. Some of them are (Pereira et al., 2009; Zgłobicki & Baran-Zgłobicka, 2013; Raeisi et al., 2022):

- The incorporation of Geomorphosites in management and development plans.
- The training of local tourist guides and interpreters.
- The management of geotourism routes to access the most interesting sites guided by interpreters and guides.
- The replanning of old routes using geotouristic resources.
- The increase of geoeducation programs and attractive activities for visitors of all conditions and ages, activities already successfully established in many NPAs in Spain.
- Dissemination through websites and social networks as the most effective promotion mechanism among tourists.
- Adaptation and promotion of entrances through non-massified areas to the NPA, stimulating visits to the villages closest to these accesses.

And it is also important to take into account the opinion of visitors through on-site surveys to better understand the assessment of geomorphosites and the knowledge of local managers and governments about the geotouristic potential of the territory (Zgłobicki & Baran-Zgłobicka, 2013; Mateos et al., 2020; Ruiz-Pedrosa & Serrano, 2023).

In the two selected study areas, the Sierras de la Paramera and la Serrota and the Cañón del Río Lobos Natural Park, the proposals and activities developed have responded to their needs for use, management, and conservation. There are two very diverse spaces in their uses and planning, but with common problems, such as poor management by the Department of Environment of the Junta de Castilla y León. On the one hand, the Sierras de la Paramera and la Serrota, proposed to be included in the Network of NPAs of Castilla y León since 1992, have not yet been included under any protection category recognized by this law. On May 5, 1992, the Order of Initiation of the Natural Resources Management Plan of the Sierras de La Paramera and Serrota Natural Space (1992) was published, but it seems that 30 years have not been enough. The inaction of the Junta de Castilla y León has had fatal consequences in this natural space, such as the Navalacruz 2022 wildfire, which devastated the entire area. The lax protection provided by a natural space being part of the Natura 2000 Network, without any type of funding, management, or planning, does not seem sufficient for the conservation of these spaces. Just as we would regret a year later with the fire in the Sierra de la Culebra, another area whose declaration has also been pending since 1991, and which is also part of the Natura 2000 Network. A Natura 2000 Network that, if not incorporated into the regional legislation through specific planning instruments, shows no effectiveness.

The Cañón del Río Lobos Natural Park would have a better fate if it had been declared in 1985, but it has still been affected by a long process of declaration and planning that, almost 40 years later, has not seen its planning approved. The Declaration Decree is the only planning instrument in force, which, as mentioned, is ridiculously brief. Despite this, it has received some attention and management. It has well-developed and recently renovated tourist infrastructure: a Park House and River Classroom, a footbridge over the Río Lobos inaugurated at the end of 2020, and a renovated viewpoint in 2024 are some of the recent actions taken. However, investment in education and dissemination is equally scarce. The Cañón del Río Lobos is part of the School Visits Program of the Natural Heritage Foundation. But these initiatives have not been enough to diversify the offer, and as mentioned, tourists continue to flock to the area around the Hermitage. Particularly shameful in the Cañón del Río Lobos is the attention given to the abiotic heritage, specifically the geological and geomorphological aspects since its declaration. A clear example is the offer of supposed georoutes that visit Inventory Sites of Geological Interest in the Park. This information is offered at the Park House, on one of the interpretive panels, as we have already pointed out. However, in 2019, we asked the Park House employees about these georoutes, and they told us that, despite the advertising, the georoutes had not been designed, and the inventory of Sites of Geological Interest had not been carried out. Six years later, we returned to the Park House, and the situation has not changed. The panel is still in the same place, offering the georoutes, but this time the excuse was different. On August 15, 2024, a national holiday with a significant influx of visitors to the Park and the House, they told us that the environmental agent in charge of the georoutes was not there, and there was no brochure or information available to the public. Without intending to blame the employees of the House, who we know work in precarious conditions, it is shameful that in an NPA like the Cañón del Río Lobos, one of the most visited in the community, georoutes are offered that do not exist. Any visitor minimally interested in Geology is met with a completely negative response when trying to learn about the geology of the Park.

The limited protection and dissemination of abiotic heritage in our community and country is especially sad after being able to conduct research in Italy and Switzerland. All the selected geomorphosites in these countries had protection measures in place, as both Switzerland and Italy have made significant efforts at the national and regional levels to recognize geomorphosites. Switzerland, for example, has a Federal Inventory of Landscapes and Natural Monuments with 162 recognized landscapes and elements, developed in collaboration with the cantons. They have also created a national inventory of Geotrails specifically designed for geotourism, based on geological and landscape characteristics, with trails, maps, and guides to promote Geology to the public (an initiative of the Swiss Academy of Sciences and the "Geology Experience" association). The Valais Canton website also provides information on the geological sites of interest in the region. On the other hand, Italy has a National Inventory of Geosites, with over 3000 geosites across the country. The Emilia-Romagna region has developed its own inventory of geosites, with nearly 700 sites categorized as regional or local interest. Each geosite has a descriptive sheet with scientific content and maps. In both Switzerland and Italy, unlike in Spain, there are numerous initiatives for dissemination and environmental education, the availability of interpretation centers, and interpretive geotrails. All of these initiatives focus on geoheritage, abiotic heritage, geotourism, and involve universities and research groups actively working, often at the behest of administrations that seek researchers to create scientific and educational materials.

In Spain, there are some similar national initiatives, although with little development and no recent activity, such as the Spanish Landscape Inventory, which was initiated in April 2011, for the development of the Spanish Inventory of Natural Heritage and Biodiversity. However, almost 15 years later, the Inventory has not been completed, with only a preliminary work, the Atlas of the Landscapes of Spain, a 2004 work at a scale of 1:200,000 for the peninsula and 1:50,000 for the islands. Since then, no national-level effort has been made, resulting in a very poor inventory that responds to morphostructural sets. There is a National Inventory of Geological Sites of Interest prepared, which distinguishes between sites of local importance, although the level of detail at the local scale is very underdeveloped. For the Cañón del Río Lobos, there are two sites of local importance whose main interest is stratigraphic, and the entire Cañón del Río is a site of main geomorphological interest. In the Sierra de la Paramera, only a site of petrological–geochemical interest has been inventoried, corresponding to the Portacho del Cuchillo diatreme. In the Serrota, the Moraines of the Serrota have been inventoried, of local importance, although the Identification sheet does not have any description type, and a dyke in Garganta del Villar, of stratigraphic interest and also without description.

At the regional level, the Natural Heritage Network of Castilla y León recognizes, within the Network of Natural Areas of Special Interest, the category of geological or paleontological sites of special interest, whose designation is, as we have already mentioned, particularly problematic. Paleontological heritage is regulated by the Cultural Heritage Law, belonging to the archaeological heritage, so this category of geological and paleontological sites is not very coherent. However, it is irrelevant, since no geological or paleontological sites have been inventoried since this figure was created with the 2015 Law on Natural Heritage of Castilla y León. Once again, the inaction and lack of interest of the Junta de Castilla y León in abiotic heritage is notable, as the rest of the categories of Natural Areas of Special Interest have seen some development (Wetlands of special interest, Notable trees, and Natural recreation areas). The regional effort in the recognition and protection of geological and geomorphological heritage is directly null.

Given all this, the proposals to be developed in each study area have necessarily been different and adapted. In the Sierras de la Paramera and la Serrota, with no tourist infrastructure, no interpretation elements (neither panels nor interpretation centers), and no recreational areas, it does not seem sensible to propose geotourism initiatives. It is a space that needs a valuation of its heritage from the bottom up, born from the local population, a process that seems to be gaining strength, especially after the wildfire. We have already seen that the municipalities of the Paramera region have agreed, at the initiative of the Provincial Council, to design tourist routes, although no interpretive or informative material has been developed about their heritage. These routes, as of today, are purely for hiking and do not serve, therefore, for a geotourism planning of the NPA. Therefore, with the dissemination route carried out in Ulaca in 2021, the aim was precisely to delve into the valuation of the abiotic heritage, specifically geomorphological, among the residents of the Paramera region. On the other hand, we also aim to disseminate the geomorphological heritage in a space where, despite its spectacular granite modeling, attention has only been focused on cultural heritage. Therefore, with the elaboration of the geomorphological map of Ulaca and the relationship between landforms and the archaeological site, the importance of Geomorphology is reinforced.

Meanwhile, in Cañón del Río Lobos, with a well-developed tourist infrastructure but not focused on geotourism, it was optimal to propose a diversification of the tourist offer based on the inventory of geomorphosites, the geotouristic map, and the design of georoutes, which could imply a change in the touristic model. Since the predominant visitors are short-stay visitors who only visit the canyon and the hermitage without staying overnight, a multiday touristic model could be promoted, offering a greater variety of sites to visit and encouraging visitors to stay longer and explore the park from a broader perspective, encompassing the natural—abiotic and biotic—and cultural heritage. This could have a positive impact on the economy of the park's municipalities and surrounding areas, promoting local development initiatives through the promotion of its natural heritage. However, this work has not enjoyed dissemination or implementation in the Park, as, unlike in Italy and Switzerland, initiatives on geoheritage issues mainly arise from university research groups, who face the disinterest of the administrations.

In the Sierras de la Paramera and la Serrota, we have seen that the geotouristic potential of the geomorphosites is low. However, this poor assessment does not stem from a lack of intrinsic interest, which have medium to high scenic, scientific, and conservation values, but rather from the underdevelopment of their additional values, which can be easily improved. With an investment in improving accessibility, installing recreational areas, and creating interpretive products, it would be enough to increase these additional values and elevate their geotouristic potential to high. The potential of geomorphosites largely depends on accessibility, infrastructure, and protection. In the Paramera and la Serrota, there is no type of planning or management, but in the Río Lobos area, there is, which enhances the potential use of geomorphosites for geotourism and geographical education. Even with low potential use, the geomorphological value and additional value, as well as didactic and touristic value, should be considered. If a geomorphosite is interesting from a didactic and touristic point of view, it should not be dismissed, but actions should be focused on improving its accessibility and facilitating its use. The didactic potential is also affected by the presence of didactic material. Despite the absence of this material in the Paramera and la Serrota, there are geomorphosites with high didactic potential derived from their high scientific and/or cultural interest. Therefore, work can be done on developing didactic material for education at all levels, as intended in this book and in future work, starting with didactic evaluation and continuing, in the future, to develop specific didactic materials. Therefore, efforts can be made to improve the didactic potential of these sites, and thus geomorphosites with medium potential would transition to high potential.

It is worth asking why, with a very similar number of inhabitants, just over 3000, the difference between La Paramera and La Serrota and the Cañón de Río Lobos is so abysmal in terms of their socioeconomic aspect. A health center in Hontoria del Pinar, none in La Paramera; four educational centers with 309 students in Río Lobos, compared to 2 centers and 53 students in La Paramera; 4 hotels, 37 rural tourism accommodations, and 19 restaurants in Río Lobos, compared to 3 hotels, 60 rural tourism accommodations, and 16 restaurants in La Paramera (Table 6.1).

The Paramera and the Serrota, with a higher number of inhabitants than the Cañón del Río Lobos, have many fewer services, which can be understandable given their proximity to the capital of Ávila, with the Río Lobos being twice as far from Soria. However, the Paramera and the Serrota have good rural tourism services, with 60 accommodations (mostly rural houses), which almost double the number of accommodations in the Río Lobos, 3 hotels, and 16 restaurants, very similar to the Río Lobos. It seems, therefore, that the influx of visitors in the Paramera is based on a type of rest tourism, with groups of friends and families staying in rural houses and seeking pool and disconnection, not learning about geoheritage, because this offer does not exist either.

If an effort were made to promote geotourism in La Paramera and La Serrota, it could be successful, based on the following reasons: (1) There is already a tourist infrastructure capable of accommodating rural tourists; (2) There is already a public that stays in these rural accommodations, with the province of Ávila being a leader in the region, so there is a target audience interested in geotourism focused on the natural and cultural heritage of La Paramera and La Serrota; (3) Specifically in La Serrota, there is already a hik-

**Table 6.1** Socioeconomic scope in the two NPAs under study

| | Population (2023) | Health centers | Educational centers | Rural tourism accommodations | Hotels | Restaurants |
|---|---|---|---|---|---|---|
| Paramera and Serrota | 3288 | 0 | 2 | 60 | 3 | 16 |
| Río Lobos | 3132 | 1 | 4 | 37 | 4 | 19 |

ing tradition; (4) The Ávila Provincial Council has recently shown interest in creating routes in La Paramera, indicating potential interest and support from this administration. Therefore, a promotion and valuation plan for geotourism would be an important territorial resource for the region.

And the declaration of an NPA, as is the case of the Cañón del Río Lobos Natural Park, involves a promotion and tourist development that can have positive effects on its economic scope. We have already seen that the CRLNP receives public funding for the improvement of its infrastructure and services, and what are some of the most recent actions developed. However, if we go back to its declaration in 1985, there has been a positive impact in all areas of the Park, as analyzed by Bachiller (2014), who conducted a study of the socioeconomic influence of the Park from its declaration until 2011. The improvement of its state of conservation, the development of the commercial and tourist sector, a slower population loss than in the rest of the province, and the provision of infrastructure and collective facilities are some of the positive impacts pointed out by the author. Therefore, if the Sierras de la Paramera and la Serrota were recognized, and a firm commitment was made to the exploitation of the potential geotouristic resources, it is not unreasonable to think that there would be a significant improvement in the services of its municipalities. However, the negative effects of the tourist development of the Park are also identified, such as overcrowding and deterioration of the most visited points. In this study by Bachiller (2014), the need to publicize other less visited but equally interesting points of the Park is already recognized.

In short, both NPA share common needs: the study and dissemination of their geomorphological heritage, a commitment to its protection and planning, and the search for a balance between the promotion of natural heritage and sustainable development that ensures, above all, its protection. These are very difficult needs to meet in the face of a neglected geographical education, a very incipient geotourism that has not seen any development in our community, and the inaction of public administrations.

# Conclusions

# 7

The geomorphological heritage, and specifically geomorphosites, are places with great didactic and geotouristic potential that lies in their enormous scientific and cultural interest and their links with the territory and the development of societies. Geomorphosites are not elements of high scientific value at an international, national, or regional scale, but they do have local value due to their relationship with the territory, environmental and cultural heritage, and their connection with the populations that inhabit them.

A potential that is diminished by the limited recognition of geographical education and geotourism, with a lack of proper integration and planning of the uses of these places in Spain and specifically in Castilla y León.

**Geographical education** and the teaching of Earth Sciences in general are currently unable to meet the learning objectives and competencies established in the recommendations of the European Union and in the curricula of countries such as Spain, Italy, and Switzerland, although the situation is similar worldwide. These subjects are often relegated to courses such as History and Biology, resulting in students lacking specific knowledge of Geology, geomorphology, landforms, and landscapes, which hinders their appreciation of Natural Heritage as something to protect and conserve. Geomorphosites, which represent not only geological and geomorphological processes of high scientific value but also a cultural and territorial legacy, can be used as educational resources that, through field trips, help reverse the neglect of Earth Sciences in school systems. The proposed method for evaluating the didactic potential of geomorphosites facilitates their incorporation into formal education, regardless of the country and legislation in which it is applied. The geomorphosites analyzed in Spain, Italy, and Switzerland have medium to high didactic potential due to their high geological–geomorphological scientific value, good accessibility in most cases, and additional values that make them highly multidisciplinary for teaching.

The use and planning of geomorphosites with a **geotouristic orientation** in Natural Parks allows for diversifying visits, relieving congestion in overcrowded areas, and connecting them to other natural environments (habitats, forests, geosites, or landscapes) through proposed uses and visitable locations linked to geomorphosites. Additionally, a geotouristic use of geomorphosites facilitates expanding the range of recreational, educational, or geotouristic activities in two ways: (1) as a tool for diversifying visited locations; and (2) as tools for local development in surrounding areas by attracting population and generating new territorial dynamics, always with the involvement of local communities in various tasks. The ideal framework for incorporating geomorphology, through geomorphosites and their potential use, as an attractive and enriching geotouristic resource, is through their inclusion in appropriate and integrative planning and management instruments, such as Special Plans, Natural Resource Management Plans, and Use and Management Plans, which do not currently exist for the Cañón del Río Lobos, much less for the Paramera and the Serrota. Therefore, it is necessary to incorporate geomorphosites into strategic, management, or promotional plans where they are used as tools for territorial planning and management to help conserve the natural environment and promote sustainable public use. It is about going beyond a simple inventory of geomorphosites and having these sites actively participate in managing visitor flows, guiding or limiting uses, and protecting geoheritage.

The **geomorphological mapping** carried out in the Cañón del Río Lobos National Park and in the Sierras de la Paramera and la Serrota has been essential for an in-depth study of the geomorphological heritage, the geomorphosites and their intrinsic values, from a spatial and territorial perspective that can be very useful for the management of National Parks and geomorphosites as educational and geotourism resources. The geomorphological map also allows establishing the rela-



R. M. Ruiz Pedrosa, *Geomorphological Heritage as Didactic and Geotouristic Resources*, Geoheritage, Geoparks and Geotourism, https://doi.org/10.1007/978-3-032-06205-5_7

tionship between landforms, their spatial distribution, and their connections with cultural elements. The connection between natural and cultural elements in culturally relevant spaces, such as the archaeological site of Ulaca, helps to better understand both current landscapes and the environment in which different societies inhabited them. It aims to understand in detail the geographical constraints and the abilities of cultures to exploit or adapt to the environment. It is a technical document useful for acquiring information that can be used by managers, interpreters, or archaeologists for a better understanding of the environment. However, it is not a dissemination tool, and its analysis and information must be qualified and transcribed using tools adapted to users (educational levels, tourists, geotourists, hikers, local residents). One way to provide this information to visitors is to incorporate it into routes and walks through interpreters, as has been done in Ulaca with local residents of the Paramera (the most effective cultural transfer system), documentation, signage, or virtual guides closely related to archaeological remains. The incorporation of geomorphological elements in the management of cultural spaces is necessary to ensure their protection against potential visitor flows and infrastructure development, as well as to guarantee the survival and use of this cultural service for local societies and visitors.

Activities and proposals have been carried out in the two study areas for the dissemination of geoheritage, based on the current uses and needs of each space. On the one hand, the **Sierras de la Paramera and la Serrota**, which are part of the Natura 2000 Network, have been waiting for over 30 years to be incorporated into the Natural Spaces Network of Castilla y León. Therefore, it lacks interpretive materials and geotouristic infrastructure, with medium accessibility values but medium to high scientific and cultural value in some of the geomorphosites. The 12 geomorphosites inventories in the Paramera and la Serrota are representative of their structural relief and the spectacular granite modeling that characterizes them, as well as the glacial traces present in La Serrota. The strategies implemented have been based on the evaluation of didactic potential, the development of the geomorphological map of Ulaca and la Serrota, and a dissemination activity with the local residents.

On the other hand, the **Cañón del Río Lobos Natural Park**, one of the first Natural Parks declared in the region, is currently in a state of planning lethargy. The park features folded relief and spectacular karstic formations represented in 14 geomorphosites, which have attracted visitors and established a certain level of natural tourism in the park. However, this tourism lacks interpretive elements related to the geomorphology and landforms, focusing mainly on a few visually striking locations while neglecting the rest of the park. To address this issue, georoutes have been designed and proposed to connect and explore the identified geomorphosites, aiming to promote the dissemination of the park's geomorphology and cultural elements and encourage sustainable public use through diversification of the tourism offer.

However, all these educational and geotourism initiatives, without effective and real collaboration between researchers, private agents, and public administrations, are doomed to fail, as seen in the Río Lobos, where georoutes have been designed but have not been promoted in the Park. Therefore, there is still much work to be done in educating and disseminating geoheritage, and geomorphological heritage must receive the recognition it deserves for its importance in understanding Earth's history, shaping societies, and supporting landscapes whose beauty should be protected, preserved, and passed on to future generations. But you cannot value what you do not know, and you cannot protect what you do not value. This is the goal of this book and of many researchers who focus their work on geoheritage: to study, raise awareness, and educate.

# References


Abad Soria, J., & García Quiroga, F. (2011). El Valle Amblés y las Sierras de Ávila, la Paramera y la Serrota. Patrimonio natural abulense. En *Cuadernos de Patrimonio Abulense* (Vol. 12). Institución Gran Duque de Ávila. Diputación de Ávila.

Abad Soria, J., Balset Izquierdo, J., Iglesias Aparicio, C., Castro, J., & Rodríguez Martín, C. T. (2022). Propuesta de Plan de Regeneración y Recuperación del Medium Natural y Landscape de la Sierra de la Paramera (Ávila) y su entorno.

Agudo Segura, L., Moreno Muñoz, D., & García Marín, R. (2022). Análisis de los cambios en el comportamiento de los consumidores Touristics en la Región de Murcia (España) tras la aparición de la COVID-19. *ROTUR. Revista de Ocio y Turismo, 16*(1), 1–13. https://doi.org/10.17979/rotur.2022.16.1.8746

Alario Trigueros, M., Molinero Hernando, F., & Morales Prieto, E. (2018). La persistencia de la dualidad rural y el valor de la nueva ruralidad en Castilla y León (España). *Investigaciones Geográficas, 70*, 9. https://www.redalyc.org/articulo.oa?id=17664421001

Almagro Gorbea, M. (2006). El "Canto de los Responsos" de Ulaca (Ávila): un rito celta del Más Allá. *Revista de Ciencias de las Religiones, 11*, 5–38.

Almagro Gorbea, M & Álvarez Sanchís, J.R (1993). La sauna de Ulaca: Saunas y baños iniciáticos en el mundo céltico. *Cuad. Arqueol. Univ. Navar. 1*, 177–253

Álvarez-Sanchís, J.R (1999). Los Vettones; Real Academia de la Historia: Madrid, Spain

Arenillas, M., & Martinez de Pison, E. (1976). La morfologia glaciar de la Serrota (Avila). *Boletin de la Real Sociedad geografica Madrid, 1*, 21–26.

Aryasa, A. M., Bambang, A. N., & Muhammad, F. (2017). The study of environmental carrying capacity for sustainable tourism in Telaga Warna Telaga Pengilon Nature Park, Dieng Plateau, Central Java. *IOP Conference Series: Earth and Environmental Science, 70*(1), 012003. https://doi.org/10.1088/17551315/70/1/012003

Bachiller, J. M. (2014). Valoración de la influencia socioeconómica y territorial de los Espacios Naturales Protegidos. El Cañón del Río Lobos. *Polígonos. Revista de geografía, 26*, 223–254.

Bollati, I., Pelfini, M., & Pellegrini, L. (2012). A geomorphosites selection method for educational purposes: A case study in Trebbia Valley (Emilia Romagna, Italy). *Geografia Fisica e Dinamica Quaternaria, 35*(1), 23–35. https://doi.org/10.4461/GFDQ.2012.35.3

Bollati, I., Crosa Lenz, B., Zanoletti, E., & Pelfini, M. (2017a). Geomorphological mapping for the valorization of the alpine environment. A methodological proposal tested in the Loana Valley (Sesia Val Grande Geopark, Western Italian Alps). *Journal of Mountain Science, 14*(6), 1023–1038. https://doi.org/10.1007/s11629-017-4427-7

Bollati, I., Pellegrini, M., Reynard, E., & Pelfini, M. (2017b). Water driven processes and landforms evolution rates in mountain geomorphosites: Examples from Swiss Alps. *Catena, 158*, 321–339. https://doi.org/10.1016/j.catena.2017.07.013

Bouzekraoui, H., Barakat, A., Mouaddine, A., El Youssi, M., Touhami, F., & Hafid, A. (2018a). Mapping geoheritage for geotourism management, a case study of Aït Bou Oulli Valley in Central High-Atlas (Morocco). *Environmental Earth Sciences, 77*(11), 413. https://doi.org/10.1007/s12665-018-7589-x

Bouzekraoui, H., Barakat, A., Touhami, F., Mouaddine, A., & El Youssi, M. (2018b). Inventory and assessment of geomorphosites for geotourism development: A case study of Aït Bou Oulli valley (Central High-Atlas, Morocco). *Area, 50*(3), 331343. https://doi.org/10.1111/area.12380

Brilha, J. (2016). Inventory and quantitative assessment of geosites and geodiversity sites: A review. *Geoheritage, 8*(2), 119–134. https://doi.org/10.1007/s12371-014-0139-3

Brilha, J., & Reynard, E. (2018). Geoheritage and geoconservation: The challenges. In *Geoheritage*. Elsevier.

Brilha, J., Gray, M., Pereira, D. I., & Pereira, P. (2018). Geodiversity: An integrative review as a contribution to the sustainable management of the whole of nature. *Environmental Science and Policy, 86*, 19–28. https://doi.org/10.1016/j.envsci.2018.05.001

Brkic-Vejmelka, J., Pejdo, A., & Segaric, N. (2018). Sustainable development from the perspective of geography education. *European Journal of Geography, 9*(1), 121–132. https://www.eurogeojournal.eu/index.php/egj/article/view/103

Bullón Mata, T., Martinez de Pison, E., Arenillas Parra, T., Arenillas Parra, M., Sanz Herráiz, C., Troitiño Vinuesa, M. A., Burgues Hoyos, J. A., & Juarez del Canto, D. (1988). Análisis del medio físico de Ávila: delimitación de unidades y estructura territorial. Junta de Castilla y León, Consejería de Fomento, Dirección General de Urbanismo, Vivienda y Medium Ambiente.

Bussard, J., Martin, S., Monbaron, M., Reynard, E., & Khalki, Y. E. (2022). Les paysages géomorphologiques du Haut Atlas central (Maroc) : potentiel éducatif et éléments pour la médiation scientifique. *Géomorphologie: relief, processus, environnement, 28*(3), 173–185. https://doi.org/10.4000/geomorphologie.17103

Calvo Peña, N., & Conde Iglesias, I. (2022). *Plan de restauración del incendio de agosto de 2021 en la Sierra de la Paramera (Ávila)*. Junta de Castilla y León.

Canovas García, F. (2021). Facilitar la enseñanza/aprendizaje de Geografía Física jugando con el ordenador: SIG de licencia gratuita. Diseño y resolución de tareas y validación científica de su eficacia didáctica a partir de una aplicación empírica. *Didáctica Geográfica, 24*, 225–227. https://didacticageografica.agegeografia.es/index.php/didacticageografica/article/view/691

Capdepón Frías, M. (2023). El turismo de naturaleza en España: de las políticas de recuperación post-pandemia a la propuesta de alternativas. *Boletín de la Asociación de Geógrafos Españoles, 99*. https://doi.org/10.21138/bage.3444

Capote, R., Diaz, M., Gabaldon, V., Gomez, J. J., Sanchez de la Torre, L., Ruiz, P., Rosell, J., Sopeña, A., & Yebenes, A. (1982). *Evolución sedimentológica y tectónica del ciclo alpino en el tercio noroc-*




R. M. Ruiz Pedrosa, *Geomorphological Heritage as Didactic and Geotouristic Resources*, Geoheritage, Geoparks and Geotourism, https://doi.org/10.1007/978-3-032-06205-5

*cidental de la rama castellana de la Cordillera Ibérica*. Instituto Geológico y Minero de España.

Carballo, R. R., León, C. J., & Carballo, M. M. (2019). Fighting overtourism in Lanzarote (Spain). *Worldwide Hospitality and Tourism Themes, 11*(5), 506–515. https://doi.org/10.1108/WHATT-06-2019-0043

Carrasco, R. M., Pedraza, J., & Palacios, D. (2022). The glaciers of the Sierra de Gredos. In *Iberia. Land of glaciers. How the mountains were shaped by glaciers* (pp. 457–484). Elsevier.

Carton, A., Coratza, P., & Marchetti, M. (2005). Guidelines for geomorphological sites mapping: Examples from Italy. *Géomorphologie : relief, processus, environnement, 11*(3), 209–218. https://doi.org/10.4000/geomorphologie.374

Casas Jericó, M., Puig Baguer, J., & Erneta Highrriba, L. (2018). El paisaje en el contexto curricular de la LOMCE: una oportunidad educativa, ¿aprovechada o desaprovechada? *Didáctica Geográfica, 18*. https://didacticageografica.agegeografia.es/index.php/didacticageografica/article/view/382

Casas Jericó, M., Erneta Highrriba, L., & Puig Baguer, J. (2019). La encuesta como herramienta para conocer las ideas previas sobre paisaje del alumnado de ESO: una experiencia piloto en Navarra. *Didáctica Geográfica, 19*, 47–76. https://doi.org/10.21138/DG.416

Castaldini, D., Valdati, J., & Ilies, D. C. (2005). The contribution of geomorphological mapping to environmental tourism in protected areas: Examples from the Apennines of Modena northern Italy. *Géomorphologie, 7*, 91–106.

Castaldini, D., Valdati, J., & Ilies, D. C. (2009). Geomorphological and geotourist maps of the Upper Tagliole Valley Modena Apennines, Northern Italy. *Memorie Descrittive della Carta Geologica d'Italia, 87*, 29–38.

Cayla, N. (2014). An overview of new technologies applied to the management of geoheritage. *Geoheritage, 6*(2), 91–102. https://doi.org/10.1007/s12371-014-0113-0

Cayla, N., & Martin, S. (2018). Digital geovisualisation technologies applied to geoheritage management. In E. Reynard & J. Brilha (Eds.), *Geoheritage* (Vol. 16, pp. 289–303). Elsevier. https://doi.org/10.1016/B978-0-12-809531-7.00016-2

Cendrero, A. (1980). Geología ambiental, bases doctrinales y metodológicas. Ponencias de la I Reunión de Geología Ambiental y Ordenación del Territorio, 162.

Cerrillo-Cuenca, E., de Sanjosé Blasco, J. J., Bueno-Ramírez, P., Pérez-Álvarez, J. A., de Balbín Behrmann, R., & Sánchez-Fernández, M. (2021). Emergent heritage: The digital conservation of archaeological sites in reservoirs and the case of the Dolmen de Guadalperal (Spain). *Heritage Science, 9*(1), 114. https://doi.org/10.1186/s40494-021-00590-5

Cocean, G., & Cocean, P. (2017). An assessment of gorges for purposes of identifying geomorphosites of geotourism value in the Apuseni Mountains (Romania). *Geoheritage, 9*(1), 71–81. https://doi.org/10.1007/s12371-016-01805

Coratza, P., & Hobléa, F. (2018). The specificities of geomorphological heritage. In E. Reynard & J. Brilha (Eds.), *Geoheritage* (Vol. 5, pp. 87–106). Elsevier. https://doi.org/10.1016/B978-0-12-809531-7.00005-8

Coratza, P., & Regolini-Bissig, G. (2009). Methods for mapping geomorphosites. In E. Reynard, P. Coratza, & G. Regolini-Bissig (Eds.), *Geomorphosites* (pp. 89–103). Pfeil Verlag.

Coratza, P., Bollati, I. M., Panizza, V., Brandolini, P., Castaldini, D., Cucchi, F., Deiana, G., Del Monte, M., Faccini, F., Finocchiaro, F., Gioia, D., Melis, R., Minopoli, C., Nesci, O., Paliaga, G., Pennetta, M., Perotti, L., Pica, A., Tognetto, F., & Pelfini, M. (2021). Advances in geoheritage mapping: Application to iconic geomorphological examples from the Italian landscape. *Sustainability, 13*(20), 11538. https://doi.org/10.3390/su132011538

Coratza, P., Vandelli, V., & Ghinoi, A. (2023). Increasing geoheritage awareness through non-formal learning. *Sustainability, 15*(1), 868. https://doi.org/10.3390/su15010868

Crespo Castellanos, J. M. (2012). Un itinerario didáctico para la interpretación de los elementos físicos de los paisajes de la Sierra de Guadarrama. *Didáctica Geográfica, 13*, 15–34. https://didacticageografica.agegeografia.es/index.php/didacticageografica/article/view/94

Crespo Castellanos, J. M., & Rodríguez de Castro, Á. V. (2020). La contribución de la educación geográfica para alcanzarlos Objetivos de Desarrollo sostenible. In F. J. Hinojo, J. Sadio Ramos, J. A. López Núñez, & J. M. R. Rodríguez (Eds.), *Experiencias e Investigaciones en Contextos Educativos* (pp. 376–385). Dykinson.

Crofts, R., Gordon, J. E., & Santucci, V. L. (2015). Geoconservation in protected areas. In G. L. Worboys, M. Lockwood, A. Kothari, S. Feary, & I. Pulsford (Eds.), *Protected area governance and management* (pp. 531–568). ANU Press.

da Silva Lopes, H., Remoaldo, P. C., Ribeiro, V., & Martín-Vide, J. (2021). Effects of the COVID-19 pandemic on tourist risk perceptions—The case study of Porto. *Sustainability, 13*(11), 6399. https://doi.org/10.3390/su13116399

de Miguel González, R., Claudino, S., & Souto, X. M. (2016). La utopía de la educación geográfica en las Declaraciones Internacionales de la UGI. In *XIV Coloquio Internacional de Geocrítica; Las utopías y la construcción de la sociedad del futuro*.

de Prado, C. (1864). *Descripción Física y Geológica de la Provincia de Madrid*. Junta General de Estadística.

de Vicente, G., Cunha, P. P., Muñoz-Martín, A., Cloetingh, S. A. P. L., Olaiz, A., & Vegas, R. (2018). The Spanish-Portuguese central system: An example of intense intraplate deformation and strain partitioning. *Tectonics, 37*(12), 4444–4469. https://doi.org/10.1029/2018TC005204

del Rosario Piñeiro Peleteiro, M. (1997). El pensamiento geográfico y el trabajo de campo en el siglo XX. *Didáctica Geográfica, 2*, 25–31. https://didacticageografica.agegeografia.es/index.php/didacticageografica/article/view/136

Delgado Huertos, E. (2007). Una década de Educación Ambiental en Castilla y León entre congresos (1994-2004). *TABANQUE. Revista Pedagógica, 20*, 185–230.

Dóniz-Páez, F. J., Becerra-Ramírez, R., García-Romero, L., & Hernández Ramos, W. (2023). Itinerario geoTouristic en el Valle de El Golfo como ejemplo de desarrollo territorial en el Geoparque Mundial Unesco de El Hierro, España. In *XXV Coloquio de Historia Canario-Americana (2022)* (p. XXV-048).

Dot Jutglà, E., Romagosa Casals, F., & Noguera Noguera, M. (2022). El incremento del turismo de proximidad en Cataluña en verano de 2020: una oportunidad para la consolidación del turismo rural como una forma de turismo sostenible y segura. *Investigaciones Turísticas, 23*, 162. https://doi.org/10.14198/INTURI2022.23.8

Dowling, R. K. (2011). Geotourism's global growth. *Geoheritage, 3*(1), 1–13. https://doi.org/10.1007/s12371-010-0024-7

Dowling, R. K., & Newsome, D. (2010). *Geotourism: The tourism of geology and landscape*. Goodfellow Publishers.

Dowling, R. K., & Newsome, D. (2017). Geotourism destinations—Visitor impacts and site management considerations. *Czech Journal of Tourism, 6*(2), 111–129. https://doi.org/10.1515/cjot-2017-0006

Dube, C. (2017). The uptake of education for sustainable development in geography curricula in South African secondary schools. In *Schooling for sustainable development in Africa* (pp. 93–105). Springer International Publishing. https://doi.org/10.1007/978-3-319-45989-9_7

Escriche, M. M. (2011). La función didáctica del geoturismo. Propuestas Para la region de Murcia. The didactic function of geotourism. Proposals for the region of Murcia. *Gran Tour, 4*, 62–93.

Fernández, C., & Vega, J. A. (2011). Erosión después de incendios forestales. *Boletín del CIDEU, 10*, 23–36.

Fernández Álvarez, R. (2019). Los geoparques como recurso para la enseñanzaaprendizaje del espacio geográfico en Educación Primaria: el paisaje de las áreas de montaña. *Didáctica Geográfica, 20*, 27–53. https://doi.org/10.21138/DG.440

Fernández Álvarez, R. (2020). Geoparks and education: UNESCO Global Geopark Villuercas-Ibores-Jara as a case study in Spain. *Geosciences (Switzerland), 10*(1), 27. https://doi.org/10.3390/geosciences10010027

Fernández Ruiz, J., Herreros Villanueva, V., Lillo Ramos, J., López Sopeña, F., & Olivé Davó, A. (2006). *Memoria de la Hoja 555 de Navalulgordo.* Mapa Geológico de España.

Floquet, M. (1991). La plateforme nord-castillane au Cretace Superieur (Espagne). In *Memoires Géologiques de l'Université de Dijon*. Centre des Sciences de la Terre.

Fuertes-Gutiérrez, I., García-Ortiz, E., & Fernández-Martínez, E. (2016). Anthropic threats to geological heritage: Characterization and management: A case study in the Dinosaur Tracksites of La Rioja (Spain). *Geoheritage, 8*(2), 135–153. https://doi.org/10.1007/s12371-015-0142-3

Gago-García, C., González-Relaño, R., Serrano Cambronero, M., & Babinger, F. (2021). Impacto de la crisis de la COVID-19 en el empleo del sector Touristic en España: perspectivas territorial y de género. *Boletín de la Asociación de Geógrafos Españoles, 91*. https://doi.org/10.21138/bage.3162

García-Álvarez, J., & Marías-Martínez, D. (2002). Nacionalismo y educación geográfica en la España del siglo XX. Una aproximación a través de los manuales de Bachillerato. *Documentos de Trabajo del IDEGA. Xeografía*, 11, 1–38.

García Martín, M., Villar Lama, A., Fraile Jurado, P., Sánchez Carnero, N., & Márquez Pérez, J. (2019). Se hace Geografía al andar: la salida de campo itinerante y senderista. *Didáctica Geográfica, 19*(19), 103–125. https://doi.org/10.21138/DG.418

García Paredes, C. (2022). Las Tecnologías de la Información Geográfica para el análisis espacial y sociodemográfico de la población extranjera en contextos escolares de la Comunidad de Madrid. *Didáctica Geográfica, 23*, 151–179. https://doi.org/10.21138/DG.656

García Ruiz, A. L. (1994). Los itinerarios didácticos una de las claves para la enseñanza y comprensión de la Geografía. *Iber: Didáctica de las ciencias sociales, geografía e historia, 1*, 117–126.

Garzón, M. G., Ubanell, A. G., & Rosales, F. (1981). Morfoestructura y sedimentación terciarias en el valle de Amblés (Sistema Central español). *Cuadernos de Geología Ibérica, 7*, 655–665.

González Amuchastegui, M. J., Serrano Cañadas, E., & González García, M. (2014). Sitees de interés geomorfológico, geoheritage y gestión de espacios naturales protegidos: el Natural Park de Valderejo (Álava, España). *Revista de Geografía Norte Grande, 59*, 45–64.

González de Amezúa, E. (1919). La Serrota. Anuario del Club Alpino Español (pp. 44–46).

González-Amuchástegui, M. J., Serrano Cañadas, E., González Trueba, J. J., & González García, M. (2013). Geomorphosites, a useful tool for environmental management in natural protected areas Álava, the Basque Country, Spain. In G. EDITEM (Ed.), *Managing geosites in protected areas* (pp. 87–94).

González-Domingo, A., Fosse, J., & Costa-Salavedra, C. (2021). *Managing overtourism in natural protected areas: Learnings from National Parks in Spain and in Europe*. NATUR Project: Conserving and Escaping Overtourism. Ecounion.

González-Trueba, J. J. (2007). *El macizo central de los Picos de Europa. Geomorphology y sus implicaciones geoecológjcas en la High montaña cantábrica* [Tesis Doctoral, Universidad de Cantabria].

Gordon, J. E. (2012). Rediscovering a sense of wonder: Geoheritage, geotourism and cultural landscape experiences. *Geoheritage, 4*(1–2), 65–77. https://doi.org/10.1007/s12371-011-0051-z

Gordon, J. E. (2018). Geoheritage, geotourism and the cultural landscape: Enhancing the visitor experience and promoting geoconservation. *Geosciences (Switzerland), 8*(4), 136. https://doi.org/10.3390/geosciences8040136

Gordon, J. E., & Barron, H. F. (2011). *Scotland's geodiversity: Development of the basis for a national framework* (Scottish Natural Heritage Commissioned Report, 417).

Gordon, J. E., & Barron, H. F. (2013). The role of geodiversity in delivering ecosystem services and benefits in Scotland. *Scottish Journal of Geology, 49*(1), 41–58. https://doi.org/10.1144/sjg2011-465

Gordon, J. E., Barron, H. F., Hansom, J. D., & Thomas, M. F. (2012). Engaging with geodiversity-why it matters. *Proceedings of the Geologists' Association, 123*(1), 1–6. https://doi.org/10.1016/j.pgeola.2011.08.002

Grande Ibarra, J. (2006). La Evolución del turismo rural en España y las nuevas oportunidades del turismo de naturaleza. *Revista de Estudios Touristics, 169–170*, 85–102. https://doi.org/10.61520/et.169-1702006.986

Grangier, L. (2019). *Le Geo Guide Lausanne: un outil de valorisation du patrimoine. Communication, motivations, experience* [Tesis Doctoral, Universidad de Lausana].

Gray, M. (2008). Geodiversity: Developing the paradigm. *Proceedings of the Geologists' Association, 119*(3–4), 287–298. https://doi.org/10.1016/S00167878(08)80307-0

Gray, M. (2011). Other nature: Geodiversity and geosystem services. *Environmental Conservation, 38*(3), 271–274. https://doi.org/10.1017/S0376892911000117

Gray, M. (2018). Geodiversity: The backbone of geoheritage and geoconservation. In E. Reynard & J. Brilha (Eds.), *Geoheritage* (pp. 13–25). Elsevier. https://doi.org/10.1016/B978-0-12-809531-7.00001-0

Gray, M. (2019). Geodiversity, geoheritage and geoconservation for society. *International Journal of Geoheritage and Parks, 7*(4), 226–236. https://doi.org/10.1016/j.ijgeop.2019.11.001

Gray, M., Gordon, J. E., & Brown, E. J. (2013). Geodiversity and the ecosystem approach: The contribution of geoscience in delivering integrated environmental management. *Proceedings of the Geologists'Association, 124*(4), 659–673. https://doi.org/10.1016/j.pgeola.2013.01.003

Gray, M., Fox, N., Gordon, J. E., Brilha, J., Charkraborty, A., Garcia, M. D. G., Hjort, J., Kubalíková, L., Seijmonsbergen, A. C., & Urban, J. (2024). Boundary of ecosystem services: A response to. *Journal of environmental management, 351*, 119666. https://doi.org/10.1016/j.jenvman.2023.119666

Gress, D. R., & Tschapka, J. M. (2017). Bridging geography and education for sustainable development: A Korean example. *Journal of Geography, 116*(1), 34–43. https://doi.org/10.1080/00221341.2015.1119874

Grindsted, T. S. (2015). The matter of geography in education for sustainable development: The case of Danish university geography. In W. Leal Filho (Ed.), *Transformative approaches to sustainable development at universities. World sustainability series* (pp. 13–24). Springer. https://doi.org/10.1007/978-3-319-088372_2

Hernández Pacheco, F., & Vidal Box, C. (1934). *El glaciarismo cuaternario de La Serrota*. Museo Nacional de Ciencias Naturales.

Herrera Anangonó, R. C., Delgado Campuzano, D. V., Moreira Espinoza, J. A., & Toala Tuarez, P. J. (2021). La reactivación turística post covid-19 de las áreas naturales protegidas y su incidencia en la mejora de la experiencia de los turistas en el Ecuador. *Siembra, 8*(2), e3071. https://doi.org/10.29166/siembra.v8i2.3071

Herrera-Franco, G., Carrión-Mero, P., Montalván-Burbano, N., Caicedo-Potosí, J., & Berrezueta, E. (2022). Geoheritage and geosites: A bibliometric analysis and literature review. *Geosciences, 12*(4), 169. https://doi.org/10.3390/geosciences12040169

Hose, T. A. (1996). Geotourism, or can tourists become casual rock hounds? In M. Bennett (Ed.), *Geology on your doorstep* (pp. 207–228). The Geological Society.

Hurtado Beltrán, M. F. (2024). De los teóricos en cómo enseñar la ciudad, a lo que aprenden los estudiantes: la ruta de la geografía urbana escolar en España y Colombia. *Didáctica Geográfica, 25*, 159–179. https://doi.org/10.21138/DG.703

Kelly, R., Fleming, A., Pecl, G. T., von Gönner, J., & Bonn, A. (2020). Citizen science and marine conservation: A global review. *Philosophical Transactions of the Royal Society B: Biological Sciences, 375*(1814), 20190461. https://doi.org/10.1098/rstb.2019.0461

Kozlowski, S. (2004). Geodiversity. The concept and scope of geodiversity. *Polish Geological Review (Przeglad geologiczny), 52*(8/2), 833–839.

Kubalikoba, L. (2013). Geomorphosite assessment for geotourism purposes. *Czech Journal of Tourism, 2*(2), 80–104. https://doi.org/10.2478/cjot-2013-0005

Kubalíková, L. (2019). Assessing geotourism resources on a local level: A case study from Southern Moravia (Czech Republic). *Resources, 8*(3), 150. https://doi.org/10.3390/resources8030150

Kubalíková, L. (2020). Cultural ecosystem services of geodiversity: A case study from Stranska skala (Brno, Czech Republic). *Land, 9*(4), 105. https://doi.org/10.3390/land9040105

Kubalíková, L., & Kirchner, K. (2016). Geosite and geomorphosite assessment as a tool for geoconservation and geotourism purposes: A case study from Vizovická vrchovina Highland (Eastern Part of The Czech Republic). *Geoheritage, 8*(1), 5–14. https://doi.org/10.1007/s12371-015-0143-2

Kubalíková, L., Drápela, E., Bajer, A., Zapletalová, D., Balková, M., Zágoršek, K., Kirchner, K., Kuda, F., & Roštínský, P. (2021a). Geological paths—Their use for the regional geography teaching (pp. 113–124). https://doi.org/10.5817/cz.muni.p2109694-2020-9

Kubalíková, L., Kirchner, K., & Bajer, A. (2021b). Geomorphological resources for geoeducation and geotourism. In R. B. Singh, D. Wei, & S. Anand (Eds.), *Global geographical heritage, geoparks and geotourism* (Vol. 18, pp. 343–358). Springer.

Kürüm Varolgüneş, F., Çelik, F., Del Río-Rama, M., & Álvarez-García, J. (2022). Reassessment of sustainable rural tourism strategies after COVID-19. *Frontiers in Psychology, 13*. https://doi.org/10.3389/fpsyg.2022.944412

Lambiel, C. (2021). Glacial and periglacial landscapes in the Hérens Valley. In E. Reynar (Ed.), *Landscapes and landforms of Switzerland. World geomorphological landscapes* (pp. 263–275). Springer. https://doi.org/10.1007/978-3-030-43203-4_18

Leung, Y., Spenceley, A., Hvenegaard, G., & Buckley, R. (2019). *Tourism and visitor management in protected areas: Guidelines for sustainability*. IUCN, International Union for Conservation of Nature. https://doi.org/10.2305/IUCN.CH.2018.PAG.27.en

López Fernandez, J. A., & Peral Velasco, A. J. (2018). Las vías verdes: escenario para trabajar el Medium rural en Educación Primaria. *Didáctica Geográfica, 18*, 171–192. https://didacticageografica.age-geografia.es/index.php/didacticageografica/article/view/388

Manning, R., Valliere, W., Wang, B., Lawson, S., & Newman, P. (2002). Estimating day use social carrying capacity in Yosemite national park. *Leisure/Loisir, 27*(1–2), 77–102. https://doi.org/10.1080/14927713.2002.9651296

Mao, Y., Robinson, A. M., & Dowling, R. (2009). Potential geotourists: An Australian case study. *Journal of Tourism, X*(1), 71–80.

Marino Alfonso, J. L., Poblete Piedrabuena, M. Á., Beato Bergua, S., & Herrera Arenas, D. (2021). Geotourism itineraries and augmented reality in the geomorphosites of the Arribes del Duero Natural Park (Zamora Sector, Spain). *Geoheritage, 13*(1), 16. https://doi.org/10.1007/s12371-021-00539-x

Martin, S. (2012). *Valoriser le géopatrimoine par la médiation indirecte et la visualisation des objets géomorphologiques* [Tesis Doctoral, Université de Lausanne]. https://serval.unil.ch/en/notice/serval:BIB_CA7EE5EA74AF

Martínez de Pisón, E. (1983). Cultura y ciencia del paisaje. *Agricultura y Sociedad, 27*, 9–32.

Martínez de Pisón, E. (2010). Saber ver el paisaje. *Estudios Geográficos, 71*(269), 395. https://doi.org/10.3989/estgeogr.201013

Martínez De Pisón, E. (2012). Sobre la idea y la enseñanza del paisaje. *Nimbus, 2930*, 373–380.

Martinez de Pisón, E. (2014). Ordesa: del valle perdido al símbolo patrimonial. *Ería, 94*, 145–160. https://reunido.uniovi.es/index.php/RCG/article/view/10449

Martínez-Hernández, C., & Mínguez, C. (2023). The Anthropocene and the sustainable development goals: Key elements in geography higher education? *International Journal of Sustainability in Higher Education, 24*(7), 1648–1667. https://doi.org/10.1108/IJSHE-09-2022-0316

Martín-Parra, L. M., Martínez-Salanova, J., Moreno, F., Contreras, E., Iglesias, A., & Martín Herrero, D. (2008). *Memoria de la Hoja 530 de Vadillo de la Sierra. Mapa Geológico de España. Escala 1:50.000*. Segunda serie-Primera edición.

Maté-González, M. Á., Sáez Blázquez, C., Carrasco García, P., Rodríguez-Hernández, J., Fernández Hernández, J., Vallés Iriso, J., Torres, Y., Troitiño Torralba, L., Courtenay, L. A., González-Aguilera, D., López-Cuervo, S., Aguirre de Mata, J., Velasco Gómez, J., Piras, M., di Filippo, A., Yravedra, J., Fernández Fernández, M., Chapa, T., Ruiz Zapatero, G., & Álvarez-Sanchís, J. R. (2021). Towards a combined use of geophysics and remote sensing techniques for the characterization of a singular building: "El Torreón" (the tower) at Ulaca Oppidum (Solosancho, Ávila, Spain). *Sensors, 21*(9), 2934. https://doi.org/10.3390/s21092934

Maté-González, M. Á., Rodríguez-Hernández, J., Sáez Blázquez, C., Troitiño Torralba, L., Sánchez-Aparicio, L. J., Fernández Hernández, J., Herrero Tejedor, T. R., Fabián García, J. F., Piras, M., Díaz-Sánchez, C., González-Aguilera, D., Ruiz Zapatero, G., & Álvarez-Sanchís, J. R. (2022). Challenges and possibilities of archaeological sites virtual tours: The Ulaca Oppidum (Central Spain) as a case study. *Remote Sensing, 14*(3), 524. https://doi.org/10.3390/rs14030524

Mateos, A. B., Leco, F., & Pérez, A. (2020). Visitors' perception of the overcrowding of a protected natural area: A case applied to the natural reserve "Garganta de los Infiernos" (Caceres, Spain). *Sustainability, 12*(22), 9503. https://doi.org/10.3390/su12229503

McArdell, B. W., & Sartori, M. (2021). The Illgraben torrent system. In E. Reynard (Ed.), *Landscapes and landforms of Switzerland. World geomorphological landscapes* (pp. 367–378). Springer. https://doi.org/10.1007/978-3-030-43203-4_25

McCool, S., & Kohl, J. (2017). *What is overtourism in protected areas and what can we do about it? Sustainable tourism webinars*. UN Environment, UNWTO.

Meadows, M. E. (2020). Geography education for sustainable development. *Geography and Sustainability, 1*(1), 88–92. https://doi.org/10.1016/j.geosus.2020.02.001

Méndez Gutiérrez del Valle, R. (2022). Turismo, pandemia y nuevos contrastes territoriales en España. *Ikara. Revista de Geografías iberoamericanas, 1*. https://doi.org/10.18239/Ikara.3006

Migón, P. (2006). *Granite landscapes of the world*. Oxford University Press.

Migoń, P. (2021). Granite landscapes, geodiversity and geoheritage—Global context. *Heritage, 4*(1), 198–219. https://doi.org/10.3390/heritage4010012

Migón, P., & Thomas, M. F. (2002). Grus weathering mantles. Problems of interpretation. *Catena, 49*, 5–24.

Migoń, P., & Vieira, G. (2014). Granite geomorphology and its geological controls, Serra da Estrela, Portugal. *Geomorphology, 226*, 1–14. https://doi.org/10.1016/j.geomorph.2014.07.027

Mishra, S. (2020). *Tools and resources for nature-based tourism*. The World Bank Group.

Molina, E. (1991). *Geomorfologia y Geoquímica del Paisaje: Dos Ejemplos en el Interior de la Meseta Ibérica*. Ediciones Universidad de Salamanca.

Molina, E., Blanco, J. A., Pellitero, E., & Cantano, M. (1987). Weathering processes and morphological evolution of the Spanish Hercynian massif. In V. Gardiner (Ed.), *International geomorphology* (pp. 957–977). Wiley.

Morard, S. (2011). Evaluation du paysage géomorphologique du projet de Parc Naturel Régional (PNR) Pfyn-Finges (Valais, Suisse) par la méthode de l'OFEV. *Geovisions, 36*, 49–60.

Moreno, F. (2004). Concepciones de los alumnos sobre la historia. *Enseñanza de las ciencias sociales: revista de investigación, 3*, 75–83. https://raco.cat/index.php/EnsenanzaCS/article/view/126194.

Moreno, F. (2008). *Geomorphology. Memoria del Mapa Geomorfológico 1/50.000. Vadillo de la Sierra. No 530*. Instituto Geológico y Minero.

Morón Monge, M. C., Morón Monge, H., & Trabajo Rite, M. (2023). El Proyecto de innovación docente los "Cabezos de Huelva". Educando en el activismo ciudadano: algunos resultados. *Didáctica Geográfica, 24*, 83–112. https://doi.org/10.21138/DG.659

Morote Seguido, Á. F. (2019). Las salidas de campo en España como recurso didáctico para la enseñanza de la Geografía. Una revisión bibliográfica. *Geographicalia, 71*, 27–49.

Mucivuna, V. C., Reynard, E., & da Glória Motta Garcia, M. (2019). Geomorphosites assessment methods: Comparative analysis and typology. *Geoheritage, 11*(4), 1799–1815. https://doi.org/10.1007/s12371-019-00394-x

Muñoz, V., Isabel, M., & Alfaro, C. (2010). Investigación didáctica concepciones del alumnado de secundaria sobre la comprensión y el aprendizaje conceptos de geografía. *Enseñanza de las ciencias sociales: revista de investigación, 9*, 3–16. https://raco.cat/index.php/EnsenanzaCS/article/view/191354

Németh, B., Németh, K., Procter, J. N., & Farrelly, T. (2021). Geoheritage conservation: Systematic mapping study for conceptual synthesis. *Geoheritage, 13*(2), 45. https://doi.org/10.1007/s12371-021-00561-z

Nesci, O., & Valentini, L. (2020). Science, poetry, and music for landscapes of the Marche region, Italy: Communicating the conservation of natural heritage. *Geoscience Communication, 3*(2), 393–406. https://doi.org/10.5194/gc-3-3932020

Newsome, D., & Dowling, R. (2018). Geoheritage and geotourism. In *Geoheritage*. Elsevier. https://doi.org/10.1016/B978-0-12-809531-7.00017-4

Newsome, D., Dowling, R., & Leung, Y.-F. (2012). The nature and management of geotourism: A case study of two established iconic geotourism destinations. *Tourism Management Perspectives, 2–3*, 19–27. https://doi.org/10.1016/j.tmp.2011.12.009

Nieto, L. M. (2001). Geodiversidad: propuesta de una definición integradora. *Boletín Geológico y Minero, 112*(2), 3–12.

Notari Llorens, M., Pardo Fabregat, F., & Jordán Vidal, M. (2016). Estudio preliminar de la adquisición de conocimientos de la concienciación ecológica en la educación secundaria obligatoria: la educación ambiental entre los adolescentes del nordeste de España. *Revista Internacional de Aprendizaje en Ciencia, Matemáticas y Tecnología, 3*, 131. http://sobrelaeducacion.com

Ólafsdóttir, R., & Dowling, R. (2014). Geotourism and geoparks—A tool for geoconservation and rural development in vulnerable environments: A case study from Iceland. *Geoheritage, 6*(1), 71–87. https://doi.org/10.1007/s12371013-0095-3

Orion, N. (1993). A model for the development and implementation of field trips as an integral part of the science curriculum. *School Science and Mathematics, 93*(6), 325–331. https://doi.org/10.1111/j.1949-8594.1993.tb12254.x

Orion, N. (2007). A holistic approach for science education for all. *Eurasia Journal of Mathematics, 3*(2), 111–118.

Orion, N., & Hofstein, A. (1994). Factors that influence learning during a scientific field trip in a natural environment. *Journal of Research in Science Teaching, 31*(10), 1097–1119. https://doi.org/10.1002/tea.3660311005

Orion, N., and Fortner, W. R. (2003). Mediterranean models for integrating environmental education and earth sciences through earth systems education. *Mediterranean Journal of Educational Studies*, 8(1), 97–111.

Otto, J. C., & Smith, M. (2013). Geomorphological mapping. In *Geomorphological techniques (online edition)* (Vol. 2.6). British Society for Geomorphology.

Páez, J. D., Ramírez, R. B., & Yanes, E. B. (2021). Geoheritage and geotourism in volcanic Tenerife's natural protected areas (Canary Islands, Spain). *Cuadernos Geograficos, 60*(2), 52–71. https://doi.org/10.30827/CUADGEO.V60I2.15572

Panizza, M. (2001). Geomorphosites: Concepts, methods and examples of geomorphological survey. *Chinese Science Bulletin, 46*(S1), 4–5. https://doi.org/10.1007/BF03187227

Panizza, M (2003). Géomorphologie et tourisme dans un paysage culturel integré. In Géomorphologie et Tourisme; Reynard, E., Holzmann, C., Guex, D., Summermater, N., Eds.; IGUL, Université de Lausanne: Lausanne, Switzerland; pp. 11–18.

Panizza, M., & Piacente, S. (1993). Geomorphological assets evaluation. *Zeitschrift für Geomorphologie, 87*, 13–18.

Panizza, M., & Piacente, S. (2003). *Geomorphologia Culturale*. Pitagora.

Panizza, M., & Piacente. (2009). Cultural geomorphology and geodiversity. In *Gemorphosites* (pp. 35–48). Pfeil.

Pasquaré Mariotto, F., Drymoni, K., Bonali, F. L., Tibaldi, A., Corti, N., & Oppizzi, P. (2023). Geosite assessment and communication: A review. *Resources, 12*(2), 29. https://doi.org/10.3390/resources12020029

Pedraza, J. (1989). La morfoGenesis del Sistema Central y su relación con la morfología granítica. *Cadernos do Laboratorio Xeolóxico de Laxe, 13*, 31–46.

Pedraza, J., Sanz, M. A., & Martín, A. (1989). Formas graníticas de la Pedriza. In *Cuadernos madrileños del Medium Ambiente*. Agencia de Medium Ambiente de la Comunidad de Madrid.

Pellitero Ondicol, R. (2012). Una puesta en valor del relieve como elemento patrimonial: georutas por el Alto Carrión (Montaña Palentina). *Polígonos Revista de Geografía, 17*, 113–132. https://doi.org/10.18002/pol.v0i17.400

Peña Monné, J. L. (1997). *Cartografía geomorfológica básica y aplicada*. Geoforma Ediciones.

Pereira, P., Pereira, D. I., & Caetano, M. I. (2009). The geomorphological heritage approach in protected areas: Geoconservation vs. geotourism in Portuguese natural parks. *Memoria Descriptiva della Carta Geologica d'Italia, 87*, 135–144.

Pijet-Migoń, E., & Migoń, P. (2019). Promoting and interpreting geoheritage at the local level—Bottom-up approach in the land of extinct volcanoes, Sudetes, SW Poland. *Geoheritage, 11*(4), 1227–1236. https://doi.org/10.1007/s12371-01900357-2

Piñeiro-Peleteiro, M. del R. (1997). El pensamiento geográfico y el trabajo de campo en el siglo XX. *Didáctica Geográfica*, 2, 25–31. https://didacticageografica.agegeografia.es/index.php/didacticageografica/article/view/136

Polukhina, A., Sheresheva, M., Efremova, M., Suranova, O., Agalakova, O., & Antonov-Ovseenko, A. (2021). The concept of sustainable rural tourism development in the face of COVID-19 crisis: Evidence from Russia. *Journal of Risk and Financial Management, 14*(1), 38. https://doi.org/10.3390/jrfm14010038

Pralong, J.-P. (2005). A method for assessing tourist potential and use of geomorphological sites. *Géomorphologie : relief, processus, environnement, 11*(3), 189–196. https://doi.org/10.4000/geomorphologie.350

Prosser, C. D. (2019). Communities, quarries and geoheritage—Making the connections. *Geoheritage, 11*(4), 1277–1289. https://doi.org/10.1007/s12371019-00355-4

Raeisi, R., Dincă, I., Almodaresi, S. A., (Nellie) Swart, M. P, & Boloor, A. (2022). An assessment of geosites and geomorphosites in the Lut Desert of Shahdad region for potential geotourism development. *Land, 11*(5), 736. https://doi.org/10.3390/land11050736

Rebelo, D., Marques, L., & Costa, N. (2011). Actividades en ambientes exteriores al aula en la Educación en Ciencias: contribuciones para su operatividad. *Enseñanza de las Ciencias de la Tierra, 19*(1), 15–25.

Regolini-Bissig, G. (2008). Mapping geomorphosites: An analysis of geotourist maps. *Geotourism/Geoturystyka, 14*(1), 3. https://doi.org/10.7494/geotour.2008.14.3

Regolini-Bissig, G. (2010). Mapping geoheritage for interpretive purpose: Definition and interdisciplinary approach. In G. Regolini-Bissig & E. Reynard (Eds.), *Mapping geoheritage* (pp. 1–13). IGUL-Université de Lausanne.

Regolini-Bissig, G. (2011). Cartographier les géomorphosites: objectifs, publics et propositions méthodologiques. *Geovisions, Université de Lausanne, Institut de géographie, 38*, 383–384.

Regolini-Bissig, G. (2012). Cartographier les géomorphosites : objectifs, publics et propositions méthodologiques. *Géomorphologie : relief, processus, environnement, 18*(3), 383–384. https://doi.org/10.4000/geomorphologie.10076

Reynard, E. (2003). Geomorphologie et tourisme. Quelles relations? In E. Reynard, C. Holzmann, D. Guex, & N. Summermatter (Eds.), *Géomorphologie et Tourisme* (pp. 1–10). Institute de Geographie IGUL-Université de Lausanne.

Reynard, E. (2008). Scientific research and tourist promotion of geomorphological heritage. *Geografía Física e Dinamica Quaternaria, 31*, 225–230.

Reynard, E. (2009). Geomorphosites: Definitions and characteristics. In E. Reynard, P. Coratza, & G. Regolini-Bissig (Eds.), *Geomorphosites* (pp. 9–20). Pfeil.

Reynard, E., & Coratza, P. (2013). Scientific research on geomorphosites. A review of the activities of the IAG working group on geomorphosites over the last twelve years. *Geografia Fisica e Dinamica Quaternaria, 36*(1), 159–168. https://doi.org/10.4461/GFDQ.2013.36.13

Reynard, E., & Coratza, P. (2016). The importance of mountain geomorphosites for environmental education: Examples from the Italian dolomites and the Swiss alps. *Acta Geographica Slovenica, 56*(2), 291–303. https://doi.org/10.3986/AGS.1684

Reynard, E., & Giusti, C. (2018). The landscape and the cultural value of geoheritage. In E. Reynard & J. Brilha (Eds.), *Geoheritage* (Vol. 8, pp. 147–166). Elsevier. https://doi.org/10.1016/B978-0-12-809531-7.00008-3

Reynard, E., & Panizza, M. (2005). Geomorphosites: Definition, assessment and mapping. *Géomorphologie: relief, processus, environnement, 11*(3), 177–180. https://doi.org/10.4000/geomorphologie.337

Reynard, E., Fontana, G., Kozlik, L., & Scapozza, C. (2007). A method for assessing «scientific» and «additional values» of geomorphosites. *Geographica Helvetica, 62*(3), 148–158. https://doi.org/10.5194/gh-62-148-2007

Reynard, E., Coratza, P., & Regolini-Bissig, G. (2009). *Geomorphosites*. Dr. Friedrich Pfeil.

Reynard, E., Clivaz, M., Keller, R., & Backhaus, N. (2021). L'approche par les prestations paysagères, un cadre analytique et un outil de gestion des géopatrimoines à forte composante paysagère. *Géo-Regards, 14*(1), 35–53. https://doi.org/10.33055/GEOREGARDS.2021.014.01.35

Ríos Rodríguez, N., Nieto Masot, A., & Cárdenas Alonso, G. (2022). Los efectos de la COVID-19 en el sector Touristic de las Comunidades Autónomas españolas. *Boletín de la Asociación de Geógrafos Españoles, 94*, 1–36. https://doi.org/10.21138/bage.3279

Rodrigues, J., Costa e Silva, E., & Pereira, D. I. (2023). How can geoscience communication Foster public engagement with geoconservation? *Geoheritage, 15*(1), 32. https://doi.org/10.1007/s12371-023-00800-5

Ruiz Zapatero, G. (2005). Castro de Ulaca. Solosancho, Ávila. In *Cuadernos de Patrimonio Abulense* (Vol. 3). Institución Gran Duque de Alba. Diputación de Ávila.

Ruiz-Pedrosa, R. M. (2019). *Sitees de interés geomorfológico y su valoración didáctica en el Natural Park cañón del río Lobos*. Universidad de Valladolid.

Ruiz-Pedrosa, R. M. (2024). Patrimonio geomorfológico como recurso didáctico y geoturístico en espacios naturales protegidos. Estudio de casos en las Sierras de la Paramera y la Serrota y en el Cañón del Río Lobos. Tesis Doctoral, Universidad de Valladolid: https://uvadoc.uva.es/handle/10324/72353

Ruiz-Pedrosa, R. M., & Serrano, E. (2023). Granite landscapes and landforms in the Castro de Ulaca Site (Ávila, Spain): A narrow relationship between natural and cultural heritage. *Sustainability, 15*(13), 10470. https://doi.org/10.3390/su151310470

Ruiz-Pedrosa, R. M., & Serrano, E. (2024). Dissemination and interpretation of natural heritage in Sierra de la Paramera (Ávila, Spain). An experimental activity on geomorphosites, cultural heritage and landscape. *Geoheritage, 16*(3), 86. https://doi.org/10.1007/s12371-024-00990-6

Ruiz-Pedrosa, R. M., González-Amuchástegui, M. J., & Serrano, E. (2024). Geomorphosites as geotouristic resources: Assessment of geomorphological heritage for local development in the Río Lobos Natural Park. *Land, 13*(2), 128. https://doi.org/10.3390/land13020128

Salas, R., Guimerá, J., Mas, R., Martín Closas, C., Meléndez, A., & Alonso, A. (2001). Evolution of the Mesozoic Central Iberian Rift System and its Cenozoic inversion (Iberian chain). In W. Cavazza, A. H. F. R. Roberston, P. Ziegler, & P. Crasquin (Eds.), *Peri-Tethyan rift/wrench basins and passive margins* (Vol. 186, pp. 145–185). Mémoires Museum Nationale de Histoire Naturelle.

Sánchez Almodóvar, E., & Olcina Cantos, J. (2024). Enseñando la complejidad: clima, cambio climático y extremos atmosféricos en Educación Secundaria. *Didáctica Geográfica, 25*, 125–158. https://doi.org/10.21138/DG.704

Sánchez Ogallar, A. (1995). El trabajo de campo y las excursiones. En M. J. Marrón Gaite, & A. Moreno Jiménez (Eds.), *Enseñar geografía : de la teoría a la practica* (pp. 159–184).

Santos, I., Henriques, R., Mariano, G., & Pereira, D. I. (2018). Methodologies to represent and promote the geoheritage using unmanned aerial vehicles, multimedia technologies, and augmented reality. *Geoheritage, 10*(2), 143. https://doi.org/10.1007/s12371-018-0305-0

Sanz Herráiz, C. (1988). *El relieve del Guadarrama oriental*. Comunidad de Madrid, Consejería de Política Territorial.

Schmieder, O. (1915). Die Sierra de Gredos. *Mitteilungen der Geographischen Gesellschaft in Manchen, X*(1).

Segovia, R. (2008). *El Drenaje Subterráneo en el Acuífero Kárstico del Cañón del rio Lobos (Soria-Burgos)* [Tesis Doctoral, Universidad Politécnica de Madrid].

Seijmonsbergen, A. C. (2013). 14.4 The modern geomorphological map. In J. Shroder, A. D. Switzer, & D. M. Kennedy (Eds.), *Treatise on geomorphology* (Vol. 14, pp. 35–52). Elsevier. https://doi.org/10.1016/B978-0-12-374739-6.00371-7

Sequeira, M. A., Paquet, H., & Begonha, A. (2002). Weathering of granites in a temperate climate (NW Portugal): Granitic saprolites and arenization. *Catena, 49*, 41–56.

Serantes-Pazos, A. (2011). Os equipamentos para a Educación ambiental en Galicia: Análise da realidade e propostas de mellora da calidade. Tesis Doctoral, Universidade da Coruña

Serantes, A., Oliver, M. F., & Escudero, C. (2015). *La profesionalización en los equipamientos de educación ambiental en España vista desde dentro*. Seminario de Equipamientos de Educación

Ambiental del CENEAM. http://www.magrama.gob.es/es/cene-am/grupos-de-trabajo-y-seminarios/

Serrano, E. (2002). Geomorphology, natural heritage and protected areas. Lines of research in Spain. In P. Coratza & M. Marchetti (Eds.), *Geomorphological sites: Research, assessment and improvement. Workshop proceedings* (pp. 27–33). Università degli studi de Modena.

Serrano, E. (2012). Montañas, paisaje y patrimonio. *Nimbus, 29*, 701–718.

Serrano, E. (2023). *Glaciares. Cultura y Patrimonio. La huella cultural de los glaciares pirenaicos*. Ediciones Universidad de Valladolid.

Serrano, E., & González Trueba, J. J. (2011). Environmental education and landscape leisure. Geotourist map and geomorphosites in the Picos de Europa National Park. *GeoJournal of Tourism and Geosites, IV*(2), 295–308.

Serrano, E., & González-Amuchastegui, M. J. (2020). Cultural heritage, landforms, and integrated territorial heritage: The close relationship between tufas, cultural remains, and landscape in the Upper Ebro Basin (Cantabrian Mountains, Spain). *Geoheritage, 12*(4), 86. https://doi.org/10.1007/s12371-02000513-z

Serrano, E., & González-Trueba, J. J. (2005). Assessment of geomorphosites in natural protected areas: The Picos de Europa National Park (Spain). *Géomorphologie : relief, processus, environnement, 3*, 197–208.

Serrano, E., & Ruiz-Flaño, P. (2007). Geodiversity. A theoretical and applied concept. *Geographica Helvetica, 62*, 140–147.

Serrano, E., González-Trueba, J. J., Pellitero, R., González-García, M., & Gómez-Lende, M. (2013). Quaternary glacial evolution in the Central Cantabrian Mountains (Northern Spain). *Geomorphology, 196*, 65–82. https://doi.org/10.1016/j.geomorph.2012.05.001

Serrano, E., González Amuchastegui, M. J., & Ruiz Pedrosa, R. M. (2020). Patrimonio natural y Geomorphology. In *Los Sitees de interés geomorfológico del Natural Park del Cañón del Río Lobos*. Ediciones Universidad de Valladolid.

Serrano Gil, Ó. (2021). Hitos del paisaje urbano en una ciudad declarada Patrimonio Mundial de la Humanidad (Cuenca): itinerario didáctico y trabajo de campo con alumnos universitarios. *Didáctica Geográfica, 22*, 145–169. https://doi.org/10.21138/DG.607

Sharples, C. (1995). Geoconservation in forest management—Principles and practice. *Tasforests, 7*, 37–50.

Smith, M., Griffiths, J. S., & Paron, P. (2011). *Geomorphological mapping. Methods and applications*. Elsevier.

Solís Solís, J. M., Albet i Mas, A., & Garcia Ramon, M. D. (2019). *Pensamiento e historia de la geografía*. http://hdl.handle.net/10609/148304

Sopeña, A., Gutiérrez Marco, J. C., Sánchez Moya, Y., Gómez, J. J., Más, R., García, A., & Lago, M. (2004). Cordilleras Ibérica y catalana. In J. A. Vera (Ed.), *Geología de España* (pp. 67–527). SGE-IGME.

Spenceley, A., & Snyman, S. (2017). Protected area tourism: Progress, innovation and sustainability. *Tourism and Hospitality Research, 17*(1), 3–7. https://doi.org/10.1177/1467358416646646

Stanley, M. (2001). Geodiversity strategy. *Progeo News, 1*, 6–9.

Stepišnik, U., & Trenchovska, A. (2018). A new quantitative model for comprehensive geodiversity evaluation: The Škocjan Caves Regional Park, Slovenia. *Geoheritage, 10*(1), 39–48. https://doi.org/10.1007/s12371-017-0216-5

Stepišnik, U., Ilc Klun, M., & Repe, B. (2017). Assessment of educational potential of geodiversity on example of Cerknica Polje, Slovenia. *Dela, 2017*(47), 5–21. https://doi.org/10.4312/dela.47.1.5-39

Tilden, F. (1957). *Interpreting our heritage*. University of North Carolina Press.

Timmons, A. L. (2019). Too much of a good thing: Overcrowding at America's National Parks. *Notre Dame Law Review, 94*(2), 986–1018.

Torabi Farsani, N., Coelho, C., & Costa, C. (2012). Geotourism and geoparks as gateways to socio-cultural sustainability in Qeshm Rural Areas, Iran. *Asia Pacific Journal of Tourism Research, 17*(1), 30–48. https://doi.org/10.1080/10941665.2011.610145

Tormey, D. (2019). New approaches to communication and education through geoheritage. *International Journal of Geoheritage and Parks, 7*(4), 192–198. KeAi Communications Co. https://doi.org/10.1016/j.ijgeop.2020.01.001

Tricart, J. (1970). Legende pour la carte géomorphologique de la France au 1/50.000: Vol. RCP 77 (CNRS).

Twidale, C. R. (1982). *Granite landforms*. Elsevier.

Twidale, C. R., & Vidal-Romani, J. R. (2005). *Landforms and geology of granite terrains*. CRP Press.

Vaishar, A., & Šťastná, M. (2022). Impact of the COVID-19 pandemic on rural tourism in Czechia preliminary considerations. *Current Issues in Tourism, 25*(2), 187–191. https://doi.org/10.1080/13683500.2020.1839027

Vandelli, V., Migoń, P., Palmgren, Y., Spyrou, E., Saitis, G., Andrikopoulou, M. E., Coratza, P., Medjkane, M., Prieto, C., Kalovrektis, K., Lissak, C., Papadopoulos, A., Papastamatiou, N., Evelpidou, N., Maquaire, O., Psycharis, S., Stroeven, A. P., & Soldati, M. (2024). Towards enhanced understanding and experience of landforms, geohazards, and geoheritage through virtual reality technologies in education: Lessons from the GeoVT project. *Geosciences, 14*(5), 127. https://doi.org/10.3390/geosciences14050127

Vasconcelos, C., Ribeiro, T., Cardoso, A., Orion, N., & Ben-Shalom, R. (2020). Educational theoretical framework underpinning geoethical educational resources. In C. Vasconcelos, N. Orion, & R. Ben-Shalom (Eds.), *Teaching resources for higher education geoethics* (pp. 20–32). U.Porto Ediçoes. https://doi.org/10.24840/978-989-746-254-2

Verstappen, H. T. (2011). Old and new trends in geomorphological and landform mapping. In M. J. Smith, P. Paron, & J. S. Griffiths (Eds.), *Developments in earth surface processes* (Vol. 15, pp. 13–38). Elsevier. https://doi.org/10.1016/B978-0-444-53446-0.00002-1

Wilson, E. O. (1992). *The diversity of life*. Belknap Press. https://doi.org/10.1177/027046769401400114

Wood, M. E. (2002). *Ecotourism: Principles, practices and policies for sustainability, United Nations Environment Programme*. United Nations Publication.

Worton, G. J., & Gillard, R. (2013). Local communities and young people—The future of geoconservation. *Proceedings of the Geologists' Association, 124*(4), 681–690. https://doi.org/10.1016/j.pgeola.2013.01.006

Zaragoza Sáez, Á., & Morote Seguido, Á.-F. (2024). Conocimiento del alumnado de Educación Secundaria de Orihuela (España) sobre el riesgo de inundación. *Didáctica Geográfica, 25*, 35–60. https://doi.org/10.21138/DG.698

Zgłobicki, W., & Baran-Zgłobicka, B. (2013). Geomorphological heritage as a tourist attraction. A case study in Lubelskie Province, SE Poland. *Geoheritage, 5*(2), 137–149. https://doi.org/10.1007/s12371-013-0076-6

Zouros, N. C. (2007). Geomorphosite assessment and management in protected areas of Greece case study of the Lesvos Island—Coastal geomorphosites. *Geographica Helvetica, 62*(3), 169–180. https://doi.org/10.5194/gh-62-169-2007

# Index




R. M. Ruiz Pedrosa, *Geomorphological Heritage as Didactic and Geotouristic Resources*, Geoheritage, Geoparks and Geotourism, https://doi.org/10.1007/978-3-032-06205-5

## S



## T

MIX
Papier aus verantwortungsvollen Quellen
Paper from responsible sources
FSC® C105338

If you have any concerns about our products,
you can contact us on
**ProductSafety@springernature.com**

In case Publisher is established outside the EU,
the EU authorized representative is:
**Springer Nature Customer Service Center GmbH**
**Europaplatz 3, 69115 Heidelberg, Germany**

Printed by Libri Plureos GmbH
in Hamburg, Germany